Nanostructure Science and Technology

Nanostructure science and technology now forms a common thread that runs through all physical and materials sciences and is emerging in industrial applications as nanotechnology. The breadth of the subject material is demonstrated by the fact that it covers and intertwines many of the traditional areas of physics, chemistry, biology, and medicine. Within each main topic in this field there can be many subfields. For example, the electrical properties of nanostructured materials is a topic that can cover electron transport in semiconductor quantum dots, self-assembled molecular nanostructures, carbon nanotubes, chemically tailored hybrid magnetic-semiconductor nanostructures, colloidal quantum dots, nanostructured superconductors, nanocrystalline electronic junctions, etc. Obviously, no one book can cope with such a diversity of subject matter. The nanostructured material system is, however, of increasing significance in our technology-dominated economy and this suggests the need for a series of books to cover recent developments.

The scope of the series is designed to cover as much of the subject matter as possible – from physics and chemistry to biology and medicine, and from basic science to applications. At present, the most significant subject areas are concentrated in basic science and mainly within physics and chemistry, but as time goes by more importance will inevitably be given to subjects in applied science and will also include biology and medicine. The series will naturally accommodate this flow of developments in the sciences and technology of nanostructures and maintain its topicality by virtue of its broad emphasis. It is important that emerging areas in the biological and medical sciences, for example, not be ignored as, despite their diversity, developments in this field are often interlinked. The series will maintain the required cohesiveness from a judicious mix of edited volumes and monographs that while covering subfields in depth will also contain more general and interdisciplinary texts.

Thus the series is planned to cover in a coherent fashion the developments in basic research from the distinct viewpoints of physics, chemistry, biology, and materials science and also the engineering technologies emerging from this research. Each volume will also reflect this flow from science to technology. As time goes by, the earlier series volumes will then serve as reference texts to subsequent volumes.

Shikha Gulati
Editor

Functionalized Magnetic Nanomaterials

Advanced Biomedical and Industrial Applications

Springer

Editor
Shikha Gulati
Department of Chemistry
Sri Venkateswara College
University of Delhi
New Delhi, Delhi, India

ISSN 1571-5744 ISSN 2197-7976 (electronic)
Nanostructure Science and Technology
ISBN 978-3-031-97198-3 ISBN 978-3-031-97199-0 (eBook)
https://doi.org/10.1007/978-3-031-97199-0

This Springer imprint is published by the registered company Springer Nature Switzerland AG
The registered company address is: Gewerbestrasse 11, 6330 Cham, Switzerland

Dedicated to My Beloved Daughter and Parents

Preface

The field of Functionalized Magnetic Nanomaterials (FMNs) has witnessed remarkable advancements in recent years, offering groundbreaking solutions in biomedicine, industry, and environmental sustainability. These nanomaterials possess distinctive magnetic, catalytic, and biocompatible properties, making them invaluable for applications such as targeted drug delivery, non-invasive diagnostics, water purification, renewable energy technologies, and industrial waste management. Their versatility and multifunctionality have led to rapid integration into cutting-edge scientific and technological developments, addressing critical global challenges.

This book, Functionalized Magnetic Nanomaterials (FMNs): Advanced Biomedical, Industrial, and Technological Applications, builds upon the foundational principles covered in Volume 1, extending the discussion to advanced applications, novel synthesis strategies, and real-world implementations of FMNs. The 13 chapters in this volume explore how FMNs have transformed biomedical research, particularly in drug delivery, hyperthermia-based cancer therapy, biosensing, and medical imaging. Their ability to enhance imaging contrast, improve therapeutic efficacy, and enable precise disease targeting makes them invaluable in modern healthcare. Beyond biomedicine, FMNs have also made a significant impact in environmental and industrial applications. Their role in water purification, industrial waste management, and pollution control is particularly noteworthy, as their surface functionalization enables them to adsorb, degrade, or catalytically transform contaminants, making them highly effective in sustainable environmental remediation.

The book also focuses on the growing importance of FMNs in renewable energy technologies, where they contribute to advancements in fuel cells, supercapacitors, and electrocatalysis, thereby enhancing energy efficiency and storage capabilities. The final sections of this book discuss safety regulations, environmental impact, and recent studies showcasing the successful deployment of FMNs in various industries. The book concludes with a forward-looking perspective on emerging trends, interdisciplinary research opportunities, and the future of FMN-based technologies.

With a multidisciplinary approach, this book serves as a valuable resource for scientists, engineers, industrialists, healthcare professionals, and policymakers working on advanced nanomaterial applications. By bridging the gap between fundamental research and practical implementation, it aims to inspire new innovations and drive future breakthroughs in nanotechnology, materials science, and applied engineering. This volume is designed to be a valuable resource for researchers, academicians, industry professionals, and students engaged in interdisciplinary studies involving nanotechnology, materials science, biomedical engineering, environmental chemistry, and sustainable energy solutions.

We hope that this book serves as an informative guide and a source of inspiration for further exploration and innovation in this rapidly evolving domain.

New Delhi, India Dr. Shikha Gulati

Acknowledgments

First and foremost, I am profoundly grateful for the blessings, guidance, and strength that have supported me throughout this journey. Without this unwavering source of inspiration, the successful completion of this book would not have been possible.

I would like to express my sincere appreciation to Prof. C. Sheela Reddy, former principal of Sri Venkateswara College, for her invaluable mentorship and encouragement during my time at the institution. Her leadership and vision have greatly influenced my academic and professional growth, and I am deeply thankful for her support.

I am also immensely grateful to Prof. V. Ravi, Principal of Sri Venkateswara College, University of Delhi, for fostering a research-driven and intellectually stimulating environment. His dedication to academic excellence and encouragement of scholarly pursuits have been instrumental in shaping my work and facilitating the completion of this volume.

My heartfelt thanks go to my parents, my daughter, and my dear friends, whose unwavering love, patience, and encouragement have been my pillars of strength. Their support has been essential in allowing me to devote myself to this endeavor with dedication and perseverance.

This book is the result of the collective expertise and efforts of numerous distinguished scholars and researchers. I extend my deepest gratitude to all the contributing authors, whose knowledge, commitment, and scholarly contributions have been invaluable in shaping the content of this volume. Their dedication has significantly enriched the book, making it a valuable resource for the field of functionalized magnetic nanomaterials.

I also wish to acknowledge the assistance and support of many individuals who have directly or indirectly contributed to the completion of this book. Their thoughtful insights, constructive suggestions, and continuous encouragement have been truly appreciated throughout the process.

A special note of appreciation is due to the Springer team, whose professionalism, guidance, and meticulous attention to detail have been crucial in ensuring the smooth execution of this project. Their support has been instrumental in bringing this book to publication.

Finally, I extend my sincere gratitude to Springer Nature for the opportunity to publish and share this work. Their commitment to academic publishing and global dissemination of scientific knowledge is commendable, and I am honored to be a part of their scholarly endeavors.

This book represents a shared commitment to advancing knowledge, fostering collaboration, and pushing the boundaries of scientific research. I hope it serves as a valuable reference for researchers, academicians, and industry professionals engaged in the study and application of functionalized magnetic nanomaterials.

Dr. Shikha Gulati

About This Book

This volume focuses on the advanced applications of Functionalized Magnetic Nanomaterials (FMNs) across biomedical, industrial, and technological domains, showcasing their versatility and transformative impact. The book presents cutting-edge research and innovative approaches in fields such as biomedicine, waste management, and energy storage, emphasizing FMNs' unique properties and potential for interdisciplinary applications.

The book begins with an overview of FMNs in advanced applications, highlighting their role in revolutionizing biomedicine and industry. Subsequent chapters explore their diverse biomedical uses, including targeted drug delivery systems, diagnostic imaging, and biosensing. These sections provide an in-depth understanding of FMNs' role in enhancing therapeutic strategies, improving imaging accuracy, and enabling early disease detection.

In the industrial sector, FMNs are discussed in the context of water purification, industrial waste management, and renewable energy applications. The book details their efficacy in removing contaminants from water, addressing hazardous waste challenges, and optimizing energy storage solutions in batteries and supercapacitors. Additionally, FMNs' applications in electrocatalysis and fuel cells are explored, demonstrating their contribution to advancing clean energy technologies.

The volume also investigates the integration of FMNs into nanocomposites, where they enhance material properties for industrial and biomedical uses. A dedicated chapter examines the synergistic effects of FMNs with quantum dots, opening new frontiers in imaging, electronics, and energy applications. Furthermore, regulatory and safety aspects are addressed, providing insights into environmental and health considerations surrounding FMNs' usage.

The final chapter provides a comprehensive outlook on future research directions, emphasizing FMNs' potential to reshape various scientific and technological fields.

This volume serves as an essential resource for researchers, academicians, and industry professionals, providing a thorough exploration of FMNs' advanced applications and paving the way for future innovations in nanotechnology.

Contents

1 **Introduction to Advanced Applications of Functionalized Magnetic Nanomaterials (FMNs)** 1
Shikha Gulati, Mahalaxmi Chitranjan, Rashmi Khohal, and Rakshita Yadav

2 **Biomedical Applications of Functionalized Magnetic Nanomaterials (FMNs)** 39
Murali Mohan Achari Kamsali and Vasantha Thoti

3 **Functionalized Magnetic Nanomaterials (FMNs) in Drug Delivery Systems** 59
Jyoti Jain, Mahipal Bishnoi, and Deepak Sharma

4 **Magnetic Hyperthermia Using Magnetic Nanoparticles: A Promising Strategy in Cancer Treatment** 79
Shefali Shukla, Sreya Jayan, P. K. Hasna, and Shikha Gulati

5 **Diagnostic Imaging with Functionalized Magnetic Nanomaterials (FMNs)** 105
Pooja, Aman Maurya, and Laishram Saya

6 **Biosensing with Functionalized Magnetic Nanomaterials (FMNs)** 119
Vivek Chintada and Narasimha Golla

7 **Functionalized Magnetic Nanomaterials (FMNs) in Water Treatment and Purification** 137
Mohd Tahoor, Kritika Chaturvedi, and Laishram Saya

8 **Industrial Waste Management Using Functionalized Magnetic Nanomaterials (FMNs)** 159
Vasantha Thoti and Murali Mohan Achari Kamsali

9 Functionalized Magnetic Nanomaterials (FMNs) in Renewable Energy Applications 175
Parul Pant, Prerita Bahl, and Himanshi Agrawal

10 Nanocomposites Incorporating Functionalized Magnetic Nanomaterials (FMNs) 199
Parul, Vivek Das, and Kriti Bhandari

11 Quantum Dots and Functionalized Magnetic Nanomaterials (FMNs): Synergistic Effects in Advanced Technologies 219
Roopali Sharma, Sakshi Singh, and Jaya Tuteja

12 Regulatory and Safety Aspects of Functionalized Magnetic Nanomaterials (FMNs) 249
Sini Kuriakose and Yukti Monga

13 Conclusions and Future Outlook of Functionalized Magnetic Nanomaterials (FMNs) in Advanced Applications 265
Rakshita Yadav and Shikha Gulati

Index 289

About the Editor

Dr. Shikha Gulati (M.Sc., Ph.D.) is a distinguished Assistant Professor of Chemistry at Sri Venkateswara College, University of Delhi. With a strong dedication to academia and research, she has made remarkable contributions to the fields of Inorganic Chemistry, Green Chemistry, Environmental Remediation, Nanomaterials, Catalysis, and the integration of Artificial Intelligence in Chemistry.

Her research primarily focuses on the development of advanced nanomaterials for catalytic applications and the implementation of AI-driven approaches in chemical sciences. She is particularly committed to designing sustainable and eco-friendly solutions for environmental challenges, such as wastewater treatment using nanotechnology and promoting green synthetic methodologies.

Dr. Gulati has an extensive publication record, with numerous research articles in esteemed international journals, showcasing her commitment to pioneering scientific progress. She has also authored and contributed to several academic books, and her work is frequently referenced in undergraduate and postgraduate curricula, underlining its academic impact.

Beyond her research endeavors, she is deeply involved in mentoring young scholars and guiding students in innovative scientific projects, fostering a research-driven learning environment. She has played a key role in organizing national and international conferences, seminars, and workshops, encouraging interdisciplinary collaboration among researchers, academicians, and industry experts.

With her expertise in chemistry, nanotechnology, artificial intelligence, and sustainable environmental solutions, Dr. Gulati has meticulously curated this book to provide a well-structured and insightful exploration of contemporary advancements. Her editorial expertise ensures that this book serves as a valuable resource for students, researchers, and professionals in the field.

Abbreviations

3-D	3-Dimensional
AA	Amino Acid
AD	Alzheimer's Disease
AES	Auger Electron Spectroscopy
AI	Artificial Intelligence
AMF	Alternating Magnetic Field
AMLs	Antibody-Conjugated Magnetoliposomes
AOP	Advanced Oxidation Processes
APBA	*m*-Aminophenylboronic Acid
APS	Ammonium Persulfate
APTES	Aminopropyl Triethoxysilane
BBB	Blood-Brain Barrier
BNNS	Boron Nitride NanoSheets
BPA	Bisphenol A
BSA	Bovine Serum Albumin
CAGR	Compound Annual Growth Rate
CAs	Contrast Agents
CDPA	Cyclodextrin Grafted on Poly(isophthalamide)
CFO-BFO	Cobalt Ferrite-Bismuth Ferrite
CIP	Ciprofloxacin
CKD	Chronic Kidney Disease
CMNTs	Carbon Dot- and Magnetite-Modified Magnetic Carbon Nanotubes
CNS	Central Nervous System
CNTs	Carbon Nanotubes
CP	Calcium Phosphate
CR	Congo Red
CT	Computed Tomography
CTAB	Cetyltrimethylammonium Bromide
CVD	Chemical Vapor Deposition
DC	Dendritic Cells

DDT	Dichlorodiphenyltrichloroethane
DEX	Dexamethasone
DL	Detection Limit
DOPA	Dopamine
DOX	Doxorubicin
DSSCs	Dye-Sensitized Solar Cells
DTPA	Diethylenetriamine Pentaacetic Acid
DTX	Docetaxel
EBRT	External Beam Radiation Therapy
ECM	Extracellular Matrix
EDLCs	Electrochemical Double-Layer Capacitors
EDS	Energy-Dispersive Spectroscopy
EDX	Energy-Dispersive X-Ray Spectroscopy
EDXRF	Energy-Dispersive X-Ray Fluorescence
EEI	Electrode Electrolyte Interface
EGF	Epidermal-Derived Growth Factor
ELIME	Enzyme-Linked Immuno-Magnetic Electrochemical Method
EMF	Electromotive Force
FAAS	Flame Atomic Absorption Spectroscopy
FCV	Feline Calicivirus
$Fe_2O_3@SiO_2$	Silica-Coated Fe_2O_3 Nanoparticles
$Fe_3O_4@TiO_2$	Fe_3O_4 Coated with TiO_2
FESEM	Field Emission Scanning Electron Microscopy
FMFN	Fullerene Nanocomposite
FMNs	Functionalized Magnetic Nanomaterials
f-MWCNTs	functionalized MWCNTs
FTIR	Fourier Transform InfraRed spectroscopy
GEM	GemcitabineGO—Graphene Oxide
GO	Graphene Oxide
GQDs	Graphene Quantum Dots
HER	Hydrogen Evolution Reaction
HER2 Receptor	Human Epidermal Growth Factor 2 Receptor
HGMS	High Gradient Magnetic Separation
HOR	Hydrogen Oxidation Reaction
HPLC	High-Performance Liquid Chromatography
HSA	Human Serum Albumin
HSAB	Hard and Soft Acids and Bases
HT	Hyperthermia
IARC	International Agency for Research on Cancer
IDA	Iron Deficiency Anemia
IMRT	Intensity-Modulated Radiation Therapy
IMSiNTs	Improved Mesoporous Silica Nanotubes
IONPs	Iron Oxide Nanoparticles
IO-OA/PLGA	Iron Oxide-Oleic Acid/Poly(lactic-co-glycolic acid)
LBL	Layer By Layer

LDA	Layered Double Hydroxide
LHT	Local Hyperthermia
LMB	LeucoMethylene Blue
LOD	Limits of Detection
MAGO	Magnetic Amine-Functionalized Graphene Oxide
MB	Methylene Blue
MBs	Micro-Bundle Structure
MCPMs	Magnetic Crosslinked Copolymer Microspheres
MCs	Magnetic Clusters
MDSC	Myeloid-Derived Suppressor Cells
MePEG	Methoxy Poly(ethylene glycol)
MGO	Magnetic Graphene Oxide
MHN	Magnetic Hybrid Nanomaterials
MHT	Magnetic Hyperthermia Therapy
MIONCs	Magnetic Iron Oxide Nanocomposites
MIONs	Magnetic Iron Oxide Nanoparticles
miRNA	micro Ribonucleic acid
ML	Machine Learning
MLDH	Magnesium–Aluminum-Layered Double Hydroxide
MMOF	Magnetic Copper-Based Metal Framework
MNMs	Magnetic Nanomaterials
MNP	Magnetic Nanoparticle
MOF	Metal-Organic Framework
MONPs	Magnetic Oxide Nanoparticles
MPI	Magnetic Particle Imaging
MPNC	Magnetic Polymer Nanocomposite
MRA	MR Angiography
MRAM	Magnetoresistive Random-Access Memory
MRI	Magnetic Resonance Imaging
MS	Multiple Sclerosis
MSNPs	Mesoporous Silica Nanoparticles
MTT	3-(4,5-Dimethylthiazol-2-yl)-2,5-diphenyltetrazolium Bromide
MTX	Methotrexate
MWCNT/Pe	Multiwalled Carbon Nanotubes/Pectin
MWCNTs	Multiwalled Carbon Nanotubes
NDDs	Novel Drug Delivery System
NMR	Nuclear Magnetic Resonance
NP	Nanoparticle
NPs	Nanoparticles
NW	Nanowires
OG	Orange G
OH	Hydroxyl Radicals
ORR	Oxygen Reduction Reaction
PA	Photoacoustic Imaging
PAA	Polyacrylic Acid

PAAH	Poly-Allylamine-Hydrochloride
PAAMA	Co-Polymer made of Equimolar Acrylic and Maleic Acid Monomers
PAHs	Polycyclic Aromatic Hydrocarbons
PAI	Photoacoustic Imaging
PANI	Polyaniline
PCBs	Poly Chlorinated Biphenyls
PCL	Poly(ε-caprolactone)
pDA	Plasmid Vector DNA
PDA	Poly(dopamine)
PEA	2-phenylethylamine
PEG	Polyethylene Glycol
PEI	Polyethyleneimine
PELA	Copolymer of poly(ethylene glycol) and poly(d,l-lactide)
PEMFC	Proton Exchange Membrane Fuel Cells
PEO-b-Pγ MPS	Poly(ethylene oxide)-block-poly(γ-methacryloxypropyltrimethoxysilane)
PET	Positron Emission Tomography
PFAS	Per- and Polyfluoroalkyl Substances
PFCs	Perfluorinated Compounds
PGA	Poly(glycolic acid)
PHMB	Polyhexamethylene Biguanide Hydrochloride
PIONs	Porous Iron Oxide Nano Agents
PLA	Poly(lactic acid)
PLGA	Poly(lactic-co-glycolic acid)
PMMA	Polymethyl Methacrylate
PNC	Polymer Nanocomposites
PNS	Peripheral Nervous System
PSA	Poly(Styreneacrylic) Acid
PTT	Photothermal Therapy
PTX	Paclitaxel
PVA	Polyvinyl Alcohol
PZT	Lead Zirconium Titanate
QD	Quantum Dots
QDSSCs	Quantum Dot Sensitized Solar Cells
RF	Radio Frequency
rGO	Reduced Graphene Oxide
rGO–S/Co_3O_4	Sulphur-rGO@cobalt Oxide Composites
RhB	Rhodamine B
RHT	Regional Hyperthermia
ROS	Reactive Oxygen Species
RT	Radiotherapy
SAR	Specific Absorption Rate
SAXS	Small- Angle X- Ray Scattering
SBRT	Stereotactic Body Radiotherapy

SDS	Sodium Dodecyl Sulfate
SEM	Scanning Electron Microscopy
siRNA	Small Interfering Ribonucleic Acid
SLN	Sentinel Lymph Node
SO^4	Sulphate Radicals
SPECT	Single-Photon Emission Computed Tomography
SPIO	Superparamagnetic Iron Oxide
SPIONs	Superparamagnetic Iron Oxide Nanoparticles
SRS	Stereotactic Radiosurgery
SSNPs	Solid Silica Nanoparticles
SWCNT	Single-Walled Carbon Nanotubes
SWCNTs	Single-Walled CNTs
TC	Tetracycline
TCP	Tricalcium Phosphate
TE	Thermoelectric
TEM	Transmission Electron Microscopy
TGA	Thermogravimetric Analysis
TiO_2	Titanium Dioxide
TOF	Turnover Frequency
TSS	Total Suspended Solids
VOC	Volatile Organic Compound
VOCs	Volatile Organic Compounds
WBH	Whole Body Hyperthermia
WHPCO	Wet hydrogen Peroxide Catalytic Oxidation
XRD	X-Ray Diffraction
ZIF	Zeolitic Imidazolate Frameworks

Chapter 1
Introduction to Advanced Applications of Functionalized Magnetic Nanomaterials (FMNs)

Shikha Gulati, Mahalaxmi Chitranjan, Rashmi Khohal, and Rakshita Yadav

Abstract This chapter provides an introductory overview of the advanced applications of functionalized magnetic nanomaterials (FMNs) in biomedical, industrial, and technological domains. FMNs have emerged as versatile and innovative materials, offering unique magnetic properties, high surface area, and tunable functionalities that enable their use across diverse fields. In biomedicine, FMNs are pivotal in diagnostics, drug delivery, and hyperthermia therapy, addressing critical healthcare challenges. Industrially, FMNs are driving advancements in catalysis, water purification, and environmental remediation, contributing to sustainable and efficient processes. The chapter emphasizes the interdisciplinary nature of FMNs, highlighting their integration into the health and technology sectors. It sets the foundation for the subsequent chapters by exploring current trends, challenges, and future prospects, offering insights into the transformative potential of FMNs in shaping modern science and industry.

Keywords Magnetic · Nanomaterials · Functionalized · Biomedical · Industrial · Technological

1.1 Introduction

Functionalized magnetic nanomaterials (FMNs) represent a significant advancement in nanotechnology, offering versatile applications across various scientific and industrial fields [1]. These materials are composed of magnetic nanoparticles, typically iron oxides like magnetite (Fe_3O_4) or maghemite (γ-Fe_2O_3), that are surface-functionalized with different organic or inorganic coatings to enhance their stability,

S. Gulati (✉)
Department of Chemistry, Sri Venkateswara College, University of Delhi, Delhi, India
e-mail: shikhagulati@svc.ac.in

M. Chitranjan · R. Khohal · R. Yadav
Department of Biochemistry, Sri Venkateswara College, University of Delhi, Delhi, India

S. Gulati (ed.), *Functionalized Magnetic Nanomaterials*, Nanostructure Science and Technology, https://doi.org/10.1007/978-3-031-97199-0_1

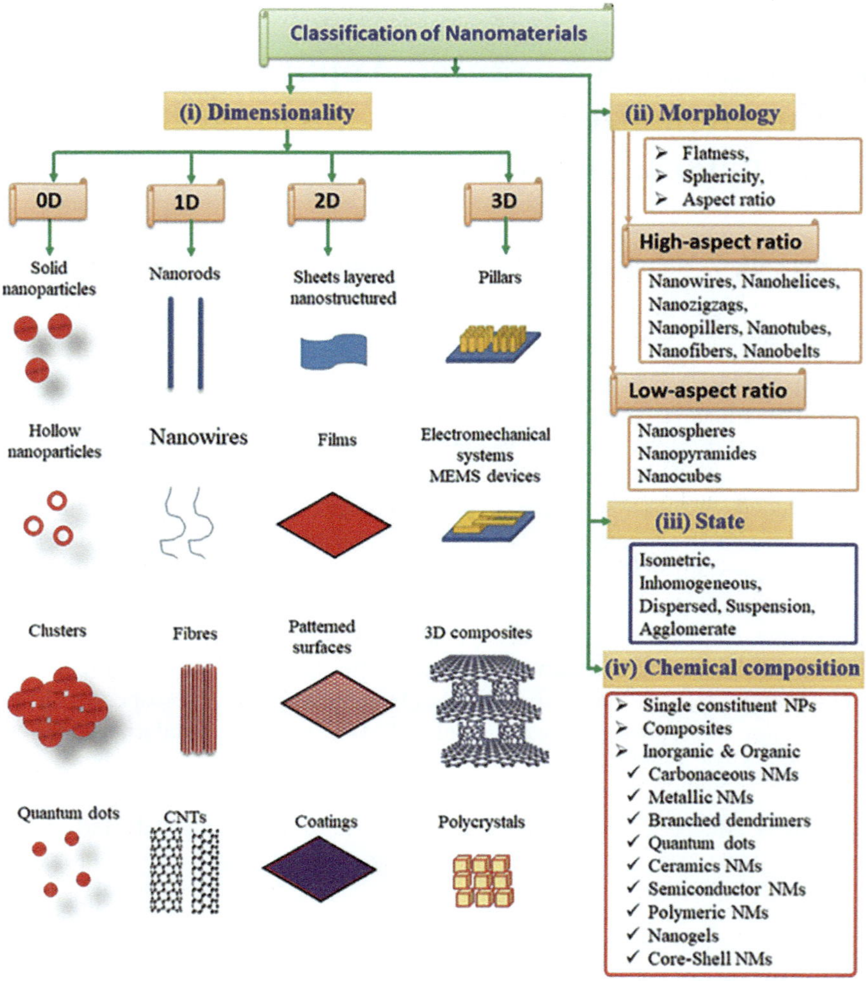

Fig. 1.1 Schematic representation of functionalized magnetic nanomaterials (FMNs) Reproduced with permission from Ref. [6]

dispersibility, and specificity for targeted applications (Fig. 1.1) [2, 3]. The functionalization process imparts unique properties that allow FMNs to be tailored for specific uses, ranging from environmental remediation to biomedical applications and catalysis. One of the most prominent areas where FMNs are applied is environmental remediation. Their high surface area, magnetic responsiveness, and functionalized surfaces make them excellent candidates for removing pollutants, such as heavy metals and organic contaminants, from wastewater [4]. These materials can be easily separated from the environment using external magnetic fields, reducing secondary pollution and improving efficiency [5]. Similarly, FMNs have shown exceptional potential in catalysis, particularly in organic synthesis and photocatalytic degradation of pollutants. Their ability to act as recyclable catalysts with high efficiency has attracted significant research interest.

Functionalized magnetic nanomaterials (FMNs) have gained significant traction in industrial and technological applications due to their enhanced catalytic properties, reusability, and ease of recovery. In the chemical and pharmaceutical industries, FMNs serve as efficient catalysts for large-scale organic transformations, including hydrogenation, oxidation, and cross-coupling reactions, reducing the need for expensive and toxic metal catalysts [7]. Acid- and base-functionalized FMNs are widely employed in petrochemical refining and fine chemical synthesis, where they facilitate selective reactions while minimizing byproduct formation [8]. Additionally, in the energy sector, FMNs play a crucial role in hydrogen production and fuel cell technologies. Their application in electrocatalysis for hydrogen evolution reactions (HER) has led to the development of cost-effective and efficient hydrogen fuel generation, supporting the transition toward renewable energy sources [9]. The ability to magnetically separate and recycle FMNs further enhances their sustainability, making them a preferred choice for various catalytic processes in industrial settings.

Beyond catalysis, FMNs are also revolutionizing environmental remediation technologies and sensor development. Photocatalytic FMNs, especially those coupled with semiconductor materials, are extensively used in wastewater treatment plants to degrade organic pollutants, including hazardous dyes, pharmaceuticals, and industrial effluents, under visible light irradiation [10]. Their high photocatalytic efficiency enables large-scale deployment in air and water purification systems, contributing to sustainable environmental management. In addition, FMNs are utilized in advanced sensing technologies, including gas sensors, biosensors, and electrochemical sensors, for real-time detection of environmental pollutants, pathogens, and biomolecules [11]. These materials are also integrated into smart coatings, self-cleaning surfaces, and antimicrobial applications, further expanding their technological significance. The multifunctional capabilities of FMNs, combined with their magnetic properties, have positioned them as key materials for cutting-edge advancements in energy, healthcare, and environmental sustainability, driving innovation across various industries.

Biomedical applications have been drastically impacted by the advancements in magnetic nanoparticle (MNP) functionalization. MNPs' low toxicity and flexible, modular design make them ideal for a wide range of technical applications in applied nanoscience. Magnetic fields can be used to alter a class of nanoparticles known as MNPs. These MNPs typically consist of two parts: functional chemical components and magnetic elements, including iron, nickel, and cobalt. Because it influences the physicochemical characteristics, stability, biocompatibility, efficiency, pharmacokinetics, separation, recovery, recyclability, and mobility in aqueous media, researchers are focusing on the precision of MNP synthesis and surface functionalization. In biomedical applications, FMNs are widely used for drug delivery, imaging, and hyperthermia therapy (Fig. 1.2) [12]. Their biocompatibility and ability to be functionalized with biomolecules allow for targeted delivery of therapeutic agents, enhancing treatment efficacy while minimizing side effects. Additionally, FMNs

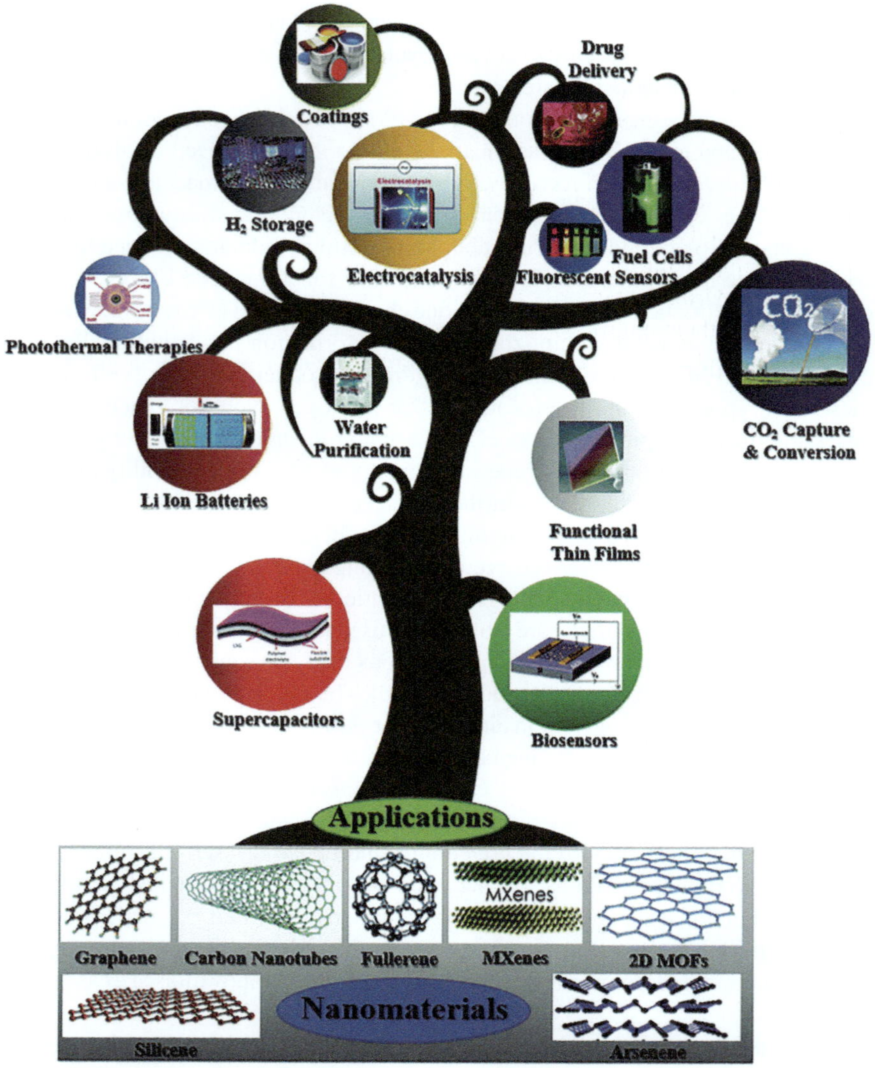

Fig. 1.2 Advanced biomedical, industrial, and technological applications of functionalized magnetic nanomaterials (FMNs). Reproduced with permission from [14]

have been utilized in biosensing applications, enabling rapid and sensitive detection of biomolecules for diagnostic purposes [13].

The following benefits of MNPs show up as a lot of fascinating prospects in the biomedical field:

1. By adjusting procedure parameters to the study's interests, the MNPs' size and shape can be managed.

2. Because MNPs can capture the signal of the magnetic moment of protons surrounding them, they are essential contrast enhancement agents in magnetic resonance imaging (MRI).
3. An external magnetic field can be used to manipulate MNPs, which offers significant benefits for a range of applications.

Despite their promising applications, challenges remain in the large-scale synthesis, stability, and potential toxicity of FMNs. Researchers are continuously working on improving their functionalization techniques to enhance biocompatibility and reduce environmental impact. Future developments in FMNs will likely focus on improving their multifunctionality, integrating artificial intelligence for smart applications, and optimizing their performance for sustainable and industrial applications [15]. These advancements will drive FMNs further into real-world applications, making them indispensable tools in science and technology.

1.2 Fundamental Properties of FMNs

Functionalized magnetic nanomaterials (FMNs) exhibit unique magnetic properties that make them highly valuable in various applications, including catalysis, environmental remediation, drug delivery, and biomedical imaging. Their behavior is primarily governed by size, composition, surface functionalization, and external magnetic field interactions [16, 17]. The fundamental magnetic characteristics of FMNs arise from their nanoscale dimensions, leading to several unique phenomena such as superparamagnetism, ferromagnetism, and antiferromagnetism (Fig. 1.3) [18]. These properties are determined by the intrinsic nature of the material as well as external factors, such as temperature and applied magnetic fields [19]. Table 1.1 compares the properties, examples, and uses of superparamagnetic, ferromagnetic, and antiferromagnetic materials in various fields.

1.2.1 Superparamagnetism

Superparamagnetism is a crucial characteristic of FMNs, particularly for nanoparticles smaller than the single-domain critical size. In this state, the nanoparticles exhibit magnetization only in the presence of an external magnetic field and lose magnetization once the field is removed. This property is highly desirable in biomedical applications, such as targeted drug delivery and magnetic resonance imaging (MRI), as it prevents unwanted aggregation of nanoparticles due to remanent magnetization [18, 19].

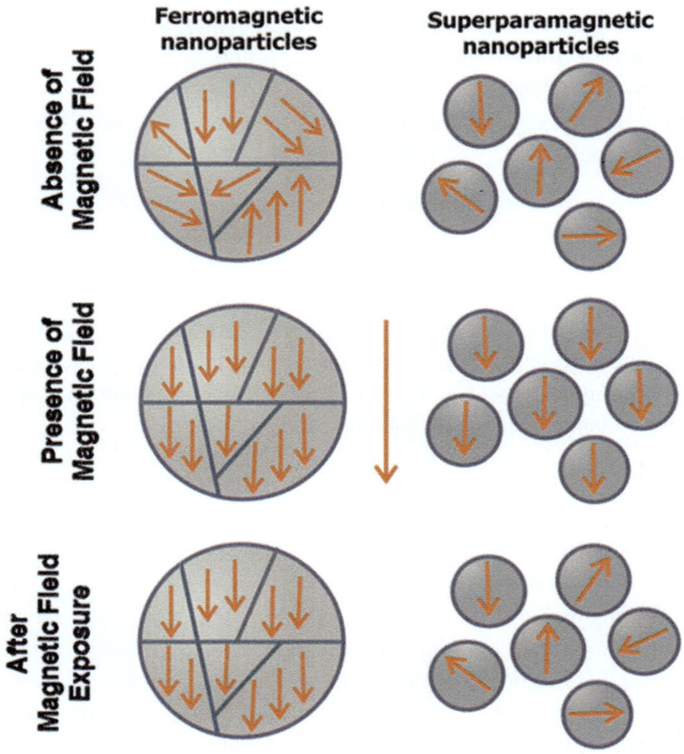

Fig. 1.3 Illustration showing superparamagnetic, ferromagnetic, and antiferromagnetic behaviors with external field interactions. Reproduced with permission from [27]

Table 1.1 Comparison of magnetic properties of FMNs

Property	Super paramagnetism	Ferro magnetism	Anti ferromagnetism	Reference
Retains magnetization after field removal	No	Yes	No	
Typical material examples	Fe3O4, γ-Fe2O3	Fe, Co, Ni	MnO, FeO	[21, 22]
Application areas	Drug delivery, MRI	Catalysis, data storage	Spintronics, magnetic sensors	[23–26]

1.2.2 Ferromagnetism and Antiferromagnetism

FMNs composed of iron, cobalt, or nickel exhibit ferromagnetic behavior, meaning they maintain spontaneous magnetization even in the absence of an external magnetic field. However, at the nanoscale, these materials may transition to superparamagnetic behavior due to thermal fluctuations. In contrast, antiferromagnetic materials exhibit opposing spin orientations, canceling out the overall magnetic moment. These properties influence the performance of FMNs in catalysis and spintronic applications [18].

1.2.3 Magnetic Anisotropy

Magnetic anisotropy refers to the directional dependence of a material's magnetization. In FMNs, shape anisotropy, surface anisotropy, and crystalline anisotropy significantly impact their magnetic behavior. Controlling anisotropy allows precise tuning of FMNs for applications such as high-density data storage and enhanced catalytic efficiency [20].

1.3 Significance of Magnetic Characteristics in FMNs

1. Enhanced Separation Efficiency

The ability of FMNs to be manipulated using external magnetic fields enables rapid and efficient separation processes in wastewater treatment, heterogeneous catalysis, and biomedical applications. Magnetic separation reduces processing times and enhances material recovery [28].

2. Improved Biocompatibility and Drug Delivery

Superparamagnetic FMNs, particularly those functionalized with biocompatible coatings, enhance targeted drug delivery by directing therapeutic agents to specific sites using external magnets [29]. This minimizes side effects and improves treatment efficacy in diseases such as cancer [22].

3. Augmented Catalytic Performance

Magnetic nanoparticles serve as highly active catalytic centers in reactions like Fenton-like degradation, hydrogen evolution reactions, and organic transformations. Their high surface-to-volume ratio and ease of magnetic recovery make them efficient and reusable catalytic agents [30].

4. Advanced Imaging and Diagnostics

FMNs play a pivotal role in MRI contrast enhancement and hyperthermia treatments. Their magnetic properties facilitate non-invasive imaging, providing improved resolution and diagnostic accuracy in biomedical applications [31].

In conclusion, the fundamental magnetic characteristics of FMNs dictate their performance across multiple fields. Understanding and tailoring these properties enable the design of advanced materials with improved functionality, making FMNs indispensable in modern scientific and industrial applications.

1.4 Functionalization Strategies: Tailoring FMNs for Specific Applications

Functionalization strategies play a crucial role in optimizing the performance of FMNs for targeted applications (Fig. 1.4). By modifying their surface chemistry, morphology, and composition, researchers can enhance their stability, selectivity, and reusability in diverse fields such as environmental remediation, biomedical applications, and catalysis.

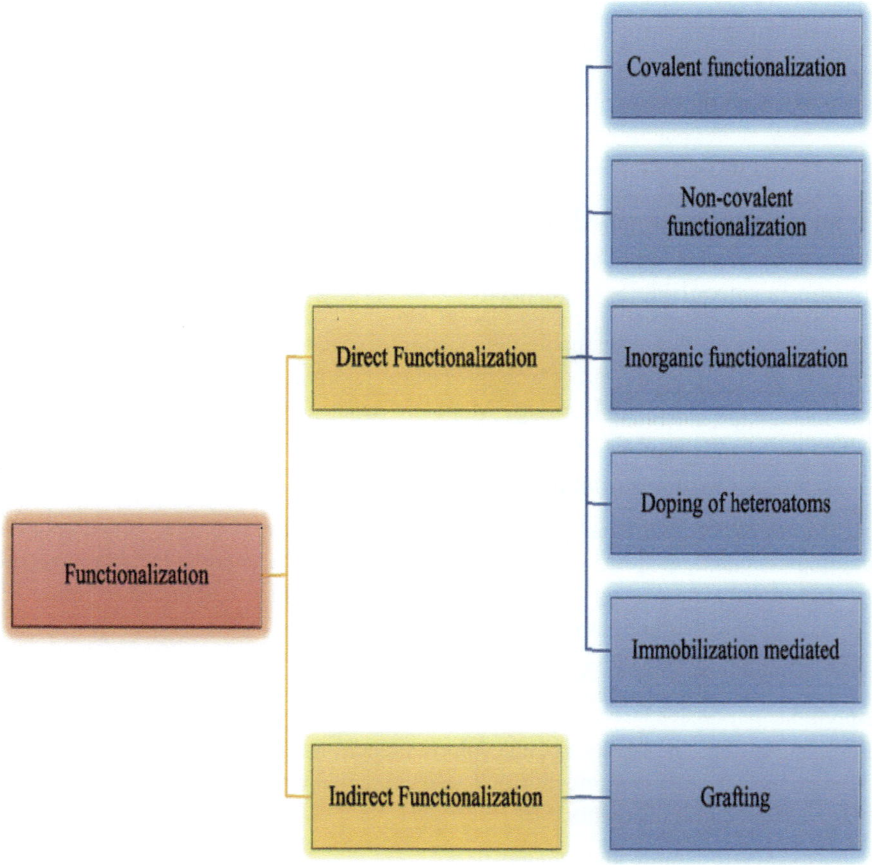

Fig. 1.4 Illustration of various functionalization strategies utilized. Reproduced with permission from [32]

1.4.1 Surface Coating and Stabilization

Surface coating of FMNs with biocompatible materials such as polymers, surfactants, or inorganic layers is a crucial functionalization strategy that prevents agglomeration and enhances their dispersibility in aqueous and biological environments [3]. This process improves the stability and efficacy of FMNs, particularly in biomedical, catalytic, and environmental applications. Proper functionalization ensures that FMNs maintain their desired physicochemical properties while interacting with complex biological or chemical systems.

A variety of stabilizing agents are commonly used to functionalize FMNs, each imparting specific advantages:

1. Polyethylene glycol (PEG): PEGylation enhances the biocompatibility of FMNs, reduces immunogenicity, and extends circulation time in biological systems. This makes PEG-coated FMNs highly suitable for drug delivery, bioimaging, and targeted therapy applications [33, 34].
2. Silica (SiO_2): Silica coating provides excellent chemical stability and facilitates the attachment of functional groups, making FMNs more versatile for catalytic and biomedical applications. The presence of a silica shell also reduces toxicity and enhances FMNs' dispersibility in physiological media [35].
3. Dopamine and Catechol Derivatives: These molecules have strong adhesive properties, forming robust coatings on FMNs [17]. Dopamine-functionalized FMNs exhibit enhanced bio-functionality, making them ideal for biosensors, medical diagnostics, and bioactive coatings. Catechol groups, inspired by mussel adhesion mechanisms, improve FMNs' interaction with biological systems and increase their stability in aqueous environments.

By selecting appropriate coating materials and functionalization methods, researchers can tailor FMNs for specific applications, optimizing their performance for drug delivery, environmental remediation, catalysis, and imaging. The choice of functionalization strategy depends on the intended application, desired surface chemistry, and required biocompatibility, ensuring FMNs exhibit enhanced efficiency and stability in real-world applications [36].

1.4.2 Functionalization with Targeting Ligands

In biomedical applications, FMNs are functionalized with specific ligands to ensure targeted delivery. These ligands enhance specificity, improving the efficiency of therapeutic interventions and reducing off-target effects [37].

1. Antibodies and Peptides: These biomolecules aid in the selective binding of FMNs to specific cellular targets, such as cancer cells [37, 38]. By conjugating FMNs with monoclonal antibodies or bioactive peptides, researchers can ensure

precise delivery of therapeutic agents to diseased tissues, minimizing adverse effects on healthy cells.
2. Folic Acid: Folic acid is widely used for targeting tumor cells that overexpress folate receptors [39]. Cancer cells often exhibit higher folate uptake due to increased metabolic demand, making folic acid-functionalized FMNs effective for drug delivery and imaging applications [40, 41].
3. Carbohydrates: Functionalization with carbohydrates facilitates cell recognition and internalization of FMNs [42]. Glycan-coated FMNs interact with specific cellular receptors, promoting uptake in targeted cells and improving biodistribution. This strategy is particularly beneficial in drug delivery, vaccine development, and immune system modulation [43].

By selecting appropriate ligands and functionalization techniques, FMNs can be engineered to optimize their performance for biomedical applications, ensuring increased efficacy, reduced toxicity, and enhanced biocompatibility.

1.4.3 Doping and Elemental Substitution

Modifying functionalized magnetic nanomaterials (FMNs) by doping them with metal ions or substituting elements is a widely used strategy to enhance their catalytic, electronic, and magnetic properties for diverse applications. By carefully selecting dopants or substituted elements, researchers can precisely tailor the physicochemical characteristics of FMNs to meet specific functional requirements. One common approach is doping FMNs with noble metals such as gold (Au) and platinum (Pt), which significantly boosts their catalytic performance. These noble metal-doped FMNs exhibit superior catalytic efficiency due to their high surface energy and enhanced electron transfer capabilities, making them ideal candidates for environmental remediation and energy conversion applications [44]. For instance, Pt-doped FMNs are extensively used in fuel cells, where they improve the oxygen reduction reaction (ORR) and hydrogen evolution reaction (HER), leading to enhanced energy efficiency. Similarly, Au-doped FMNs have demonstrated exceptional catalytic activity in the degradation of organic pollutants, making them highly effective for wastewater treatment and other environmental applications.

Another important modification strategy involves elemental substitution, such as replacing iron (Fe) ions in Fe_3O_4 with elements like manganese (Mn) or zinc (Zn). This approach significantly alters the magnetic, structural, and electronic properties of the material, expanding its applicability in biomedical and sensing technologies. Mn-substituted Fe_3O_4 enhances the magnetic response and provides improved contrast in magnetic resonance imaging (MRI), making it a valuable tool for medical diagnostics. Additionally, Zn substitution in Fe_3O_4 influences the electronic band structure and enhances the material's sensitivity in sensor applications, particularly for gas sensing and biosensing technologies. Such modifications not only optimize the fundamental properties of FMNs but also broaden their scope in various fields, from energy storage to biomedicine. By carefully tuning the composition of FMNs through doping and substitution, researchers can develop multifunctional materials with tailored properties, leading to breakthroughs in catalysis, environmental sustainability, and medical advancements.

1.4.4 Functionalization for Catalysis

FMNs used in catalysis are often functionalized with active catalytic sites to improve reaction efficiency. Table 1.2 provides an overview of functionalization strategies, their applications, and relevant references, offering insights into material modifications for diverse uses. Common strategies include:

1. Anchoring metal nanoparticles: Functionalization of functionalized magnetic nanomaterials (FMNs) plays a critical role in enhancing their catalytic efficiency, stability, and reusability. By introducing active catalytic sites onto the surface of FMNs, their performance in various chemical transformations can be significantly improved [45]. One of the most effective strategies is the anchoring of metal nanoparticles, such as gold (Au), platinum (Pt), palladium (Pd), or transition metals like nickel (Ni) and cobalt (Co). This approach increases the density of active sites, thereby accelerating oxidation and reduction reactions, which are crucial in energy production, environmental remediation, and organic synthesis [46]. The strong interaction between the metal nanoparticles and the magnetic support ensures enhanced stability, preventing catalyst deactivation over time.
2. Functionalization with acidic or basic groups: Another widely used functionalization technique involves the introduction of acidic or basic groups onto FMNs to improve their catalytic selectivity in organic transformations [1]. Acid-functionalized FMNs, often modified with sulfonic ($–SO_3H$) or carboxyl ($–COOH$) groups, are particularly useful in esterification, alkylation, and hydrolysis reactions [47]. Similarly, base-functionalized FMNs, modified with amine ($–NH_2$) or hydroxide ($–OH$) groups, are highly effective in condensation and transesterification reactions, enabling efficient and selective catalysis [48].

Table 1.2 Functionalization strategies and their applications

Functionalization method	Application area	Examples	References
PEG coating	Drug delivery	PEGylated Fe_3O_4 for cancer therapy	[49]
Metal doping	Catalysis	Au-doped Fe_3O_4 for CO oxidation	[50]
Antibody functionalization	Biomedical imaging	MRI contrast enhancement	[51]
Silica coating	Environmental remediation	SiO_2-coated Fe_3O_4 for water purification	[52]

3. Hybridization with carbon-based materials (graphene, CNTs): Additionally, hybridizing FMNs with carbon-based materials, such as graphene and carbon nanotubes (CNTs), further boosts their conductivity, stability, and catalytic performance. These carbonaceous materials provide a high surface area and excellent electron transfer properties, making them ideal for electrocatalysis and photocatalysis applications. Such functionalized FMNs not only enhance catalytic activity but also offer easy magnetic recovery and recyclability, making them highly attractive for industrial and sustainable catalytic processes.

These functionalization strategies allow FMNs to be tailored for precise applications, ensuring enhanced performance and versatility in modern technological and medical advancements.

1.5 Applications of FMNs in Biomedicine

NPs are used in drug delivery, gene therapy, tagging and tracking, hyperthermia, and medical imaging using a variety of modalities as positron emission tomography (PET), magnetic resonance imaging (MRI), and so on. It is to be mentioned that employing these for in vitro or in vivo applications, biomolecular interactions depend on the surface of the nanoparticle being chemically modified. Consequently, a variety of targeting moieties, including peptides, aptamers, and antibodies, have been used to be integrated on the surface of nanoparticles. Due to their intrinsic magnetic properties, great biocompatibility, and minimal toxicity, iron oxide nanoparticles (NPs) are a desirable family of nanostructures that have garnered a lot of interest in the medical field. For biomedical applications, it is crucial to modify the surface of iron oxide nanoparticles using different covering compounds, such as dopamine (DOPA), polyethylene glycol with a thiol end group (thiol-PEG), and polyacrylic acid (PAA) [13].

1.5.1 FMNs in Diagnostics: Magnetic Resonance Imaging (MRI) and Biosensing

MRI is a non-invasive imaging technique widely used for soft tissue visualization due to its high spatial resolution and reliance on non-ionizing radiation. It works by aligning hydrogen protons in a strong external magnetic field, followed by a brief radiofrequency (RF) pulse that excites these protons. When the pulse stops, protons emit RF signals as they return to equilibrium, which MRI scanners detect to generate images [53].

Relaxation Times and Image Contrast:

After excitation, protons undergo relaxation, categorized into two main types:

1. T1 Relaxation (Spin–Lattice): Protons realign with the magnetic field along the longitudinal axis. Tissues with shorter T1 times appear bright on T1-weighted images, useful for anatomical imaging and brain lesion assessment.
2. T2 Relaxation (Spin–Spin): Protons lose phase coherence in the transverse plane. Tissues with longer T2 times appear bright on T2-weighted images, which helps detect pathologies like inflammation or edema [53].

MRI Image Formation and Contrast.

MRI uses a grayscale system:

1. Bright (High Signal Intensity): Short T1 or T2 times.
2. Dark (Low Signal Intensity): Long T1 or T2 times.
3. Gray (Intermediate Signal Intensity): Balanced T1 and T2 times.

Water content influences relaxation times, allowing tissue differentiation in imaging [53].

1. **FunctionalizedMagnetic Nanoparticles (FMNs) in MRI**

FMNs, such as superparamagnetic iron oxide nanoparticles (SPIONs), enhance contrast by altering relaxation times. SPIONs shorten T2 relaxation, making targeted areas appear darker in T2-weighted images. Gadolinium-based FMNs enhance T1 contrast, making tissues brighter in T1-weighted images. Functionalization enables precise targeting of tumors, inflammation, or biomarkers, improving diagnostic accuracy [53].

Magnetic hybrid nanomaterials (MHNs) are magnetic nanoparticles that incorporate functional elements. Because of their biocompatibility and prospective uses as contrast agents or multimodal imaging probes, these MHNs have garnered a lot of attention. MHNs exhibit superparamagnetism as well as novel properties that can be applied to:

1. Positron emission tomography,
2. Computed tomography contrast modalities,
3. Magnetic resonance imaging, and
4. Single-photon emission computed tomography.

They do this by combining magnetic nanomaterials with functional nanocomponents like noble metals or isotopes. Higher sensitivity and spatial resolution than traditional materials are achieved by the synergistic effects of several nanoparticles combined [34].

As MRI contrast agents (CAs), magnetic nanoparticles (MNPs) have been extensively studied to aid in cancer diagnosis, therapy, and detection. Iron oxide nanoparticles (IONPs) are the most studied of these nanomaterials as CAs for MRI because of their magnetic characteristics, specifically their superparamagnetism, which results in extremely high relaxivity, their high biocompatibility because they can be integrated into iron metabolism, and their ease of surface functionalization with target molecules for molecular imaging [31].

Magnetic nanoparticles known as porous iron oxide nano agents (PIONs) can be directed to tumour tissues by a magnetic field and are employed in both in vitro and in vivo photoacoustic imaging (PAI) and magnetic resonance imaging (MRI). They have been paired with a plasmid vector (pDNA) to serve as a platform for diagnostic imaging and photothermal therapy [13].

Gadolinium-based contrast agents, widely used for T1-weighted MRI, enhance imaging by shortening T1 relaxation, producing a brighter signal. However, free Gd is toxic, posing risks such as nephrogenic systemic fibrosis and tissue accumulation, including in the brain, bones, and organs, regardless of kidney function. Concerns over toxicity and distribution have driven the search for safer alternatives, including inorganic nanoparticles (Gd oxide, fluoride, phosphate) [53].

2. **Biosensors**

To meet the growing demand for biosensors across various scientific fields, researchers are exploring nanoscale materials to enhance sensor technology. Opioids, widely used as pain relievers, present both therapeutic benefits and risks, including addiction and overdose. The global rise in opioid misuse necessitates precise detection methods for biological, environmental, and confiscated samples. Integrating advanced nanostructures into biosensors offers a promising solution for rapid and accurate opioid detection [54].

A recent review examines nanobiosensors for opioids, detailing nanomaterials used in these platforms and their detection capabilities. Over the past few decades, nanotechnology has significantly advanced, leading to the development of nanoparticles and nanomaterials that improve biosensor performance as depicted in Table 1.3 [54].

Table 1.3 Various nanomaterials employed in biosensor design, emphasizing their role in enhancing sensitivity and efficiency

Type of nanomaterial	Dimension	Confinement directions	Applications	References
Nanoparticles (solid, hollow) and quantum dots (QDS)	0 D	x, y, z	Biosensors	[55]
Nanowires, nanotubes, and CNTs (carbon nanotubes)	1 D	x, y	Photon devices	[55]
Films, plates, and sheets	2 D	x	Solar cells	[55, 56]
Nanocomposites and polycrystals	3 D	Not specified	Mobile phones	[55]

1.5.2 *Therapeutic Applications: Drug Delivery and Hyperthermia*

Inherent qualities as well as their unique features of MNPs include magnetic properties, biodegradability, biocompatibility, and the possibility for heat therapy. Furthermore, as illustrated in the figure below (Fig. 1.5), MNP applications are offered in biomedical fields that employ therapy-based hyperthermia and drug-targeted administration [12].

NPs are used in drug delivery, gene therapy, tracking and tagging, and hyperthermia. It should be mentioned that whether employing NPs for in-vitro or in-vivo applications, biomolecular interactions depend on the surface of the nanoparticle being chemically modified [13].

Drugs may be contained within the particle or absorbed at the NP surface. Moreover, it has been demonstrated that functionalization prevents NPs from clumping together and renders them biocompatible at different phases of application. The physical, chemical, and mechanical properties of NP are enhanced via functionalization, which produces synergistic effects.

1. **Carbon Nanotubes**

Carbon nanotubes' exceptional qualities, such as their high aspect ratio, improved conductivity, robustness, and biocompatibility, have made them a viable candidate for use as innovative drug and gene delivery vehicles. CNTs are effectively absorbed by a variety of cell types through a number of processes [13, 57]. Doxorubicin (DOX), methotrexate (MTX), docetaxel (DTX), paclitaxel (PTX), gemcitabine

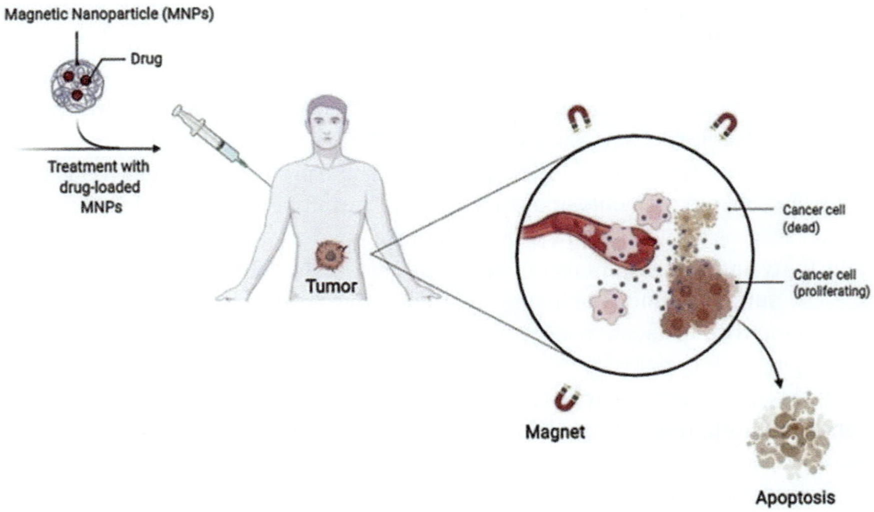

Fig. 1.5 Targeted drug delivery for cancer treatment using magnetic nanoparticles (MNPs) guided by an external magnetic field. Reproduced with permission from [12]

(GEM), anti-inflammatory medications, osteogenic dexamethasone (DEX) steroids, and other anticancer compounds have all been transported by CNTs [13]. Furthermore, CNTs' ease of surface functionalization has encouraged the use of these materials as gene delivery vectors, treating a variety of illnesses, including cancer, by delivering different genes like plasmid DNA (pDNA), microRNA (miRNA), and small interfering RNA (siRNA) [13]. Enhancing CNTs' physical characteristics, such as their solubility and dispersibility, is only one of the main goals of CNT functionalization; another is to increase their biocompatibility. They are more cytotoxic in the body due to their poor dispersibility and aggregation. One effective way to reduce their cytotoxicity and increase their drug delivery effectiveness is through surface functionalization [13]. For example, human serum albumin (HSA) and single-walled carbon nanotubes (SWCNTs) have been coupled to transport PTX intracellularly, improving PTX's anticancer efficacy in vitro. For efficient nucleic acid delivery, SWCNTs can be coupled with lipids and naturally occurring dendrimers based on amino acids. Additionally, siRNA was delivered via functionalized SWNTs for gene-based cancer treatment [13, 58].

The Need for Controlled Medication Delivery

The precise distribution of an agent to the illness site is a significant pharmacological difficulty; the majority of commonly prescribed medications, which are either orally or intravenously are not targeted. Less than 10% of the dose is thought to reach the organ of interest, and even fewer reach cellular targets as a result. To ensure adequate medication concentration at the target region, the most popular remedy is to increase the amount that is administered. This inefficiency can restrict the practical application of promising medicines by causing toxicity and off-target effects. Furthermore, adverse immunological reactions at the administration site may result from non-selective distribution [59].

Therefore, the primary goal of controlled medication delivery is to provide medicinal drugs continuously over an extended period of time. However, pulsatile medication administration is also quite appealing. In theory, zero-order or more complex (such as pulsatile) delivery profiles might be achieved with a suitable scaffold design and time/intensity control of the external field. These profiles could then replicate the physiological requirements of bioactive agents, resulting in optimal drug delivery.

Because of its wide applicability, pharmacology as a whole aims to introduce selectivity into drug administration. Using nanoscale delivery systems with cell-specific surface ligands, including liposomes and polymeric nanoparticles, is one way to improve medication selectivity.

Magnetic drug delivery technologies have the potential to significantly outperform this poor targeting performance. Early researchers imagined that magnetic nanoparticles in tissue would be captured by applied magnetic fields placed around an organism.

For instance, when a solid tumour is exposed to strong magnetic field gradients, bloodborne particles produced by an intravenous injection of magnetic nanoparticles may be trapped or gathered. Permanent magnets placed inside the target tissue or a magnetic device outside the animal could produce such gradients [60] (Fig. 1.6).

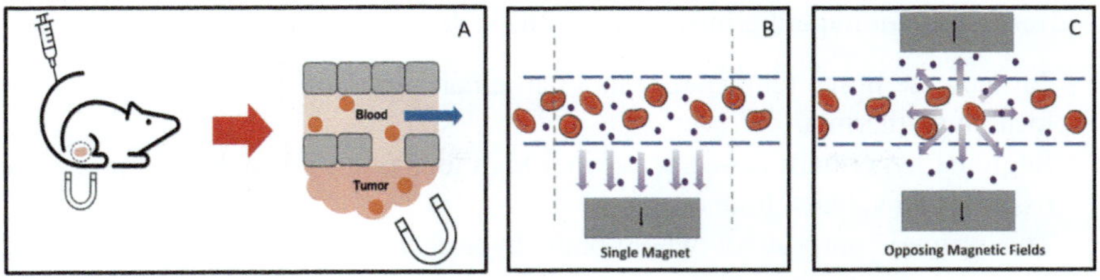

Fig. 1.6 Schematic magnetic drug delivery **a** Injected magnetic nanoparticles accumulate in regions with a strong external field gradient, with orange particles entering the extracellular space. Reproduced with permission from [61]. **b** Single magnets pull particles in one direction, unlike dual magnets, which create a stable radial force. **c** Dual-magnet system enhances drug targeting in deep tissues, overcoming the limitation of single magnets, whose magnetic gradient weakens with distance. Reproduced with permission from [60]

Magnetic Hyperthermia

The concept of hyperthermia refers to an increase in temperature inside the body. Magnetic hyperthermia has synergistic effects that enhance drug delivery systems with modified MNPs. Superparamagnetic nanoparticles in magnetic hyperthermia have a success rate in treating cancer; the death of cancer cells requires heating efficiency. Higher temperatures—up to 41–47 °C—are applied to the tissue during cancer treatment, which can destroy the cancer cells by destroying their DNA. Localized magnetic hyperthermia with reduced MNP levels can be used to treat small or deep-seated cancers in the presence of weak magnetic fields (Fig. 1.7). But the heating efficiency needs to be improved. Therefore, by using their functions, MNPs can improve the therapy of hyperthermia [12].

Fig. 1.7 Mechanisms of hyperthermia therapy using an external magnetic field for cancer treatment. Reproduced with permission from [12]

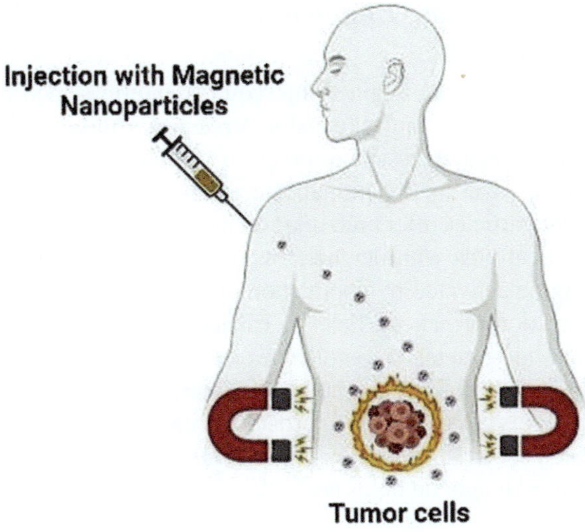

Treatments for hyperthermia fall into three categories:

1. An increase in the temperature in a particular region of the tissue is known as local hyperthermia.
2. Regional hyperthermia occurs when a high temperature is delivered to sizable tissue regions, like a limb or organ.
3. The primary purpose of whole-body hyperthermia is to treat diseases like metastatic cancer that have spread to other parts of the body.

Because of the tissue that needs to be treated, local or regional hyperthermia is the most difficult to perform of all the many forms of hyperthermia. Since the MNPs can be directed towards the cancerous tissue by an external magnetic field and then start hyperthermia therapy using magnetic pulses, magnetic hyperthermia provides a better option for the selective treatment of cancer cells in this regard. This lowers the dosages of other treatments like chemotherapy and radiotherapy [33].

The conversion of magnetic energy, as measured by the specific absorption rate (SAR), which is the rate at which electromagnetic energy is absorbed by biological tissue, is the primary cause of the temperature increase of MNPs. The amplitude, magnetic field frequency, and magnetic relaxation mechanisms of the MNPs—all of which are influenced by their size, shape, and inherent magnetic parameters—determine the SAR. By adopting this method, it is feasible to produce localized heat that can reach temperatures of 42–45 °C, which is a temperature at which cancer cells can be killed without harming healthy cells. Therefore, making sure that MNPs' heating is appropriately regulated when transmitted to the biological environment presents a major problem in hyperthermia therapy. However, this specificity is only possible if MNPs are biofunctionalized with molecules (i.e., coupled to antibodies) that target cancer cells in order to accomplish precise targeting [33].

1.5.3 Tissue Engineering and Regenerative Medicine

Effective tissue engineering requires careful selection of biomaterials, appropriate cell sourcing, and expansion strategies. Additionally, successful tissue or organ replacement necessitates precise spatial and temporal control over biological conditions, ensuring functional integration and long-term viability. Scaffolds based on nanoparticles offer both scaffold movement control and contrast for bioimaging [62].

Scaffolds with magnetic nanoparticles, which are superparamagnetic in nature, can be activated by application of an electric field [63, 64]. Creation and functionalization of nanoparticles is carried out on their surfaces using various polymers, ceramics, metals, biomolecules, etc., depending on the intended use.

Incorporating magnetic nanoparticles alongside an external magnetic field presents a promising approach to enhancing bone tissue regeneration, as it has been shown to impact cellular metabolism. In medical applications, both static and pulsed magnetic fields are widely utilized, not only as integral elements of magnetic resonance techniques but also for accelerating wound healing and promoting bone regeneration [65, 66].

Magnetic nanocomposites play a crucial role in tissue engineering by enhancing scaffold properties for bone regeneration [67]. Natural polymers like alginate and collagen, along with synthetic polymers such as poly(ε-caprolactone) (PCL), are commonly used for scaffold fabrication [62, 68]. However, PCL alone faces certain limitations, like a lack of adequate mechanical strength and cell adhesion properties [69]. To overcome these limitations, inorganic nanofillers like hydroxyapatite and tricalcium phosphate (TCP) are incorporated to improve osteogenic differentiation and mechanical performance [70, 71].

1. Periodontal Tissue Engineering

Dental tissue engineering, particularly PLGA-based composites, aids in tissue regeneration, infection control, drug delivery, cement formation, and alveolar bone protection. Derived from PGA and PLA, PLGA is an aliphatic polyester that offers excellent mechanical properties, tunable degradation, and biocompatibility, making it valuable for periodontal disease treatment [72].

Boguslavsky et al. developed a non-destructive method to graft monodisperse silica nanoparticles (~30 nm) onto polystyrene, polyethylene, and polyvinyl chloride surfaces. The addition of silica nanoparticles increased surface roughness, significantly disrupting biofilm formation. Regardless of the polymer type, bacteria failed to adhere to silica-grafted films, effectively preventing or delaying bacterial growth. Due to their distinctive antibacterial properties and ease of modification, metal nanomaterials have gained recognition in dental tissue engineering. Silver, gold, TiO_2, and ZnO are key antibacterial metals, with their effectiveness further enhanced through functional modifications [73].

2. Bone Tissue Engineering

Composites consisting of calcium phosphate (CP), IONPs, and other biopolymers (such as chitosan, bovine serum albumin, etc.) have been shown to be biocompatible scaffolds that degrade rather slowly. The biological organization of planted stem cells was expedited by composites, leading to the regeneration of bone tissue. In bone tissue engineering, scaffold materials provide mechanical support and create a conducive environment for bone regeneration. Key factors such as porosity, biocompatibility, bioabsorbability, and biological conduction must be considered.

Chitosan has been instrumental in advancing the field, with its porous structures aiding bone conduction. Carbon-based scaffolds serve as templates for bone stem cell growth, adhesion, proliferation, differentiation, and regeneration. Carbon nanotubes enhance scaffold properties due to their high tensile strength, electrical conductivity, and current transmittance [73].

3. Neural Tissue Engineering

Collagen is the only natural biopolymer approved for medical research in peripheral nerve regeneration. As a key component of connective tissue, it is widely present in human tissues, providing structural support. Gelatin, derived from collagen through acidic or alkaline hydrolysis, offers advantages such as cost-effectiveness, high availability, low toxicity, and biodegradability, making it useful in medical imaging, tissue scaffolds, and drug delivery.

Additionally, gelatin enhances cell adhesion and proliferation, serving as an effective implant material for tissue engineering when chemically modified. It can also be combined with PLA and electrospun to support axon growth and motor neuron differentiation. Electrospinning is an electro-hydrodynamic process that uses high voltage to produce fibres and particles from a polymer solution. A basic setup includes a polymer reservoir (often a syringe with a blunt needle), a pump, a high-voltage power source, and a collector. This technique is widely used for manufacturing electrospun fibers through electrostatic forces [74]. Recently, gelatin nanoparticles have been integrated into polymer scaffolds to improve biocompatibility for nerve tissue engineering.

Carbon-based nanomaterials play a crucial role in neural tissue engineering due to their strong interactions with brain tissue and neurons. Materials like graphene (G), fullerenes, and carbon nanotubes show great potential in this field. Research indicates that electrical stimulation enhances neuron regeneration, and G-based materials, with their excellent electrical conductivity, mechanical strength, and flexibility, support neural tissue engineering by promoting neuron differentiation and proliferation.

For example, incorporating functional GO nanosheets with nanofibers modified the physicochemical and biological properties of scaffolds for brain tissue repair, improving neural progenitor cell survival. Carbon nanotubes, resembling neurites in structure, are widely used in neural tissue regeneration [73].

4. Regenerative Medicine

IONP-based nano enzymes for successful tissue engineering and regenerative medicine have been reported by Michelle et al. As seen in Fig. 1.8, they have been found to have high stability, catalytic action, and an effective enzymatic nature that aid in wound healing, bone tissue engineering, and cardio-protection in angiogenesis and heart-related therapies [63].

At the nanoscale, each nanoparticle acts as a magnetic domain, generating micro-movements that affect ion channels and activate mechanotransduction pathways. However, adding magnetic nanoparticles increases scaffold rigidity, impacting cell adhesion.

In order to govern cell proliferation, differentiation, and migration and create functional tissues, nanostructured scaffolds give the cells a more convenient structural support that approximates the original bone architecture [13]. Cells within the scaffold detect compressive or tensile forces from the magnetic field, with piconewton-level forces sufficient to deform the membrane and activate mechanosensitive channels (Fig. 1.9) [76].

Magnetic particle imaging (MPI), a new method for tomographic MION-assisted imaging, which utilizes MIONS probes for in vivo monitoring of tissue regeneration. Dynamic control over delivery improves the precision and safety of biological agents, with external magnetic stimulation enabling regulated release. Real-time monitoring

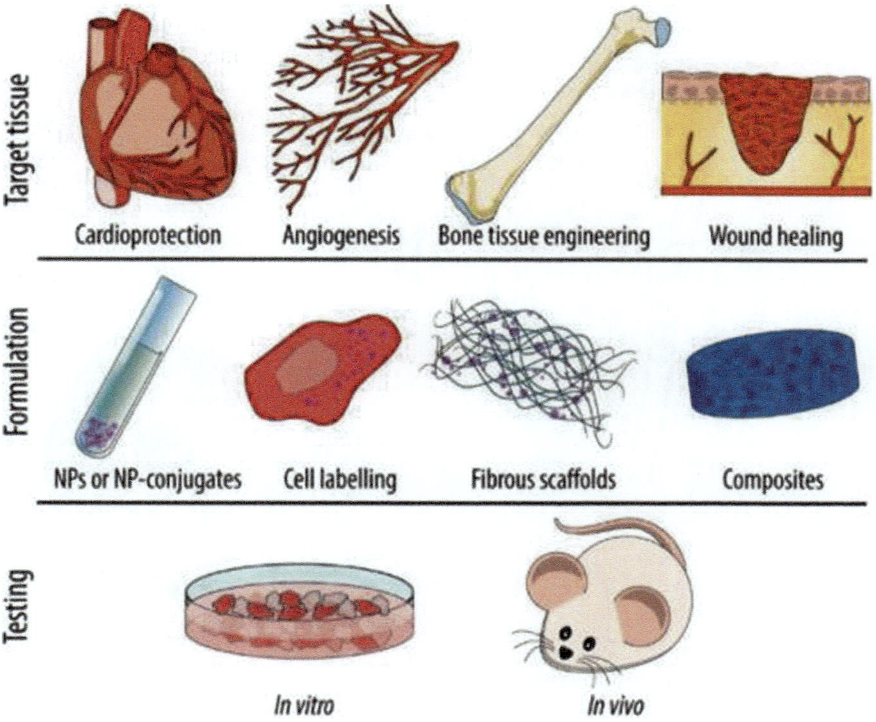

Fig. 1.8 Nanozymes are utilized either as nanoparticle (NP) solutions or as labelling agents for various mammalian cell lines. They can be integrated into scaffolds, such as fibrous networks or composite materials. Their effectiveness is assessed through both in vitro and in vivo studies. Reproduced with permission from [75]

of tissue regeneration is crucial, especially in stem cell therapies aimed at repairing the heart, brain, spinal cord, and even treating Alzheimer's disease [63].

1.6 Challenges and Biocompatibility Considerations

When using nanomaterials for organ replacement in tissue engineering, it is crucial to assess their sensitivity, immune response, toxicity, and potential effects on reproduction and foetal development. A major concern in clinical applications is the prolonged retention of functionalized nanoparticles, which may lead to toxicity and adverse effects [73].

However, long-term bioaccumulation raises concerns about potential toxicity, cancer risks, and reproductive harm. Despite widespread nanoparticle use, scientific gaps persist in understanding their hazards, and no global standards exist for nano-specific risk assessments.

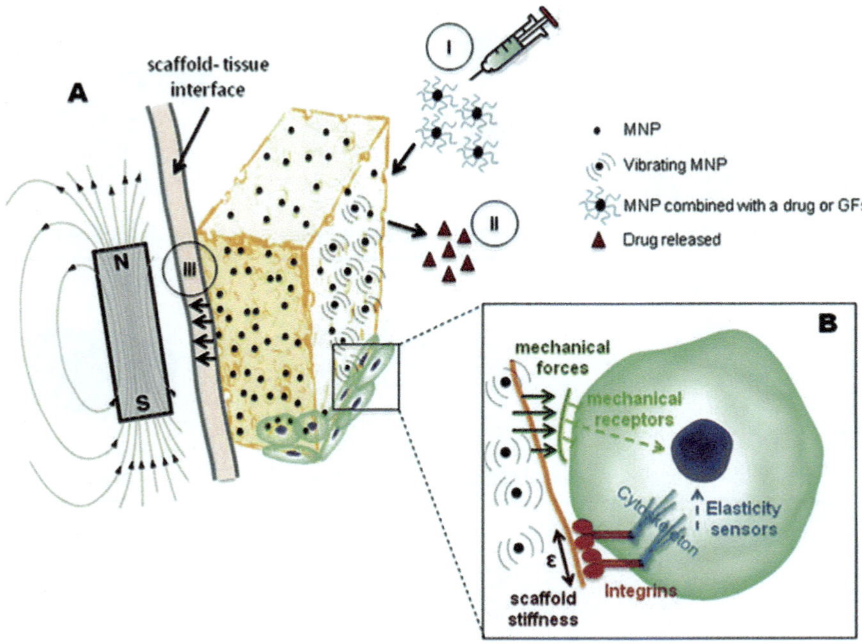

Fig. 1.9 **a** illustrates the potential benefits of magnetic nanocomposite biomaterials in tissue engineering. **I** Magnetic nanoparticles carrying drugs or growth factors are injected near the scaffold and guided by an external magnetic field. **II** Magnetic vibrations trigger drug release, with different stimuli enabling controlled release profiles. **III** Magnets can also secure the scaffold. **b** A magnified view shows how magnetic stimulation influences cellular activity. Reproduced with permission from [76]

To ensure the safe clinical translation of functionalized nanoparticles, their toxicity and adverse effects must be systematically evaluated. Key factors include metabolic pathways, acute and chronic toxicity, genotoxicity, reproductive toxicity, and organ biodistribution. Initial assessments should involve multiple animal models, followed by clinical trials if results are promising. Developing biodegradable nanoparticles enhances clinical applicability, as small NPs can be rapidly excreted via urine, necessitating precise size control. An effective strategy is designing ultrasmall, renal-clearable nanomaterials. Larger NPs primarily undergo hepatobiliary and faecal excretion, influenced by their surface properties and composition.

The biodegradability of these structures is a key factor. Inorganic nanoparticles like MnO_2, $CaCO_3$, $Ca_3(PO_4)_2$, and transition-metal dichalcogenides degrade into ions. Their biodegradability can be tailored by doping with inorganic elements (e.g., calcium-doped silica NPs), incorporating organic groups, or using biodegradable polymers. However, efficiently targeting diseased tissues remains a challenge. Local injection is being explored as an alternative to systemic drug delivery [13].

The movement of MNPs within the body is still debated, as their ability to travel through blood vessels or penetrate tissues remains uncertain. The indiscriminate nature of magnetic fields may lead to nanoparticle aggregation, with smaller MNPs potentially obstructing blood vessels and causing tissue damage, especially under strong magnetic fields.

Ensuring the safety of magnetic iron oxide nanoparticles (MIONs) in biological systems remains a significant challenge, despite their clinical approval. One major concern is the excessive release of free iron, which can lead to oxidative stress and disrupt liver metabolism. Additionally, while combining MIONs with other functional nanomaterials enhances their biomedical applications, it may introduce toxicity, making precise control over their size, composition, and surface modifications essential to reduce immunogenicity. Optimizing their degradation and clearance rates is also necessary for effective use in regenerative medicine.

Imaging techniques like MRI and MPI face limitations in differentiating between living and dead MION-labelled cells or verifying stem cell differentiation. Overcoming these challenges requires improved targeting specificity at the cellular and subcellular levels [64].

1.7 Energy Sector Applications: Hydrogen Production and Fuel Cells

Functionalized magnetic nanomaterials (FMNs) have emerged as key players in the energy sector, particularly in hydrogen production and fuel cell technologies. Hydrogen is considered a clean and sustainable fuel, and efficient methods for its generation are crucial for advancing green energy solutions. FMNs, when engineered as catalysts, significantly enhance hydrogen evolution reactions (HER) by reducing the overpotential and improving charge transfer kinetics. Doping FMNs with noble metals (such as Pt, Pd, and Ru) or transition metals (like Ni, Co, and Mo) can boost catalytic activity, enabling efficient water splitting for hydrogen generation [77]. Additionally, FMNs hybridized with conductive carbon-based materials, such as graphene and carbon nanotubes (CNTs), exhibit improved electrical conductivity and stability, further enhancing their efficiency in electrocatalytic HER processes [78]. Their magnetic properties allow for easy recovery and reuse, making them cost-effective and sustainable options for large-scale hydrogen production.

Beyond hydrogen generation, FMNs also play a crucial role in fuel cell technology. Fuel cells, which convert chemical energy into electrical energy through electrochemical reactions, require efficient electrocatalysts to facilitate reactions such as the oxygen reduction reaction (ORR) and hydrogen oxidation reaction (HOR) [9]. FMNs functionalized with metal nanoparticles or heteroatoms, such as nitrogen (N) or sulfur (S), demonstrate remarkable catalytic efficiency in these processes, as shown in Fig. 1.11. In proton exchange membrane fuel cells (PEMFCs), for example, platinum-based FMNs serve as highly active catalysts, enabling fast reaction kinetics and prolonged durability, as shown in Fig. 1.10 [79]. To reduce costs, researchers are

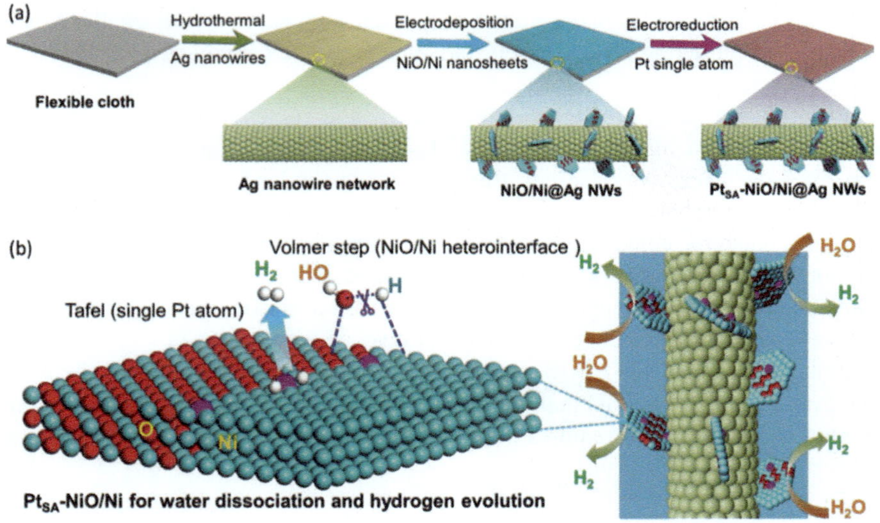

Fig. 1.10 Schematic representation of FMNs in Hydrogen Evolution Reactions (HER) Schematic illustration of the synthesis and water splitting mechanism of PtSA-NiO/Ni. **a** The synthesis process of Pt single-atom-anchored NiO/Ni heterostructure nanosheets on Ag nanowires network. **b** The mechanism of PtSA-NiO/Ni network as an efficient catalyst towards large-scale water electrolysis in alkaline media. Reproduced with permission from [79]

exploring non-noble metal FMNs, such as iron–nitrogen-carbon (Fe–N–C) composites, as potential alternatives. The magnetic recoverability of FMNs further enhances their economic viability by allowing easy separation and recycling of spent catalysts. Additionally, FMNs are being integrated into solid-state hydrogen storage systems, improving hydrogen adsorption and desorption kinetics. Table 1.4 summarizes functionalized magnetic nanomaterials (FMNs), highlighting their modification strategies, applications, advantages, and references. The combination of catalytic activity, stability, and recyclability makes FMNs highly attractive for energy storage and conversion applications, driving innovations in sustainable fuel technologies [80].

1.8 Technological Innovations: FMNs in Advanced Manufacturing and Electronics

Functionalized magnetic nanomaterials (FMNs) are driving technological advancements in manufacturing and electronics by enabling next-generation devices with enhanced efficiency, durability, and functionality [86]. Table 1.5 highlights FMNs in advanced manufacturing and electronics, covering applications and benefits. In

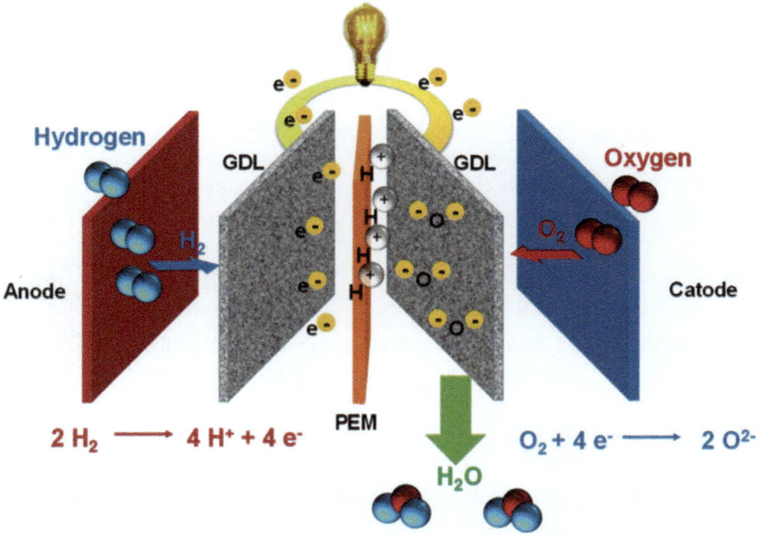

Fig. 1.11 FMNs in fuel cellsRole in oxygen reduction reaction (ORR) and hydrogen oxidation reaction (HOR). Reproduced with permission from [79]

Table 1.4 Functionalized magnetic nanomaterials for hydrogen production and fuel cells

FNM type	Modification strategy	Application	Advantages	References
Noble metal-doped FMNs	Pt, Pd, Ru doping	Hydrogen Evolution Reaction (HER)	High catalytic activity, stability	[81, 82]
Transition metal-based FMNs	Ni, Co, Mo doping	Water splitting	Cost-effective, enhanced electron transfer	[83]
Hybrid FMNs	Graphene, CNT hybridization	Fuel cell electrodes	Improved conductivity, durability	[78]
Non-noble metal FMNs	Fe–N–C composites	Oxygen Reduction Reaction (ORR)	Low-cost, high catalytic efficiency	[84]
Magnetic metal oxides	Fe_3O_4-based functionalization	Hydrogen storage	Enhanced adsorption, easy recovery	[85]

advanced manufacturing, FMNs are used in the fabrication of smart coatings, self-healing materials, and nanocomposites with superior mechanical and thermal properties [87, 88]. Their magnetic properties allow for precise manipulation in additive manufacturing (3D printing), enabling the development of highly customized components with controlled microstructures, as shown in Fig. 1.12 [89]. FMNs also play a crucial role in enhancing the performance of wear-resistant materials used

Table 1.5 FMNs in advanced manufacturing and electronics

FNM Type	Application	Key benefits	References
Magnetic metal oxides (Fe_3O_4, $CoFe_2O_4$)	Smart coatings, self-healing materials	Corrosion resistance, wear resistance	[87, 98]
Carbon-based FMNs (Graphene, CNTs)	High-speed transistors, flexible electronics	Improved conductivity, lightweight structure	[87, 94]
Spintronic FMNs (FePt, CoPt)	MRAM, data storage	Faster data transfer, lower power consumption	[92]
Hybrid FMNs	Printed circuit boards (PCBs), sensors	Enhanced thermal stability, durability	[97]
Nanocomposite FMNs	3D printing, additive manufacturing	Precise microstructure control, customizability	[87]

in aerospace, automotive, and biomedical industries. By incorporating FMNs into coatings and structural materials, manufacturers can improve corrosion resistance, electrical conductivity, and mechanical strength, leading to longer lasting and more efficient industrial components [88].

In the electronics sector, FMNs are revolutionizing the development of high-performance sensors, memory storage devices, and energy-efficient circuits [90, 91]. Magnetic nanoparticles are widely used in spintronic devices, where electron spin is exploited to achieve faster and more efficient data storage and transfer [92]. FMNs integrated into magnetoresistive random-access memory (MRAM) enhance data retention and reduce power consumption, making them ideal for next-generation computing systems [93]. Additionally, FMNs are key materials in flexible and wearable electronics, where their tunable electrical and magnetic properties enable the creation of lightweight, flexible circuits for smart textiles, medical wearables, and bendable displays (Fig. 1.13). Furthermore, FMNs functionalized with carbon-based materials such as graphene or carbon nanotubes (CNTs) exhibit excellent conductivity and thermal stability, making them ideal for high-speed transistors and printed circuit boards (PCBs). These innovations are paving the way for more sustainable and high-performance electronic devices, supporting the rapid evolution of modern technology [94–97].

1.9 Role of FMNs in Sensor Technologies and Smart Materials

Functionalized magnetic nanomaterials (FMNs) are playing a transformative role in sensor technologies and the development of smart materials, enabling highly sensitive, selective, and responsive systems for various industrial and biomedical applications. Table 1.6 summarizes FMNs in sensors and smart materials, highlighting applications, benefits, and references. FMNs are extensively used in chemical and biosensors due to their unique magnetic, electrical, and catalytic properties, as shown

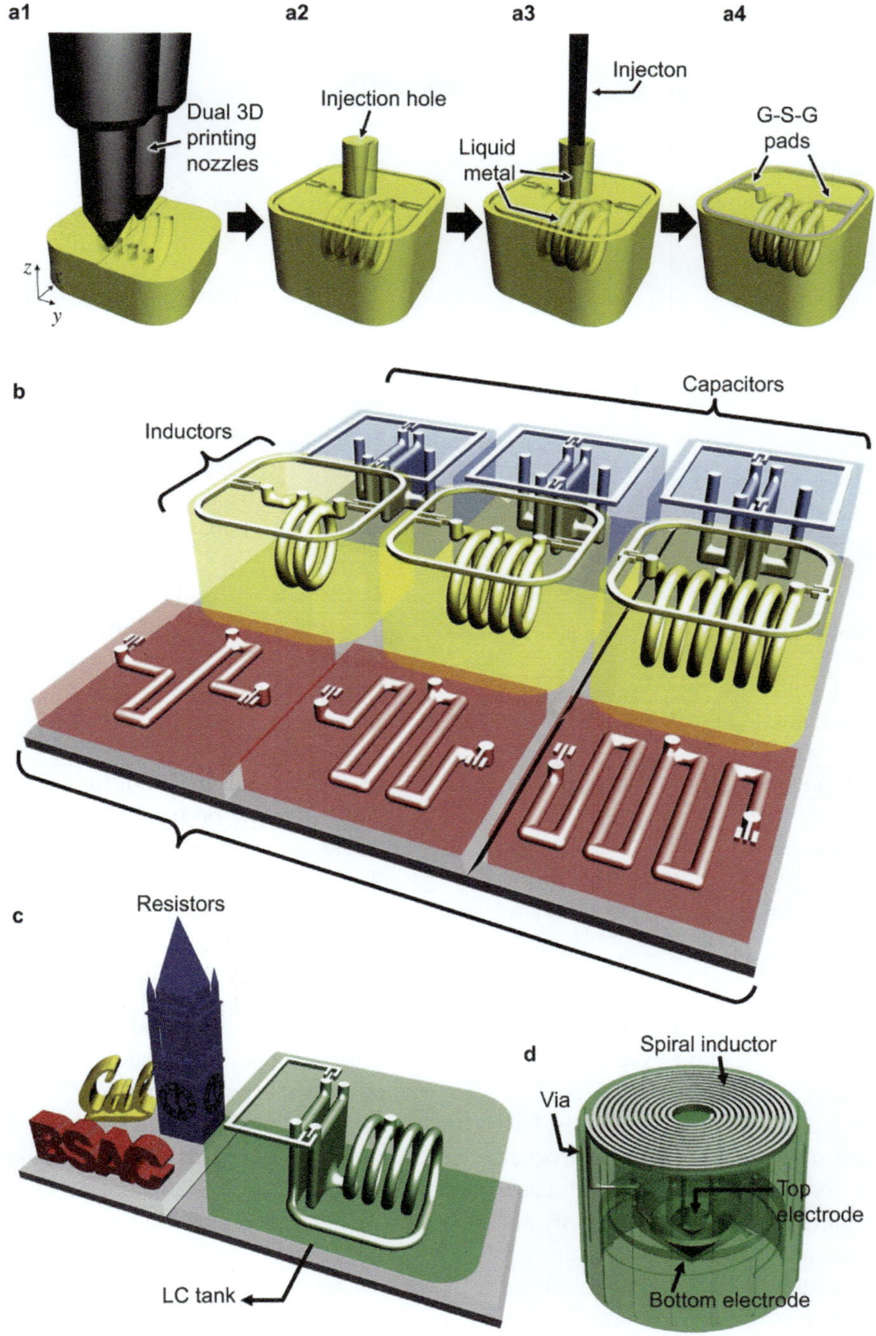

Fig. 1.12 FMNs in 3D printing and smart manufacturing. A Schematic diagram of the additive 3D manufacturing process. **a** The 3D fabrication process with embedded and electrically conductive structures. **b** 3D microelectronics components, including parallel-plate capacitors, solenoid-type inductors, and meandering-shape resistors. **c** A 3D LC tank, which is formed by combining a solenoid-type inductor and a parallel-plate capacitor. **d** A wireless passive sensor demonstration of a "smart cap," containing the 3D-printed LC-resonant circuit. Reproduced with permission from [99]

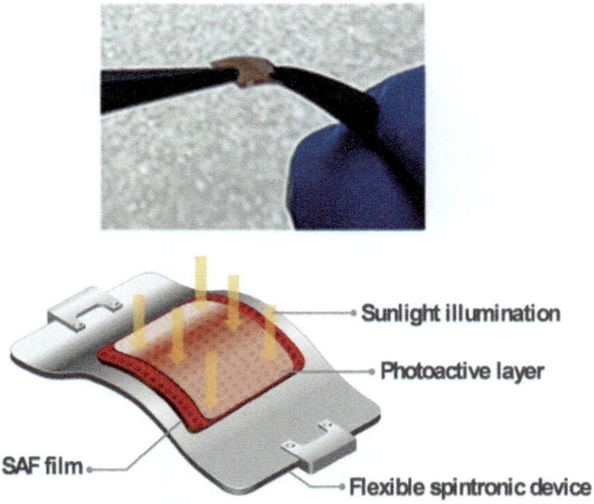

Fig. 1.13 FMNs in Spintronics and Flexible Electronics: The practical and schematic images of a flexible spintronic device. Reproduced with permission from [100]

Table 1.6 Applications of FMNs in sensors and smart materials

FNM Type	Application	Key Benefits	References
Iron oxide (Fe_3O_4) FMNs	Gas sensors (CO, VOCs)	High sensitivity, selective adsorption	[107]
Noble metal-functionalized FMNs	Electrochemical biosensors	Enhanced detection of biomolecules	[108, 109]
Magnetic hydrogels	Drug delivery, soft robotics	Controlled release, tunable responsiveness	[110]
Shape-memory FMNs	Smart coatings, self-healing materials	Damage repair, durability enhancement	[111]
Hybrid FMNs (Graphene, CNTs)	Wearable sensors, flexible electronics	Real-time monitoring, lightweight design	[112]

in Fig. 1.14 [101]. In gas sensing, FMNs such as iron oxide (Fe_3O_4) and cobalt ferrite ($CoFe_2O_4$) exhibit excellent adsorption capabilities, allowing for the detection of toxic gases like carbon monoxide (CO) and volatile organic compounds (VOCs) with high precision [102]. Similarly, FMNs functionalized with noble metals (such as gold and platinum) enhance electrochemical sensors for the detection of biomolecules, including glucose, DNA, and cancer markers [103]. These advancements have significantly improved the sensitivity and reliability of diagnostic tools used in healthcare and environmental monitoring.

In the field of smart materials, FMNs contribute to the design of adaptive and stimuli-responsive systems that change their properties in response to external signals such as magnetic fields, temperature, or pH. Magnetically responsive hydrogels and shape-memory polymers, embedded with FMNs, are being developed for drug delivery, soft robotics, and tissue engineering [104]. Additionally, FMNs are used in

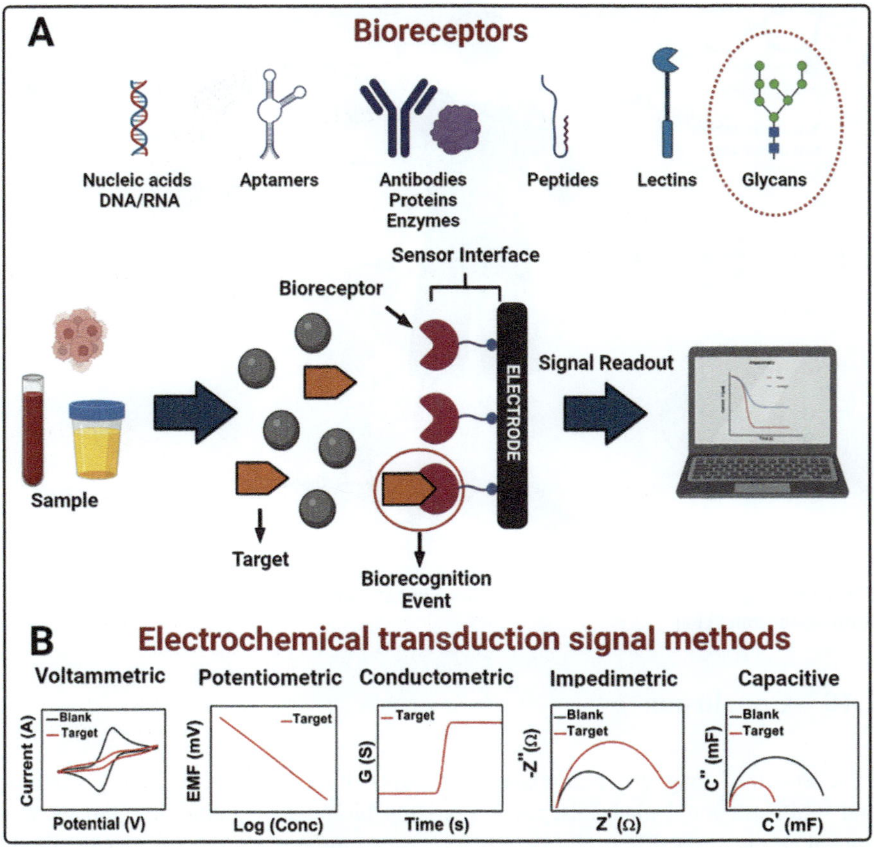

Fig. 1.14 FMNs in gas and biosensing applications. **a** Scheme of a generic electrochemical biosensor. A bioreceptor (nucleic acids, aptamers, antibodies, proteins, enzymes, peptides, lectins, glycans, etc.) attached to the electrode surface recruits the molecular target (analyte present in a sample) onto the sensor interface by an affinity reaction. After the bioreceptor binding with the target (biorecognition event), the transducer converts the binding event into a measurable signal proportional to the concentration of the target (signal readout). **b** Electrochemical transduction signal methods in biosensors: voltammetric, potentiometric, conductometric, impedimetric, and capacitive. Reproduced with permission from [113]

self-healing coatings and structural materials that can repair damage autonomously, extending the lifespan of components in aerospace, automotive, and infrastructure applications [87, 88]. Another critical area is wearable and flexible electronics, where FMNs are integrated into sensors that monitor physiological parameters such as heart rate, temperature, and blood pressure in real time (Fig. 1.15) [105]. The ability of FMNs to combine multiple functionalities—such as conductivity, magnetism, and catalytic activity—makes them ideal for next-generation sensor technologies and smart material applications, paving the way for more efficient and intelligent systems across various industries [106].

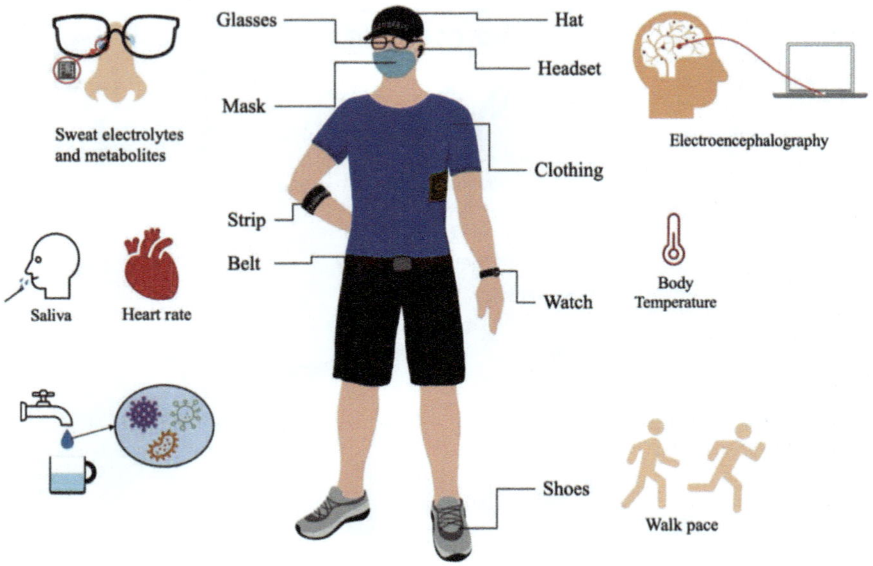

Fig. 1.15 Smart materials with FMNs in self-healing and wearable technologies. Reproduced with permission from [114]

1.10 Conclusion

Functionalized magnetic nanomaterials (FMNs) have emerged as highly versatile and multifunctional materials, revolutionizing biomedical, industrial, and technological applications. Their ability to be precisely engineered at the nanoscale allows for enhanced magnetic, catalytic, electrical, and structural properties, making them invaluable in a wide range of fields. In biomedicine, FMNs are at the forefront of next-generation drug delivery systems, diagnostic imaging, and biosensing applications, offering targeted and efficient solutions for disease detection and therapy. Their tunable surface chemistry enables biocompatibility and specificity, enhancing therapeutic efficacy while minimizing side effects.

Beyond health care, FMNs are driving significant advancements in renewable energy, environmental remediation, and industrial manufacturing. From catalyzing hydrogen evolution reactions to improving pollutant degradation through photocatalysis, FMNs are critical in sustainable energy production and environmental protection. Their role in sensor technologies, smart materials, and flexible electronics highlights their potential in developing adaptive, high-performance materials for industrial automation, wearable technology, and next-generation computing. As research continues to refine FMNs' properties and expand their applications, these materials are poised to address pressing global challenges in healthcare, energy, and technology. The interdisciplinary nature of FMNs underscores their transformative potential, ensuring that they remain central to scientific and technological progress in the coming decades.

1.11 Future Outlook

The future of functionalized magnetic nanomaterials (FMNs) holds immense potential as advancements in nanotechnology, material science, and engineering continue to push the boundaries of their applications. In biomedicine, the integration of FMNs with artificial intelligence (AI) and machine learning (ML) could enable more precise diagnostics and personalized medicine, improving early disease detection and targeted therapies. The development of biodegradable and non-toxic FMNs will further enhance their clinical applications, minimizing long-term toxicity concerns and improving patient safety. Additionally, research in magnetically controlled drug delivery systems and smart biosensors will pave the way for real-time health monitoring and remote disease management.

In industrial and technological sectors, FMNs will play a crucial role in the advancement of renewable energy technologies, particularly in hydrogen production, fuel cells, and next-generation batteries with improved efficiency and stability. Their application in smart materials, flexible electronics, and high-speed computing will continue to grow, leading to more energy-efficient and adaptive devices. Moreover, the role of FMNs in environmental sustainability, such as wastewater treatment, air purification, and carbon capture, is expected to expand, addressing critical global challenges related to pollution and climate change. Future research should focus on large-scale synthesis methods that are cost-effective, environmentally friendly, and scalable for industrial applications. With continued interdisciplinary collaboration, FMNs will remain at the forefront of innovation, significantly shaping the future of biomedicine, industry, and technology.

Abbreviations

AI	Artificial Intelligence
CAs	Contrast Agents
CNTs	Carbon Nanotubes
CP	Calcium Phosphate
DEX	Dexamethasone
DOX	Doxorubicin
DOPA	Dopamine
DTX	Docetaxel
EMF	Electromotive Force
FMNs	Functionalized Magnetic Nanomaterials
GEM	Gemcitabine
GO	Graphene Oxide
HER	Hydrogen Evolution Reaction
HOR	Hydrogen Oxidation Reaction
HSA	Human Serum Albumin

IONPs	Iron Oxide Nanoparticles
MHN	Magnetic Hybrid Nanomaterials
ML	Machine Learning
MIONs	Magnetic Iron Oxide Nanoparticles
miRNA	micro Ribonucleic acid
MNP	Magnetic Nanoparticle
MOF	Metal–Organic Framework
MPI	Magnetic Particle Imaging
MRI	Magnetic Resonance Imaging
MRAM	Magnetoresistive Random-Access Memory
MTX	Methotrexate
NW	Nanowires
ORR	Oxygen Reduction Reaction
PAA	Polyacrylic Acid
PAI	Photoacoustic Imaging
PCL	Poly(ε-caprolactone)
PCBs	Printed Circuit Boards
pDA	Plasmid Vector DNA
PEG	Polyethylene glycol
PEMFC	Proton Exchange Membrane Fuel Cells
PET	Positron Emission Tomography
PGA	Poly(glycolic acid)
PLA	Poly(lactic acid)
PLGA	Poly(lactic-co-glycolic acid)
PIONs	Porous Iron Oxide Nano Agents
PTX	Paclitaxel
QD	Quantum Dots
RF	Radio Frequency
SAR	Specific Absorption Rate
siRNA	Small Interfering Ribonucleic Acid
SPIONS	Superparamagnetic iron oxide nanoparticles
SWCNT	Single-walled carbon nanotubes
TCP	Tricalcium Phosphate
VOCs	Volatile Organic Compounds
3-D	3-Dimensional

References

1. Sharma S, Sharma H, Sharma R (2024) A review on functionalization and potential application spectrum of magnetic nanoparticles (MNPs) based systems. Chem Inorg Mater 2:100035. https://doi.org/10.1016/j.cinorg.2024.100035
2. Abdullah NH, Shameli K, Abdullah EC, Abdullah LC (2019) Solid matrices for fabrication of magnetic iron oxide nanocomposites: synthesis, properties, and application for the adsorption of heavy metal ions and dyes. Compos Part B Eng 162:538–568. https://doi.org/10.1016/j.compositesb.2018.12.075
3. Manoj GM, Shalini M, Thenmozhi K et al (2024) Recent advancements in the surface modification and functionalization of magnetic nanomaterials. Appl Surf Sci Adv 21:100608. https://doi.org/10.1016/j.apsadv.2024.100608
4. Zhou Q, Li J, Wang M, Zhao D (2016) Iron-based magnetic nanomaterials and their environmental applications. Crit Rev Environ Sci Technol 46:783–826. https://doi.org/10.1080/10643389.2016.1160815
5. Liu W-T (2006) Nanoparticles and their biological and environmental applications. J Biosci Bioeng 102:1–7. https://doi.org/10.1263/jbb.102.1
6. Saleh TA (2020) Nanomaterials: classification, properties, and environmental toxicities. Environ Technol Innov 20:101067. https://doi.org/10.1016/j.eti.2020.101067
7. Khalil M, Kadja GTM, MohM Ilmi (2021) Advanced nanomaterials for catalysis: current progress in fine chemical synthesis, hydrocarbon processing, and renewable energy. J Ind Eng Chem 93:78–100. https://doi.org/10.1016/j.jiec.2020.09.028
8. Zhou K, Zhou X, Liu J, Huang Z (2020) Application of magnetic nanoparticles in petroleum industry: a review. J Pet Sci Eng 188:106943. https://doi.org/10.1016/j.petrol.2020.106943
9. Danish MSS (2023) Exploring metal oxides for the hydrogen evolution reaction (HER) in the field of nanotechnology. RSC Sustain 1:2180–2196. https://doi.org/10.1039/D3SU00179B
10. Chakravorty A, Roy S (2024) A review of photocatalysis, basic principles, processes, and materials. Sustain Chem Environ 8:100155. https://doi.org/10.1016/j.scenv.2024.100155
11. Jung JH, Lee JH, Shinkai S (2011) Functionalized magnetic nanoparticles as chemosensors and adsorbents for toxic metal ions in environmental and biological fields. Chem Soc Rev 40:4464. https://doi.org/10.1039/c1cs15051k
12. Ullah Khan A, Chen L, Ge G (2021) Recent development for biomedical applications of magnetic nanoparticles. Inorg Chem Commun 134:108995. https://doi.org/10.1016/j.inoche.2021.108995
13. Diez-Pascual AM, Rahdar A (2022) Functional nanomaterials in biomedicine: current uses and potential applications. ChemMedChem 17:e202200142. https://doi.org/10.1002/cmdc.202200142
14. Baig N, Kammakakam I, Falath W (2021) Nanomaterials: a review of synthesis methods, properties, recent progress, and challenges. Mater Adv 2:1821–1871. https://doi.org/10.1039/D0MA00807A
15. Bai X, Zhang X (2025) Artificial intelligence-powered materials science. Nano-Micro Lett 17:135. https://doi.org/10.1007/s40820-024-01634-8
16. Gandhi AC (2022) Synthesis and characterization of functional magnetic nanomaterials. Coatings 12:857. https://doi.org/10.3390/coatings12060857
17. Hao R, Xing R, Xu Z et al (2010) Synthesis, functionalization, and biomedical applications of multifunctional magnetic nanoparticles. Adv Mater 22:2729–2742. https://doi.org/10.1002/adma.201000260
18. Ahmed MM, Abu-Elsaad N (2024) Exploring the magnetic behavior of ferrites: from diamagnetism to superparamagnetism
19. Knobel M, Nunes WC, Socolovsky LM et al (2008) Superparamagnetism and other magnetic features in granular materials: a review on ideal and real systems. J Nanosci Nanotechnol 8:2836–2857. https://doi.org/10.1166/jnn.2008.15348

20. Lisjak D, Mertelj A (2018) Anisotropic magnetic nanoparticles: a review of their properties, syntheses and potential applications. Prog Mater Sci 95:286–328. https://doi.org/10.1016/j.pmatsci.2018.03.003
21. Díez AG, Rincón-Iglesias M, Lanceros-Méndez S et al (2022) Multicomponent magnetic nanoparticle engineering: the role of structure-property relationship in advanced applications. Mater Today Chem 26:101220. https://doi.org/10.1016/j.mtchem.2022.101220
22. Mahmoudi M, Sant S, Wang B et al (2011) Superparamagnetic iron oxide nanoparticles (SPIONs): development, surface modification and applications in chemotherapy. Adv Drug Deliv Rev 63:24–46. https://doi.org/10.1016/j.addr.2010.05.006
23. Sun C, Lee JSH, Zhang M (2008) Magnetic nanoparticles in MR imaging and drug delivery. Adv Drug Deliv Rev 60:1252–1265. https://doi.org/10.1016/j.addr.2008.03.018
24. Ioannou M (2023) The role of ferromagnets and antiferromagnets for spintronic memory applications and their impact in data storage: spintronic memory applications. Emerg Minds J Stud Res 1:1–6. https://doi.org/10.59973/emjsr.6
25. Ren Z, Gao B, Xu S, Xu Q (2024) CO_2-induced strong room-temperature ferromagnetism in $BiFeO_3$. Adv Electron Mater 10:2300626. https://doi.org/10.1002/aelm.202300626
26. Khan MA, Sun J, Li B et al (2021) Magnetic sensors—a review and recent technologies. Eng Res Express 3:022005. https://doi.org/10.1088/2631-8695/ac0838
27. Belyanina I, Kolovskaya O, Zamay S et al (2017) Targeted magnetic nanotheranostics of cancer. Molecules 22:975. https://doi.org/10.3390/molecules22060975
28. Mohammed L, Gomaa HG, Ragab D, Zhu J (2017) Magnetic nanoparticles for environmental and biomedical applications: a review. Particuology 30:1–14. https://doi.org/10.1016/j.partic.2016.06.001
29. Mosbach K, Schröder U (1979) Preparation and application of magnetic polymers for targeting of drugs. FEBS Lett 102:112–116. https://doi.org/10.1016/0014-5793(79)80940-0
30. Ali A, Shah T, Ullah R et al (2021) Review on recent progress in magnetic nanoparticles: synthesis, characterization, and diverse applications. Front Chem 9:629054. https://doi.org/10.3389/fchem.2021.629054
31. Avasthi A, Caro C, Pozo-Torres E et al (2020) Magnetic nanoparticles as MRI contrast agents. Top Curr Chem 378:40. https://doi.org/10.1007/s41061-020-00302-w
32. Mahajan G, Kaur M, Gupta R (2022) Green functionalized nanomaterials: fundamentals and future opportunities. In: Green functionalized nanomaterials for environmental applications. Elsevier, pp 21–41
33. Flores-Rojas GG, López-Saucedo F, Vera-Graziano R et al (2022) Magnetic nanoparticles for medical applications: updated review. Macromol 2:374–390. https://doi.org/10.3390/macromol2030024
34. Tian X, Zhang L, Yang M et al (2018) Functional magnetic hybrid nanomaterials for biomedical diagnosis and treatment. WIREs Nanomed Nanobiotechnol 10:e1476. https://doi.org/10.1002/wnan.1476
35. Hwang J, Son J, Seo Y et al (2018) Functional silica nanoparticles conjugated with beta-glucan to deliver anti-tuberculosis drug molecules. J Ind Eng Chem 58:376–385. https://doi.org/10.1016/j.jiec.2017.09.051
36. Wieszczycka K, Staszak K, Woźniak-Budych MJ et al (2021) Surface functionalization—the way for advanced applications of smart materials. Coord Chem Rev 436:213846. https://doi.org/10.1016/j.ccr.2021.213846
37. Singh R, Srinivas SP, Kumawat M, Daima HK (2024) Ligand-based surface engineering of nanomaterials: trends, challenges, and biomedical perspectives. OpenNano 15:100194. https://doi.org/10.1016/j.onano.2023.100194
38. Marega R, Karmani L, Flamant L et al (2012) Antibody-functionalized polymer-coated gold nanoparticles targeting cancer cells: an in vitro and in vivo study. J Mater Chem 22:21305. https://doi.org/10.1039/c2jm33482h
39. Xu L, Bai Q, Zhang X, Yang H (2017) Folate-mediated chemotherapy and diagnostics: an updated review and outlook. J Control Release Off J Control Release Soc 252:73–82. https://doi.org/10.1016/j.jconrel.2017.02.023

40. Chanphai P, Thomas TJ, Tajmir-Riahi HA (2021) Application and biomolecular study of functionalized folic acid-dendrimer nanoparticles in drug delivery. J Biomol Struct Dyn 39:787–794. https://doi.org/10.1080/07391102.2020.1717994
41. Cheng J, Gu Y-J, Cheng SH, Wong W-T (2013) Surface functionalized gold nanoparticles for drug delivery. J Biomed Nanotechnol 9:1362–1369. https://doi.org/10.1166/jbn.2013.1536
42. Huang Y-F, Wang Y-F, Yan X-P (2010) Amine-functionalized magnetic nanoparticles for rapid capture and removal of bacterial pathogens. Environ Sci Technol 44:7908–7913. https://doi.org/10.1021/es102285n
43. Sampaolesi S, Nicotra F, Russo L (2019) Glycans in nanomedicine, impact and perspectives. Future Med Chem 11:43–60. https://doi.org/10.4155/fmc-2018-0368
44. Pathak TK, Kroon RE, Swart HC (2018) Photocatalytic and biological applications of Ag and Au doped ZnO nanomaterial synthesized by combustion. Vacuum 157:508–513. https://doi.org/10.1016/j.vacuum.2018.09.020
45. Gao Y, Li Z, Deng M et al (2025) Anchoring Ni nanoparticles on MoO_2 nanorods for enhancing electrocatalytic oxidation of glycerol to formate. Colloids Surf Physicochem Eng Asp 136661. https://doi.org/10.1016/j.colsurfa.2025.136661
46. Pattaweepaiboon S, Samarungkasut P, Iamprasertkun P et al (2025) Single-atom to Ultrasmall Au nanoparticles anchored on NiFe layered double hydroxide as catalyst for oxygen and hydrogen evolution reactions. ACS Appl Nano Mater 8:438–448. https://doi.org/10.1021/acsanm.4c05830
47. Mahmoudi H, Jafari AA, Saeedi S, Firouzabadi H (2015) Sulfonic acid-functionalized magnetic nanoparticles as a recyclable and eco-friendly catalyst for atom economical Michael addition reactions and bis indolyl methane synthesis. RSC Adv 5:3023–3030. https://doi.org/10.1039/C4RA11605D
48. Gong Y, Lin J, Yu C et al (2024) Pd nanoparticles anchored on amine-functionalized boron nitride Co-doped with carbon and oxygen as catalysts for formic acid dehydrogenation. ACS Appl Nano Mater 7:16245–16255. https://doi.org/10.1021/acsanm.4c02173
49. Yuan G, Yuan Y, Xu K, Luo Q (2014) Biocompatible PEGylated Fe3O4 nanoparticles as photothermal agents for near-infrared light modulated cancer therapy. Int J Mol Sci 15:18776–18788. https://doi.org/10.3390/ijms151018776
50. Kung HH, Kung MC, Costello CK (2003) Supported Au catalysts for low temperature CO oxidation. J Catal 216:425–432. https://doi.org/10.1016/S0021-9517(02)00111-2
51. Ren J, Zhang Z, Wang F et al (2012) MRI of prostate stem cell antigen expression in prostate tumors. Nanomed 7:691–703. https://doi.org/10.2217/nnm.11.147
52. Wang N, Zhou L, Guo J et al (2014) Adsorption of environmental pollutants using magnetic hybrid nanoparticles modified with β-cyclodextrin. Appl Surf Sci 305:267–273. https://doi.org/10.1016/j.apsusc.2014.03.054
53. Lapusan R, Borlan R, Focsan M (2024) Advancing MRI with magnetic nanoparticles: a comprehensive review of translational research and clinical trials. Nanoscale Adv 6:2234–2259. https://doi.org/10.1039/D3NA01064C
54. Malik S, Singh J, Goyat R et al (2023) Nanomaterials-based biosensor and their applications: a review. Heliyon 9:e19929. https://doi.org/10.1016/j.heliyon.2023.e19929
55. Joudeh N, Linke D (2022) Nanoparticle classification, physicochemical properties, characterization, and applications: a comprehensive review for biologists. J Nanobiotechnol 20:262. https://doi.org/10.1186/s12951-022-01477-8
56. Paras, Yadav K, Kumar P et al (2022) A review on low-dimensional nanomaterials: nanofabrication characterization and applications. Nanomater Basel Switz 13:160. https://doi.org/10.3390/nano13010160
57. Pillin I, Castro M, Nag Chowdhury S, Feller J-F (2016) Robustness of carbon nanotube-based sensor to probe composites' interfacial damage in situ. J Compos Mater 50:109–113. https://doi.org/10.1177/0021998315571029
58. Rathinavel S, Priyadharshini K, Panda D (2021) A review on carbon nanotube: an overview of synthesis, properties, functionalization, characterization, and the application. Mater Sci Eng B 268:115095. https://doi.org/10.1016/j.mseb.2021.115095

59. Adepu S, Ramakrishna S (2021) Controlled drug delivery systems: current status and future directions. Molecules 26:5905. https://doi.org/10.3390/molecules26195905
60. Stueber DD, Villanova J, Aponte I et al (2021) Magnetic nanoparticles in biology and medicine: past, present, and future trends. Pharmaceutics 13:943. https://doi.org/10.3390/pharmaceutics13070943
61. Al-Jamal KT, Bai J, Wang JT-W et al (2016) Magnetic drug targeting: preclinical in vivo studies, mathematical modeling, and extrapolation to humans. Nano Lett 16:5652–5660. https://doi.org/10.1021/acs.nanolett.6b02261
62. Krishani M, Shin WY, Suhaimi H, Sambudi NS (2023) Development of scaffolds from bio-based natural materials for tissue regeneration applications: a review. Gels 9:100. https://doi.org/10.3390/gels9020100
63. Liu X, Chen S, Zhang H et al (2019) Magnetic nanomaterials for advanced regenerative medicine: the promise and challenges. Adv Mater 31:1804922. https://doi.org/10.1002/adma.201804922
64. Bock N, Riminucci A, Dionigi C et al (2010) A novel route in bone tissue engineering: magnetic biomimetic scaffolds. Acta Biomater 6:786–796. https://doi.org/10.1016/j.actbio.2009.09.017
65. Russo T, Peluso V, Gloria A et al (2020) Combination design of time-dependent magnetic field and magnetic nanocomposites to guide cell behavior. Nanomaterials 10:577. https://doi.org/10.3390/nano10030577
66. Marycz K, Kornicka K, Röcken M (2018) Static Magnetic Field (SMF) as a regulator of stem cell fate—new perspectives in regenerative medicine arising from an underestimated tool. Stem Cell Rev Rep 14:785–792. https://doi.org/10.1007/s12015-018-9847-4
67. Goranov V (2024) Biomaterials functionalized with magnetic nanoparticles for tissue engineering: between advantages and challenges. Biomater Biosyst 15:100100. https://doi.org/10.1016/j.bbiosy.2024.100100
68. Reddy MSB, Ponnamma D, Choudhary R, Sadasivuni KK (2021) A comparative review of natural and synthetic biopolymer composite scaffolds. Polymers 13:1105. https://doi.org/10.3390/polym13071105
69. Guarino V, Causa F, Ambrosio L (2007) Bioactive scaffolds for bone and ligament tissue. Expert Rev Med Devices 4:405–418. https://doi.org/10.1586/17434440.4.3.405
70. Gharibshahian M, Salehi M, Beheshtizadeh N et al (2023) Recent advances on 3D-printed PCL-based composite scaffolds for bone tissue engineering. Front Bioeng Biotechnol 11:1168504. https://doi.org/10.3389/fbioe.2023.1168504
71. Fazeli N, Arefian E, Irani S et al (2021) 3D-printed PCL scaffolds coated with nanobioceramics enhance osteogenic differentiation of stem cells. ACS Omega 6:35284–35296. https://doi.org/10.1021/acsomega.1c04015
72. Lu Y, Cheng D, Niu B et al (2023) Properties of poly (Lactic-co-glycolic acid) and progress of poly (lactic-co-glycolic acid)-based biodegradable materials in biomedical research. Pharmaceuticals 16:454. https://doi.org/10.3390/ph16030454
73. Zheng X, Zhang P, Fu Z et al (2021) Applications of nanomaterials in tissue engineering. RSC Adv 11:19041–19058. https://doi.org/10.1039/D1RA01849C
74. Das SK, Chakraborty S, Naskar S, Rajabalaya R (2021) Techniques and methods used for the fabrication of bionanocomposites. In: Bionanocomposites in tissue engineering and regenerative medicine. Elsevier, pp 17–43
75. Jansman MMT, Hosta-Rigau L (2019) Cerium- and iron-oxide-based nanozymes in tissue engineering and regenerative medicine. Catalysts 9:691. https://doi.org/10.3390/catal9080691
76. Gil S, Mano JF (2014) Magnetic composite biomaterials for tissue engineering. Biomater Sci 2:812–818. https://doi.org/10.1039/C4BM00041B
77. Mishra P, Patnaik S, Parida K (2019) An overview of recent progress on noble metal modified magnetic Fe_3O_4 for photocatalytic pollutant degradation and H_2 evolution. Catal Sci Technol 9:916–941. https://doi.org/10.1039/C8CY02462F
78. Saha MS, Kundu A (2010) Functionalizing carbon nanotubes for proton exchange membrane fuel cells electrode. J Power Sources 195:6255–6261. https://doi.org/10.1016/j.jpowsour.2010.04.015

79. Tellez-Cruz MM, Escorihuela J, Solorza-Feria O, Compañ V (2021) Proton exchange membrane fuel cells (PEMFCs): advances and challenges. Polymers 13:3064. https://doi.org/10.3390/polym13183064
80. Kalibek MR, Ospanova AD, Suleimenova B et al (2024) Solid-state hydrogen storage materials. Discov Nano 19:195. https://doi.org/10.1186/s11671-024-04137-y
81. Wang J, Fang W, Hu Y et al (2021) Single atom Ru doping 2H-MoS2 as highly efficient hydrogen evolution reaction electrocatalyst in a wide pH range. Appl Catal B Environ 298:120490. https://doi.org/10.1016/j.apcatb.2021.120490
82. Chen J, Xia G, Jiang P et al (2016) Active and durable hydrogen evolution reaction catalyst derived from Pd-doped metal-organic frameworks. ACS Appl Mater Interfaces 8:13378–13383. https://doi.org/10.1021/acsami.6b01266
83. Paul SC, Dey SC, Molla MdAI et al (2021) Nanomaterials as electrocatalyst for hydrogen and oxygen evolution reaction: exploitation of challenges and current progressions. Polyhedron 193:114871. https://doi.org/10.1016/j.poly.2020.114871
84. Chen Y, Gokhale R, Serov A et al (2017) Novel highly active and selective Fe-N-C oxygen reduction electrocatalysts derived from in-situ polymerization pyrolysis. Nano Energy 38:201–209. https://doi.org/10.1016/j.nanoen.2017.05.059
85. Xu G, Zhang W, Zhang Y et al (2018) Fe3O4 nanoclusters highly dispersed on a porous graphene support as an additive for improving the hydrogen storage properties of LiBH4. RSC Adv 8:19353–19361. https://doi.org/10.1039/c8ra02762e
86. Jia J, Lan Y (2023) Synthesis, characterization, and applications of nanomaterials for energy conversion and storage. Mol Basel Switz 28:7383. https://doi.org/10.3390/molecules28217383
87. Cerdan K, Moya C, Van Puyvelde P et al (2022) Magnetic self-healing composites: synthesis and applications. Molecules 27:3796. https://doi.org/10.3390/molecules27123796
88. Romero-Fierro D, Bustamante-Torres M, Bravo-Plascencia F et al (2022) Recent trends in magnetic polymer nanocomposites for aerospace applications: a review. Polymers 14:4084. https://doi.org/10.3390/polym14194084
89. Kokkinis D, Schaffner M, Studart AR (2015) Multimaterial magnetically assisted 3D printing of composite materials. Nat Commun 6:8643. https://doi.org/10.1038/ncomms9643
90. Fu Y, Liu T, Wang H et al (2024) Applications of nanomaterial technology in biosensing. J Sci Adv Mater Devices 9:100694. https://doi.org/10.1016/j.jsamd.2024.100694
91. Dai D, Zhang Y, Yang S et al (2024) Recent advances in functional materials for optical data storage. Mol Basel Switz 29:254. https://doi.org/10.3390/molecules29010254
92. Barla P, Joshi VK, Bhat S (2021) Spintronic devices: a promising alternative to CMOS devices. J Comput Electron 20:805–837. https://doi.org/10.1007/s10825-020-01648-6
93. Xiao Z, Naik VB, Lim JH, et al (2024) Adapting magnetoresistive memory devices for accurate and on-chip-training-free in-memory computing. Sci Adv 10:eadp3710. https://doi.org/10.1126/sciadv.adp3710
94. Gazzato L, Frasconi M (2025) Carbon nanotubes and their composites for flexible electrochemical biosensors. Anal Sens 5:e202400038. https://doi.org/10.1002/anse.202400038
95. Zhang S, Liu H, Yu J et al (2020) Multi-functional flexible 2D carbon nanostructured networks. Nat Commun 11:5134. https://doi.org/10.1038/s41467-020-18977-6
96. Kamran U, Heo Y-J, Lee J, Park S-J (2019) Functionalized carbon materials for electronic devices: a review. Micromachines 10:234. https://doi.org/10.3390/mi10040234
97. Kawabata R, Araki T, Akiyama M et al (2022) Stretchable printed circuit board integrated with Ag-nanowire-based electrodes and organic transistors toward imperceptible electrophysiological sensing. Flex Print Electron 7:044002. https://doi.org/10.1088/2058-8585/ac968c
98. Comanescu C (2023) Recent advances in surface functionalization of magnetic nanoparticles. Coatings 13:1772. https://doi.org/10.3390/coatings13101772
99. Wu S-Y, Yang C, Hsu W, Lin L (2015) 3D-printed microelectronics for integrated circuitry and passive wireless sensors. Microsyst Nanoeng 1:15013. https://doi.org/10.1038/micronano.2015.13

100. Wang C, Du Y, Zhao Y et al (2023) Solar-powered switch of antiferromagnetism/ferromagnetism in flexible spintronics. Nanomaterials 13:3158. https://doi.org/10.3390/nano13243158
101. Tripathy A, Nine MJ, Silva FS (2021) Biosensing platform on ferrite magnetic nanoparticles: synthesis, functionalization, mechanism and applications. Adv Colloid Interface Sci 290:102380. https://doi.org/10.1016/j.cis.2021.102380
102. Liu X, Cheng S, Liu H et al (2012) A survey on gas sensing technology. Sensors 12:9635–9665. https://doi.org/10.3390/s120709635
103. Mehmandoust M, Erk N, Karaman C et al (2021) Sensitive and selective electrochemical detection of epirubicin as anticancer drug based on nickel ferrite decorated with gold nanoparticles. Micromachines 12:1334. https://doi.org/10.3390/mi12111334
104. Jalili NA, Muscarello M, Gaharwar AK (2016) Nanoengineered thermoresponsive magnetic hydrogels for biomedical applications. Bioeng Transl Med 1:297–305. https://doi.org/10.1002/btm2.10034
105. Yao S, Swetha P, Zhu Y (2018) Nanomaterial-enabled wearable sensors for healthcare. Adv Healthc Mater 7:1700889. https://doi.org/10.1002/adhm.201700889
106. Govindan B, Sabri MA, Hai A et al (2023) A review of advanced multifunctional magnetic nanostructures for cancer diagnosis and therapy integrated into an artificial intelligence approach. Pharmaceutics 15:868. https://doi.org/10.3390/pharmaceutics15030868
107. Zhai Q, Du B, Feng R et al (2014) A highly sensitive gas sensor based on Pd-doped $Fe_3 O_4$ nanoparticles for volatile organic compounds detection. Anal Methods 6:886–892. https://doi.org/10.1039/C3AY41867G
108. Chattopadhyay S, Sabharwal PK, Jain S et al (2019) Functionalized polymeric magnetic nanoparticle assisted SERS immunosensor for the sensitive detection of S. typhimurium. Anal Chim Acta 1067:98–106. https://doi.org/10.1016/j.aca.2019.03.050
109. Kim MS, Kweon SH, Cho S et al (2017) Pt-decorated magnetic nanozymes for facile and sensitive point-of-care bioassay. ACS Appl Mater Interfaces 9:35133–35140. https://doi.org/10.1021/acsami.7b12326
110. Kulshrestha A, Sharma S, Singh K, Kumar A (2022) Magnetoresponsive biocomposite hydrogels comprising gelatin and valine based magnetic ionic liquid surfactant as controlled release nanocarrier for drug delivery. Mater Adv 3:484–492. https://doi.org/10.1039/D1MA00758K
111. Huang Y, Deng L, Ju P et al (2018) Triple-action self-healing protective coatings based on shape memory polymers containing dual-function microspheres. ACS Appl Mater Interfaces 10:23369–23379. https://doi.org/10.1021/acsami.8b06985
112. Pyo S, Eun Y, Sim J et al (2022) Carbon nanotube-graphene hybrids for soft electronics, sensors, and actuators. Micro Nano Syst Lett 10:9. https://doi.org/10.1186/s40486-022-00151-w
113. Echeverri D, Orozco J (2022) Glycan-based electrochemical biosensors: promising tools for the detection of infectious diseases and cancer biomarkers. Molecules 27:8533. https://doi.org/10.3390/molecules27238533
114. Rong G, Zheng Y, Sawan M (2021) Energy solutions for wearable sensors: a review. Sensors 21:3806. https://doi.org/10.3390/s21113806

Chapter 2
Biomedical Applications of Functionalized Magnetic Nanomaterials (FMNs)

Murali Mohan Achari Kamsali and Vasantha Thoti

Abstract Over the past few decades, nanomaterials have been extensively used in various fields, especially in biomedicine and healthcare. Their unique physicochemical properties, such as high surface area and nanoscale size, make them highly desirable. Recently, there has been significant interest in developing therapies based on magnetic nanoparticles for biomedical applications, owing to their unique advantages. Magnetic particles, which respond to magnetic forces, are useful in drug targeting and bioseparation, including cell sorting. To be used in biomedicine, these magnetic nanomaterials often need to be modified or functionalized with other materials. This functionalization allows them to be used in various biomedical applications, such as biosensing, drug delivery, imaging, medical implants, cancer treatment, antibacterial and antiviral applications, and tissue engineering. Additionally, functionalized magnetic nanomaterials (FMNs) have applications in hyperthermic treatments. For instance, magnetite nanoparticles conjugated with antibodies, known as antibody-conjugated magneto-liposomes (AMLs), are used in hyperthermia and enhance tumor-specific contrast in MRI via systemic administration. This chapter provides a critical overview of recent research on functionalized magnetic nanomaterials in biomedical applications. It also discusses the role of the magnetic properties of these nanomaterials in targeted therapeutic approaches and improved diagnostic tools.

Keywords Biomedicine · Functionalized magnetic nanomaterials (FMNs) · Drug delivery · Tissue engineering · Hyperthermia · Applications

M. M. A. Kamsali (✉) · V. Thoti
Department of Chemistry, Sri Venkateswara College, University of Delhi, Delhi, India
e-mail: mmakamsali@svc.ac.in

S. Gulati (ed.), *Functionalized Magnetic Nanomaterials*, Nanostructure Science and Technology, https://doi.org/10.1007/978-3-031-97199-0_2

2.1 Introduction

Functionalized magnetic nanomaterials (FMNs) are an evolving area in nanotechnology aimed at enhancing the properties and functionalities of nanoscale materials through modification. These materials, typically between 1 and 100 nm, possess unique physical, chemical, and biological traits due to quantum confinement effects and high surface area-to-volume ratios. Nanomaterials (NMs) with at least one dimension in the nanoscale range exhibit novel properties such as improved mechanical strength, electrical conductivity, and catalytic activity, making them valuable in medicine, electronics, and environmental science. Functionalization, which involves attaching various functional groups, molecules, or coatings to the surface of nanoparticles (NPs), tailors their properties for specific uses, enhancing compatibility, reactivity, and stability for diverse applications [1]. This process allows NMs to interact more effectively with biological systems, electronic devices, or environmental contexts.

FMNs are notable due to their high stability, large magnetic moments, and excellent response to moderate magnetic fields. Their unique properties enable them to cross biological barriers, protect drugs from rapid degradation, and provide a large surface area for conjugating targeting ligands. Additionally, FMNs offer low production costs and superparamagnetism for external magnetic field-guided targeting. Over the past decade, research on FMNs has surged, particularly in high-interest biomedical fields such as biomaterial science, diagnostics, magnetic drug and gene delivery, hyperthermia, magnetic resonance imaging (MRI), and theranostics. FMNs have shown utility in drug delivery, medical imaging, and tissue engineering. Magnetic nanoparticles, for instance, have been extensively studied for targeted drug delivery and hyperthermia treatment in cancer therapy. Cancer remains a major global health issue, responsible for one out of every eight deaths worldwide. The International Agency for Research on Cancer (IARC) projects approximately 13.1 million cancer-related deaths by 2030 [2]. The rise in cancer cases is attributed to genetic predisposition, immune deficiencies, industrialization, population growth, and lifestyle changes. Current trends suggest a potential 70% increase in cancer cases, necessitating innovative therapeutic approaches. Magnetic nanoparticle-mediated hyperthermia has emerged as a promising cancer treatment modality, utilizing the heat-generating properties of superparamagnetic iron oxide nanoparticles (SPIONs) under an alternating magnetic field. Hyperthermia raises tissue temperatures to 40–45 °C, leading to cancer cell destruction or growth inhibition. Studies show that hyperthermia synergistically enhances radiotherapy and chemotherapy, offering a promising strategy to improve cancer treatment outcomes.

Beyond cancer therapy, FMNs are significant in cell therapy, tissue engineering, and regenerative medicine by modulating stem cell functions. Functionalized nanoparticles can be modified with bioactive molecules to serve as vehicles for drug or gene transport, aiding in controlled stem cell differentiation and regeneration under magnetic field influence. They can also be incorporated into biocompatible scaffolds with cells or bioactive substances to support tissue regeneration.

Developing diverse scaffold materials with optimized surface properties has been crucial in biomedical research to promote cell adhesion and proliferation for specific applications. Among the promising nanoscale materials are iron oxide nanoparticles (IONPs), which can be functionalized with bioactive substances, embedded in composites, and internalized by cells. IONPs are widely used in drug delivery systems for targeted delivery through specific binding proteins like antibodies or external magnetic fields. Additionally, primary cells and cell lines can be magnetically labeled with IONPs, facilitating non-invasive in vivo monitoring of cell therapy efficacy using MRI [3].

Apart from biomedical applications, FMNs have extensive utility in environmental remediation, including water purification, pollutant degradation, and air filtration. Functionalizing nanoparticles with catalytic or adsorptive properties effectively removes contaminants from water and air, contributing to environmental sustainability. In energy storage, FMNs improve the performance of batteries, supercapacitors, and fuel cells due to their superior electrical conductivity, chemical stability, and surface area, enhancing efficiency and longevity. FMNs also have potential in electronics, with applications in sensors, transistors, flexible electronics, and photodetectors. By tailoring the electronic and optical properties of nanomaterials, researchers develop devices with enhanced performance and novel functionalities [4].

This chapter focuses on the biomedical applications of FMNs, focusing on their roles in drug delivery, imaging, and tissue engineering. It highlights recent advancements and future prospects, emphasizing how the magnetic properties of FMNs enable targeted therapeutic approaches and improve diagnostic tools.

2.2 Functionalized Magnetic Nanomaterials (FMNs) in Drug Delivery

2.2.1 Targeted Drug Delivery

Functionalized nanomaterials (FMNs) have revolutionized drug delivery systems by enhancing precision and effectiveness through their magnetic properties. Advanced research has developed multicomponent nanoparticles that combine magnetic and other functional elements, enabling combination therapies targeting multiple disease pathways. These nanoparticles, known as magnetic targeted carriers, deliver particles precisely to target tissues. Surface coatings on these particles improve medicinal properties, preventing pharmacokinetic effects and toxicity from cell or protein interactions, thus enhancing biocompatibility. Functionalizing nanoparticles with ligands or antibodies ensures accurate targeting, reducing harm to healthy tissues. This targeted approach has shown benefits in cancer therapy, with folic acid-functionalized nanoparticles effectively targeting cancer cells. Additionally, nanoparticles functionalized with chemotherapeutic drugs and RNA molecules have increased treatment efficiency and reduced side effects [5]. Their size, remote control capability, and resonance response to magnetic fields make magnetic nanoparticles valuable in biomedicine.

2.2.2 *Classification of Magnetic Nanoparticles*

To select the most suitable magnetic nanoparticles (MNPs) for specific biomedical applications, it is essential to clearly classify them based on their nature, composition, and size, considering the synthesis methods and surface functionalization. Several types of MNPs have been developed and researched over time. Generally, these MNPs are grouped into three main categories: (i) Magnetic pure metals (Fe, Co, Ni), (ii) Magnetic metal oxides (Fe_2O_3, Fe_3O_4) or ferrites (MFe_2O_4, where M = Fe, Co, Zn), (iii) Multicomponent magnetic nanoparticles as core/shell MNPs or magnetic nanoclusters (Fig. 2.1). Each category has advantages and disadvantages, but its properties can be tailored for specific applications. The following sections will discuss and exemplify each category.

2.2.2.1 Magnetic Pure Metals

Pure metal materials possess distinctive properties shaped by electron distribution, particularly their magnetic behavior. Transition metals such as Fe, Ni, Co, and Mn are frequently utilized for their exceptional magnetic performance in biomedical applications. Iron nanoparticles are extensively employed due to their superior magnetic properties. Although cobalt has higher toxicity, it is also used in MRI imaging and hyperthermia treatments for tumors [6].

2.2.2.2 Magnetic Metal Oxides

Iron oxide is distinguished by its excellent magnetic properties and lower toxicity, making it ideal for nanotechnology applications. Metal oxides are preferred in biomedical fields due to their stability, low toxicity, and biocompatibility, particularly for magnetic bio-separation, hyperthermia, targeted drug delivery, and biosensing. Coating iron oxide nanoparticles (NPs) with biocompatible layers improves stability and reduces toxicity. Superparamagnetic iron oxide nanoparticles (SPIONs), 10–30 nm in diameter, are beneficial for controlled drug delivery under an external magnetic field. Spinel ferrites ($MeFe_2O_4$), doped with divalent transition metals, are tailored for specific biomedical applications, combining electrical and magnetic properties. Coated ferrite NPs with stabilizing molecules like PEG ensure biocompatibility and safe drug release without adverse toxicity effects [7].

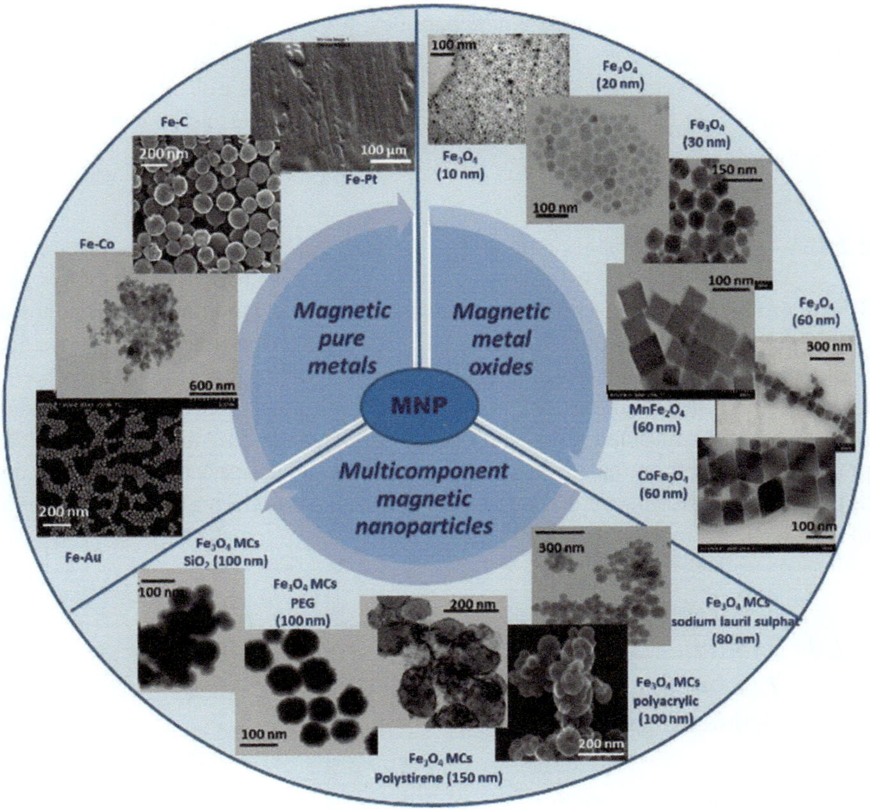

Fig. 2.1 Categorization of magnetic nanomaterials (All images are original and credited to the authors) Ref. [7] © 2023 by the authors. Licensee MDPI, Basel, Switzerland. This article is an open-access article distributed under the terms and conditions of the Creative Commons Attribution (CC BY-4.0) license (https://creativecommons.org/licenses/by/4.0/)

2.2.3 Multicomponent Magnetic Nanoparticles

Multicomponent MNPs have evolved to offer multiple functionalities by combining two or more components, providing new features unavailable in single-component materials. The two main categories of multicomponent magnetic structures with potential biomedical applications are:

2.2.3.1 Core/Shell-Type MNPs

These are the most common and widely studied multicomponent nanoparticles. Initially used in semiconductors to modify electrical and optical properties, the concept was extended to magnetic systems to protect and enhance metallic MNP properties, especially reactive Fe nanoparticles. Some core/shell magnetic systems feature a bimagnetic core, combining two magnetic metallic species covered with an oxide layer. Examples include FePt/Fe_3O_4, FePt/ZnO, Fe_3O_4/ZnO, and Fe_3O_4/Au, which benefit from the SPION core's magnetic properties and the noble metals' plasmon resonance properties [8].

2.2.3.2 Magnetic Clusters (MCs)

These are created by the controlled self-assembly of superparamagnetic nanoparticles into three-dimensional clusters. MCs have gained popularity in biomedical applications due to their innovative collective properties. Depending on the preparation method and the nature of functionalizing polymers, MNPs can form clusters of various shapes and sizes, such as spheres, rods, and nanorings. MCs with a polymer coating show promise for theragnostic applications, enhancing MRI and hyperthermia performance and allowing for the encapsulation and targeted release of drugs [9].

2.2.4 Limitations and Solutions to Overcome These Limitations in the Use of MNPs in Biomedical Applications

For effective biomedical applications, identifying ideal properties for the functionalization of bio-MNPs is crucial to enhance performance. Key factors include colloidal stability in aqueous environments to prevent agglomeration and potential negative effects on human health and the environment. Proper surface functionalization ensures biocompatibility and biospecificity. Research focuses on surface modifications to improve colloidal stability, prevent aggregation, ensure non-toxicity in physiological conditions, and introduce functional groups for targeting specific molecules. MNPs must have low toxicity and high biocompatibility for biomedical applications. MNPs can generate reactive oxygen species that cause cellular-level toxicity, including mitochondrial destruction and DNA damage. Factors like shape, size, and surface properties affect biocompatibility. Enhancing biocompatibility involves functionalizing MNPs with agents like polymers, proteins, or inorganic compounds. Proteins like albumin and silica coatings improve biocompatibility, and noble metal coatings, such as gold, further reduce toxicity [10].

2.3 Functionalized Magnetic Nanomaterials (FMNs)—Magnetic Resonance Imaging (MRI) and Diagnostics

MRI is a widely used non-invasive diagnostic technique in medicine for generating detailed images of soft tissues. When a magnetic field (B_0) is applied, hydrogen protons align with the field. MRI measures the net magnetization (M) of these protons. A radiofrequency pulse disrupts M_0 (equilibrium magnetization), causing it to spin out of equilibrium. Upon stopping the pulse, M returns to equilibrium through relaxation. This process creates detailed images. There are two relaxation processes: longitudinal (T1—recovery) and transverse (T2—decay). T1 relaxation involves energy exchange with the surroundings, producing bright signals (positive contrast). T2 relaxation involves energy transfer between spins, creating dark signals (negative contrast). T2* is an apparent transverse relaxation time affected by magnetic field inhomogeneities, making it shorter than T2. MRI can differentiate between ischemic and hemorrhagic strokes, assess damaged areas, locate metastasis, determine tumor size and stage, and diagnose demyelinating diseases, Alzheimer's dementia, and congenital heart diseases [11].

2.3.1 Overcoming MRI Limitations with FMNs

MRI's drawbacks include low specificity and high cost compared to other imaging methods like CT. This is particularly evident in breast cancer screening, where MRI's expense limits its use to high-risk groups and difficult-to-detect lesions. Consequently, professionals often prefer cheaper, faster tests such as ultrasound and mammography. Additionally, MRI's low specificity can lead to false positives and unnecessary treatments. FMNs offer a solution by enhancing MRI sensitivity and specificity, as they accumulate in target tissues, shortening proton relaxation times and improving MR images. Some FMNs shorten both T1 and T2 relaxation times, leading to dual-mode T1- and T2-weighted MRI for increased diagnostic accuracy. Dual-mode imaging produces matching T1- and T2-weighted images using a single system, providing complementary information for a more complete diagnosis. Dual-mode agents can be created through metal doping of IONPs or synthesizing MNPs with a T2-weighted core and T1-weighted shell, potentially enhancing MRI's diagnostic accuracy. MNPs can also serve as multimodal imaging platforms, such as PET-MRI, combining PET's high sensitivity with MRI's high resolution for a comprehensive examination. Many FMN formulations have completed clinical trials and received approval as MRI contrast agents [12].

2.3.2 IONPs' Clinical Uses

Numerous iron oxide nanoparticle (IONP) formulations have been approved for clinical use by the EMA and FDA, such as ferumoxytol, ferumoxide, ferumoxsil, and ferristene. Ferucarbotran and Sienna+® are approved in the EU but not in the USA. Ferumoxytol is user-friendly, administered as an IV bolus for MRA and dynamic MR, and treats IDA in adults with CKD. Ferumoxide enhances T2-weighted MRI for liver lesion detection. Ferumoxsil and ferristene are used orally for gastrointestinal imaging. Ferucarbotran is an IV liver imaging contrast agent in Germany. Sienna+® is used for sentinel lymph node detection in breast cancer. Ferumoxtran is researched for CNS visualization and MRI contrast enhancement in MS patients. Feruglose is used for coronary angiography. Mn-based agents like Mangafodipir and LumenHance™ were approved for liver imaging but withdrawn due to toxicity concerns. Mn_xO_y nanoparticles are promising for T1-weighted MRI with better toxicity profiles than Gd-based agents. Holmium (Ho) nanoparticles show significant potential in ultra-high MRI and are non-toxic at low concentrations. These advancements highlight the diverse and promising applications of IONPs in clinical imaging and diagnostics. Ni et al. reported r2 relaxation values of 222.6 mM^{-1} s^{-1} at 7 T for $NaHoF_4$ nanoparticles, whereas Atabaev et al. created PEGylated Ho_2O_3 nanoparticles with an r2 of 23.47 mM^{-1} s^{-1} at 1.5 T and green fluorescence resulting from intra-4f-transitions in Ho ions. Cytotoxicity studies showed these PEG-Ho_2O_3 nanoparticles were non-toxic at concentrations below 16 μg/mL [13]. Lin et al. demonstrated that Gd chelates loaded into MSNPs had better T_1 and T_2 relaxivity values compared to SSNPs with similar Gd agents and successfully used these nanoparticles for imaging the aorta and liver in mice [14]. Torresan et al. developed FeAu nanoparticles coated with thiolated PEG for use as CT/MRI contrast agents in mice, showing low particle accumulation in the liver and spleen while CT measurements outperformed iopromide [15]. Other alloy MNPs, such as FePt and FeNi, also exhibited high superparamagnetic properties with low toxicities as MRI contrast agents [16].

2.4 FMNs in Diagnostics

Early diagnosis is essential for effective treatment, as many critical conditions are detected too late, leading to preventable deaths. Accurate diagnostic assays are lacking, with MRI offering high resolution but low sensitivity and specificity, and PCR tests being highly sensitive and specific but time-consuming [17].

FMNs provide promising solutions for molecular diagnosis, enabling faster and simpler tests through magnetic separation, as shown in Fig. 2.2. They also enhance the sensitivity and specificity of MRI.

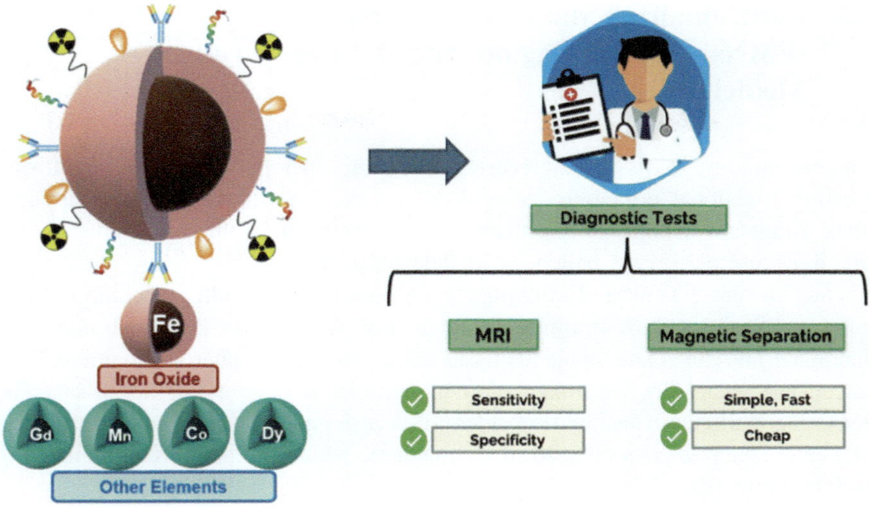

Fig. 2.2 Magnetic nanoparticles: a versatile tool for diagnostic tests in various imaging techniques Ref. [17] © 2021 by the authors. Licensee MDPI, Basel, Switzerland. This article is an open-access article distributed under the terms and conditions of the Creative Commons Attribution (CC BY) license (https://creativecommons.org/licenses/by/4.0/). (All images are original and credited to the authors.)

Chen et al. developed a highly sensitive immunosensor that employs $CoFe_2O_4$ nanoparticles conjugated with an anti-CEA antibody [18]. These $CoFe_2O_4$ nanoparticles can be incorporated into biosensors to enhance sensitivity and specificity, speeding up the process and lowering the detection limit. Furthermore, $CoFe_2O_4$ nanoparticles can detect specific DNA strands, identifying point mutations and single-nucleotide polymorphisms without the need for time-consuming PCR methods. In addition, He et al. developed an immunosensor for detecting N-terminal pro-brain natriuretic peptide (NT-proBNP), a vital diagnostic and prognostic marker for heart failure. This immunosensor showcased a wide detection range, high sensitivity, and good reproducibility [19].

Long and colleagues have investigated methods to effectively isolate and enrich phosphopeptides from complex samples. FMNs have been examined as tools to enhance phosphopeptide signals and minimize impurity interferences, crucial for early disease diagnosis. Studies showed $CuFeMnO_4$ nanoparticles were effective in neutral and acidic conditions, and Mn-doped Fe_2O_4 microspheres also exhibited good reusability and selectivity for phosphopeptides [20]. Bai et al. developed amino-functionalized MNPs for DNA separation, achieving 4–5 times higher DNA adsorption efficiency compared to silica-coated iron oxide nanoparticles (IONPs). SiO_2 nanoparticles coated with IONPs and silica showed improved accumulation and separation yield under a magnetic field, enhancing stability and water solubility [21].

2.5 Functionalized Magnetic Nanomaterials (FMNs)—Tissue Engineering and Regenerative Medicine

Nanotechnology in tissue engineering integrates nanoparticles into nanofibers, creating functional organoids for molecular screening and regenerative applications. Magnetic nanoparticles enhance tissue regeneration through guided bioactive molecule delivery. Challenges include replicating in vivo tissue characteristics and current 2D culture limitations. A universal 3D cell culture technique using magnetic particles shows promise. Koudan et al. showed that cells with non-toxic nanomaterials can create viable 3D tissue spheroids. These spheroids maintain fluorescent microcapsules with magnetic nanoparticles without cytotoxic effects. The dense nanoparticles in the microcapsules preserve the spheroids' magnetic properties. Magnetic spheroids can be utilized for magnetic patterning and tissue-engineering biofabrication (Fig. 2.3) [22].

Magnetic particles hold promise for assembling complex tissues without the need for chemical cell alteration or cell-specific materials. However, several challenges need to be addressed: (i) *Biocompatibility*: Issues with metallic cores can be mitigated using polymeric coatings, hydroxyapatite shells, and choosing biocompatible cores like Fe_2O_3 or Fe_3O_4. (ii) *Immunological Reactions*: Surface charge alterations, functionalization, and neutralization with PEG or acetylation can help mitigate immune responses. (iii) *Accumulation and Biodegradability*: Strategies like functional group coatings or lipids can improve macrophagic uptake and renal clearance to address inefficient biodegradation and clearance, which can otherwise cause organ toxicity [23]. Despite these strategies, significant challenges remain for effectively using magnetic nanoparticles in tissue regeneration therapies.

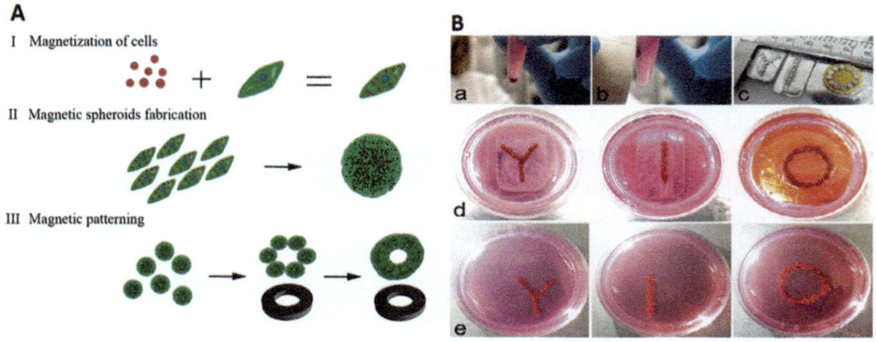

Fig. 2.3 Magnetic nanoparticle-based spheroid patterning involves the following steps: **A** The technique used to create magnetic spheroids. **B** Arrangement of tissue spheroids that are labeled with 12 magnetic microcapsules per cell under the influence of an applied magnetic field. (All images are original and belong to the authors.) Reprinted with permission from Ref. [22], 2021 Copyright © 2021, American Chemical Society

2.5.1 *FMNs as Tissue Engineering Scaffolds*

Tissue engineering aims to restore functional tissue at damaged sites using seed cells, growth factors, and 3D biodegradable scaffolds. While this method is essential for tissue regeneration and healing, a significant challenge is that cells often remain on the scaffold's surface and do not penetrate it. Recent research, however, has shown that a magneto-mechanical drive can direct cells to the scaffold's center, potentially overcoming this limitation. Parfenov et al. developed a technique using magnetic force to seed cells onto a porcine acellular carotid artery stent. The stent was submerged in a suspension of magnetically labeled cells, leading to increased cell adhesion due to iron oxide nanoparticle absorption. Additionally, chitosan-coated iron oxide nanoparticles enhanced the penetration of human osteoblasts into 3D scaffolds, improving cell interaction and accelerating the cell proliferation cycle. This magnetic approach shows significant potential for advancing bone tissue repair and regeneration, providing a valuable theoretical foundation for further research and clinical applications [24].

Fernandes et al. created magnetic nanocomposite scaffolds made of iron oxide nanoparticles and polycaprolactone. These scaffolds showed improved cell adhesion, increased alkaline phosphatase activity, and higher expression of osteogenesis-related genes compared to pure polycaprolactone scaffolds [25]. Gao et al. explored the nano-deformation of IO-OA/PLGA nanocomposites under static magnetic fields, discovering that magnetic mechanical stimulation significantly enhanced the osteogenic differentiation of MC3T3-E1 cells [26]. Cojocaru et al. investigated the combined effects of calcium phosphate composites with magnetic nanoparticle inclusions for bone tissue engineering [27]. Bigham et al. utilized a Mg_2SiO_4-$CoFe_2O_4$ nanocomposite scaffold as a multifunctional magnetic scaffold to eliminate residual bone cancerous tissues post-surgery [28].

2.5.2 *Advancements in Stem Cell Therapy with FMNs*

Recently, stem cell therapy has been proposed for repairing large bone defects. While successful in animal models, traditional stem cell transplantations have not shown comparable clinical results. Effective treatment requires maintaining stem cells' long-term characteristics, ensuring stable cell phenotypes, and reducing cell death at defect sites post-transplantation.

2.5.2.1 Materials Science and Chemical Biology Advances

Researchers are investigating the use of iron oxide nanoparticles with external magnetic fields to control and study stem cells. These nanoparticles affect cell adhesion, proliferation, movement, and osteogenic differentiation and are useful for in vivo tracking and monitoring. Liu et al. demonstrated that Fe_3O_4/BSA particles enhance osteogenesis under magnetic fields, which is promising for stem cell-based bone regeneration therapies [29].

2.5.2.2 Cardiac Tissue Regeneration

Nanotechnology is advancing cardiovascular regeneration, focusing on alternative therapies for blood vessels, heart valves, and myocardium. Nanoparticle probes enable noninvasive imaging of cardiovascular targets, aiding diagnosis and treatment monitoring [30].

2.5.2.3 Bone Regeneration

Stem cell therapy for bone defects is effective in animal models but faces clinical challenges. Iron oxide nanoparticles improve stem cell characteristics and support adhesion, proliferation, and differentiation. They also aid in the precise delivery of labeled cells or medications and visualization via MRI [3].

2.5.2.4 Nervous System Regeneration

Stem cells, progenitor cells, and differentiated cells treat neurodegenerative diseases and brain injuries. Iron oxide nanoparticles enhance brain tissue therapy and drug delivery, with labeling techniques enabling tracking and monitoring through imaging methods like MRI, PET, and multiphoton microscopy [3].

2.6 Functionalized Magnetic Nanomaterials (FMNs)—Magnetic Hyperthermia

Magnetic nanoparticles (MNPs), including SPIONs, are important in biomedicine due to their low toxicity and good biocompatibility. They are used for cell separation, MRI contrast, targeted drug delivery, and localized magnetic hyperthermia (HT) for cancer treatment. HT heats MNPs to induce apoptosis in tumor tissues, enhancing immune responses and improving tumor blood flow, though thermal ablation at higher temperatures is less suitable. While precise control methods are still needed, biodegradable NPs are being developed for more effective HT, safely delivering thermo-therapeutic substances within tumors [31].

2.6.1 *Magnetic Hyperthermia for Cancer Treatment*

FMNs are being explored for hyperthermia treatment, using an external magnetic field to generate localized heat that targets cancer cells while sparing healthy tissues. Hyperthermia (HT) selectively targets tumor cells, enhancing their sensitivity to radiation therapy (RT) and chemotherapy (CT), with minimal impact on normal tissues. Cancer accounted for nearly 10 million deaths worldwide in 2020 [32]. Current treatments include surgery, CT, and RT, but are limited by toxicity, drug resistance, and low efficacy. HT has gained interest as a non-invasive cancer therapy, with early clinical studies dating back to 1891 by William Coley using "Coley's toxin". Hyperthermia (HT) has been used in clinical cancer treatment, though maintaining precise local temperature control is difficult. HT is categorized into three types: whole-body, regional, and local. Whole-body HT employs heating blankets and thermal chambers, often as an additional treatment for metastatic diseases. Regional HT involves using heated fluids and anticancer drugs to perfuse the peritoneal cavity. However, both whole-body and regional HT have limitations in clinical use due to severe side effects like gastrointestinal symptoms and cardiac complications. Hyperthermia (HT) induces cellular changes that disrupt homeostasis and lead to cell death, with the extent of damage dependent on temperature. Moderate temperatures (39–42 °C) are generally non-lethal, while temperatures above 42 °C can kill cells. HT causes cell damage by unfolding proteins, leading to aggregation and affecting crucial processes such as DNA synthesis, repair, transcription, RNA processing, and translation, ultimately causing cell cycle arrest and death. Additionally, HT disrupts the cytoskeleton, alters membrane permeability, and induces metabolic changes, resulting in decreased energy production and increased levels of intracellular sodium (Na^+), hydrogen (H^+), and calcium (Ca^{2+}) [33].

2.6.2 *The Application of Functionalized Magnetic Nanomaterials (FMNs) for Locally Induced Hyperthermia (HT)*

Magnetic hyperthermia (MHT) utilizes FMNs and an alternating magnetic field (AMF) to produce localized heat, effectively targeting cancer cells while preserving healthy tissues. MNPs can be functionalized to target tumor cells specifically, and their properties can be adjusted for improved outcomes. FMNs enhance hyperthermia (HT) effectiveness by creating temperature gradients in tumors, avoiding damage to healthy tissues due to their small size and high reactivity. They absorb heat from external sources, concentrating it on tumors in a process called inside-out HT. Examples include iron oxide, gold nanoparticles, and carbon nanotubes. FMNs use magnetic coupling to absorb and release heat, safely targeting cancer cells without significant energy deposition in healthy tissues [34] (Fig. 2.4).

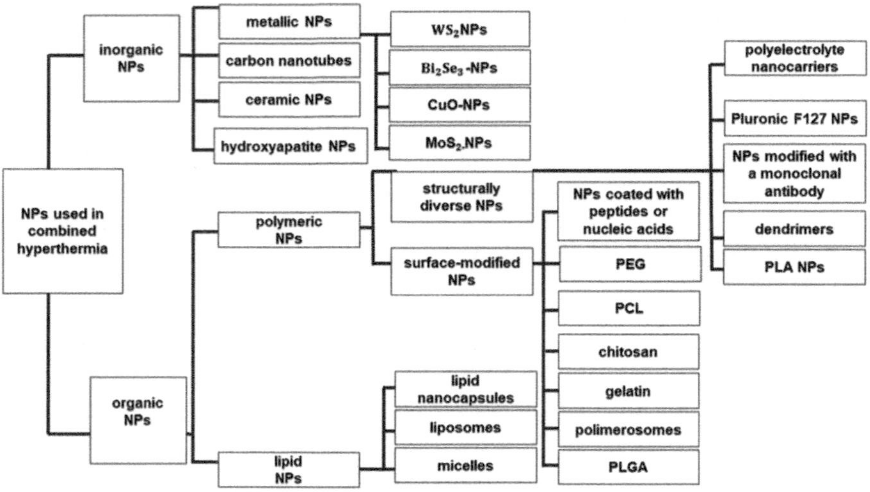

Fig. 2.4 Comprehensive classification of nanosystems for HT (all images are original and belong to the authors) Ref. [34] © 2024 by the authors. Licensee MDPI, Basel, Switzerland. This article is an open-access article distributed under the terms and conditions of the Creative Commons Attribution (CC BY-4.0) license (https://creativecommons.org/licenses/by/4.0/)

2.6.3 NP-Based MHT—Combination Therapies

One key clinical consideration in magnetic hyperthermia therapy (MHT) is the impact of increased temperature on blood flow within tumors and surrounding tissues. Combining MHT with treatments like chemotherapy (CT), radiotherapy (RT), photothermal therapy, or gene therapy enhances effectiveness. MHT combined with CT increases drug concentrations, inhibits DNA repair, and reduces apoptosis-resistant cancer cells. FMNs with MHT improve blood flow, accelerating drug delivery and release in cancer tissues. Combining MHT with RT has proven effective in treating prostate cancer, enhancing treatment efficacy and survival rates. Polymeric FMNs target genetic material delivery, restoring cellular functions and inducing programmed cell death. Increasing internal temperature through photothermal therapy shows synergistic effects when combined with FMNs and laser irradiation [35].

2.7 Functionalized Magnetic Nanomaterials (FMNs)—Biosensing and Detection

Biosensors detect and measure molecules and microorganisms by converting biological interactions into electrical signals using bioreceptors. Electrochemical and optical transductions offer high sensitivity and miniaturization but face challenges in maintaining sensitivity and selectivity over time. FMNs and beads address these issues, enhancing specificity and sensitivity. FMNs combined with electrochemical read-out methods, like ELIME (enzyme-linked immuno-magnetic electrochemical), improve analyte isolation, concentration, and detection. MNPs are also significant in microfluidic systems, enhancing analytical performance and cell separation efficiency [36].

Peng et al. developed a chitosan-functionalized nanocomposite for glucose determination, featuring magnetic nanoparticles, chitosan, β-cyclodextrin (Fe3O4-CS-CD), and multi-walled carbon nanotubes. This sensor showed excellent biocompatibility, nontoxicity, high mechanical strength, low detection limit, broad linear range, high sensitivity, excellent selectivity, and long-term stability [37]. Di Tocco et al. developed a functionalized electrochemical biosensor for triglyceride (TG) detection, using lipase immobilized on chitosan-coated magnetic nanoparticles (MNPs) dispersed in a matrix of multiwalled carbon nanotubes/pectin (MWCNT/Pe) modified with copper oxide on a glassy carbon electrode (GCE). The sensor demonstrated good sensitivity for determining TGs in human serum samples [38]. Wang et al. developed a PEG-mediated APBA-functionalized magnetic nanomaterial (APBA-PEG-MN) for detecting Staphylococcus aureus using magnetic separation and fluorescence analysis, with a limit of detection (LOD) of 270 CFU/mL in various matrices [39].

Polypyrrole (Ppy) and chitosan-coated Fe_3O_4 were used to create a platform for immobilizing glucose-6-phosphate dehydrogenase on pencil graphite electrodes. The resulting biosensor showed good linear response, high selectivity, and recovery for glucose-6-phosphate (G6P) detection [40]. Polyaniline PANI/magnetic graphene (MG) composites were synthesized to immobilize laccase for detecting phenols in various samples, exhibiting high sensitivity and a detection limit of 2.94 μM [41].

Sardaremelli et al. developed an electrocatalytic reduction and enzymatic biosensing strategy for hydrogen peroxide (H_2O_2) using polydopamine-modified magnetic nanoparticles (PDA-MNPs) and a poly(l-arginine/toluidine blue) film-modified glassy carbon electrode, achieving a detection range of 0.5–30 μM and a low limit of quantification at 0.23 μM in human plasma samples [42]. 3D hierarchically porous magnetic molybdenum trioxide–pDA-gold functionalized nanospheres (3D mag-MoO3–pDA@Au) were used in a Raman scattering sensing platform for ultrasensitive detection of SARS-CoV-2 [43]. A biosensor with gold nanoparticle-loaded magnetic reduced graphene oxide (MrGO@AuNPs) was developed for detecting bisphenol A (BPA) with high sensitivity and a detection limit as low as 0.141 pg/mL [44]. Quantum dots (QDs) integrated into magnetic nanomaterials were also used in a prototype biosensor for detecting *E. coli* in water samples, showing potential for food sample applications [45]. The main benefits of magnetic nanomaterials are their enhanced sensitivity and detection speed, affordability, and user-friendliness, even for those with minimal training. Moreover, magnetic nanoparticles boast an almost endless shelf-life, unlike biosensing methods that rely on fluorescent or radioactive materials, and they are easier to assemble and integrate with microfluidics. Magnetic nanoparticles are transforming biosensor technology, enhancing sensitivity, speed, and cost-effectiveness. They are used in medical diagnostics and bioterrorism detection. Advances include biologically relevant coatings for targeted drug delivery and magneto-optic biosensors that detect chemical binding events in real time.

2.8 Functionalized Magnetic Nanomaterials (FMNs)—Toxicology

Toxicology studies determine the harmful effects of agents on animals, humans, and the environment. Cellular toxicity manifests as altered morphology, reduced mitochondrial activity, and membrane leakage, which negatively impact cell survival and therapies [46]. In treatments, FMNs raise concerns about their potential migration, entry, and accumulation within organs, which may trigger inflammatory or immune responses. Absorbed FMNs can cause cytotoxic reactions, and their physical characteristics, such as shape and size, can lead to hazardous reactions. FMNs pose risks due to oxidative stress resulting from an imbalance of antioxidants and oxidants [47]. Oxidative stress, the main cause of reactive oxygen species (ROS) production, damages cellular proteins, lipids, and nucleic acids, leading to disease and cell death [48]. ROS sources include the MNP surface, metal ion loss, and oxidant release during enzymatic breakdown [49]. Isolated iron oxide MNPs can promote ROS production, affecting cell metabolism and enhancing apoptosis, but ROS can also be used in cancer treatment [50].

2.9 Concluding Remarks and Future Perspectives

FMNs have shown significant potential in biomedical applications such as drug delivery, imaging, tissue engineering, and cancer treatment. Their magnetic properties enable targeted therapies, improved diagnostic accuracy, and minimally invasive procedures. FMNs have advanced hyperthermic treatments and biosensing, addressing contemporary healthcare challenges. However, concerns about biocompatibility, stability, and potential cytotoxicity need further investigation.

Future research should optimize FMNs' functionalization techniques to enhance specificity and efficiency in biomedical applications. Developing biodegradable, non-toxic magnetic nanomaterials is crucial for reducing side effects and improving patient safety. Integrating FMNs with technologies like artificial intelligence, personalized medicine, and nanorobotics could revolutionize clinical applications. Rigorous preclinical and clinical trials are necessary to validate efficacy and secure regulatory approval. Addressing these challenges will enable FMNs to transform modern healthcare, enhancing patient outcomes and therapeutic precision.

Abbreviations

AMF	Alternating Magnetic Field
AMLs	Antibody-Conjugated Magnetoliposomes
APBA	*m*-Aminophenylboronic Acid
BSA	Bovine Serum Albumin
CKD	Chronic Kidney Disease
CNS	Central Nervous System
CT	Chemotherapy
DTPA	Diethylenetriamine Pentaacetic Acid
ECM	Extracellular Matrix
ELIME	Enzyme-Linked Immuno-Magnetic Electrochemical method
FMNs	Functionalized Magnetic Nanomaterials
HT	Hyperthermia
IARC	International Agency for Research on Cancer
IDA	Iron Deficiency Anemia
IONPs	Iron Oxide Nanoparticles
IO-OA/PLGA	Iron Oxide-Oleic Acid/Poly(lactic-co-glycolic acid)
MCs	Magnetic Clusters
MHT	Magnetic Hyperthermia Therapy
MNPs	Magnetic Nanoparticles
MRA	MR Angiography
MRI	Magnetic Resonance Imaging
MS	Multiple Sclerosis
MWCNT/Pe	Multiwalled Carbon Nanotubes/Pectin
NPs	Nanoparticles
PEG	Polyethylene Glycol
PNS	Peripheral Nervous System
RT	Radiotherapy
SLN	Sentinel Lymph Node
SPIONs	Superparamagnetic Iron Oxide Nanoparticles
SSNPs	Solid Silica Nanoparticles

References

1. Sharma PK, Dorlikar S, Rawat P, Malik V, Vats N, Sharma M, Rhyee JS, Kaushik AK (2021) Nanotechnology and its application: a review. In: Khondakar KR, Kaushik AK (eds) Nanotechnology in cancer management. Elsevier, pp 1–33
2. Sung H, Ferlay J, Siegel RL, Laversanne M, Soerjomataram I, Jemal A, Bray F (2021) Global cancer statistics 2020: GLOBOCAN estimates of incidence and mortality worldwide for 36 cancers in 185 countries. CA Cancer J Clin 71:209–249

3. Kermanian M, Sadighian S, Ramazani A, Naghibi M, Khoshkam M, Ghezelbash P (2021) Inulin-coated iron oxide nanoparticles: a theranostic platform for contrast-enhanced MR imaging of acute hepatic failure. ACS Biomater Sci Eng 7(6):2701–2715
4. Cai W (2024) Functional nanomaterials for sensing and detection. Nanomaterials 14(1):128
5. Ghazal H, Waqar A, Yaseen F, Shahid M, Sultana M, Tariq M, Bashir MK, Tahseen H, Raza T, Ahmad F (2024) Role of nanoparticles in enhancing chemotherapy efficacy for cancer treatment. Next Mater 2:100128
6. Nosheen S, Irfan M, Abidi SH, Syed Q, Habib F, Asghar A, Waseem B, Soomro B, Butt H (2021) A review: development of magnetic nano vectors for biomedical applications. GSC Adv Res Rev 8:85–110
7. Pusta A, Tertis M, Crăciunescu I, Turcu R, Mirel S, Cristea C (2023) Recent advances in the development of drug delivery applications of magnetic nanomaterials. Pharmaceutics 15:1872
8. Tian Q, Hu J, Zhu Y, Zou R, Chen Z, Yang S, Li R, Su Q, Han Y, Liu X (2013) Sub-10 nm Fe3O4@Cu(2–x)S core-shell nanoparticles for dual-modal imaging and photothermal therapy. J Am Chem Soc 135(23):8571–8577
9. Storozhuk L, Besenhard MO, Mourdikoudis S, LaGrow AP, Lees MR, Tung LD, Gavriilidis A, Thanh NTK (2021) Stable iron oxide nanoflowers with exceptional magnetic heating efficiency: simple and fast polyol synthesis. ACS Appl Mater Interfaces 13:45870–45880
10. Malhotra N, Lee JS, Liman RAD, Ruallo JMS, Villaflore OB, Ger TR, Hsiao CD (2020) Potential toxicity of iron oxide magnetic nanoparticles: a review. Molecules 25:3159
11. Colombo M, Carregal-Romero S, Casula MF, Gutiérrez L, Morales MP, Böhm IB, Heverhagen JT, Prosperi D, Parak WJ (2012) Biological applications of magnetic nanoparticles. Chem Soc Rev 41:4306–4334
12. Cheng KK, Chan PS, Fan S, Kwan SM, Yeung KL, Wáng YXJ, Chow AHL, Wu EX, Baum L (2015) Curcumin-conjugated magnetic nanoparticles for detecting amyloid plaques in Alzheimer's disease mice using magnetic resonance imaging (MRI). Biomaterials 44:155–172
13. Ni D, Bu W, Ehlerding EB, Cai W, Shi J (2017) Engineering of inorganic nanoparticles as magnetic resonance imaging contrast agents. Physiol Behav 46:7438–7468
14. Taylor KM, Kim JS, Rieter WJ, An H, Lin W, Lin W (2008) Mesoporous silica nanospheres as highly efficient MRI contrast agents. J Am Chem Soc
15. Amendola V, Torresan V, Forrer D, Guadagnini A, Badocco D, Pastore P, Casarin M, Selloni A, Coral D, Ceolin M et al (2020) 4D multimodal nanomedicines made of nonequilibrium Au-Fe alloy nanoparticles. ACS Nano 14:12840–12853
16. Huynh KH, Pham XH, Kim J, Lee SH, Chang H, Rho WY, Jun BH (2020) Synthesis, properties, and biological applications of metallic alloy nanoparticles. Int. J. Mol. Sci. 21:5174
17. Farinha P, Coelho JMP, Reis CP, Gaspar MM (2021) A comprehensive updated review on magnetic nanoparticles in diagnostics. Nanomaterials 11:3432
18. Chen ZG, Tang DY (2007) Antigen-antibody interaction from quartz crystal microbalance immunosensors based on magnetic $CoFe_2O_4/SiO_2$ composite nanoparticle-functionalized biomimetic interface. Bioprocess Biosyst. Eng. 30:243–249
19. He Y, Wang Y, Yang X, Xie S, Yuan R, Chai Y (2016) Metal organic frameworks combining $CoFe_2O_4$ magnetic nanoparticles as highly efficient SERS sensing platform for ultrasensitive detection of N-terminal pro-brain natriuretic peptide. ACS Appl. Mater. Interfaces 8:7683–7690
20. Long XY, Li JY, Sheng D, Lian HZ (2017) Spinel-type manganese ferrite ($MnFe_2O_4$) microspheres: a novel affinity probe for selective and fast enrichment of phosphopeptides. Talanta 166:36–45
21. Bai Y, Song M, Cui Y, Shi C, Wang D, Paoli GC, Shi X (2013) A rapid method for the detection of foodborne pathogens by extraction of a trace amount of DNA from raw milk based on amino-modified silica-coated magnetic nanoparticles and polymerase chain reaction. Anal Chim Acta 787:93–101
22. Koudan EV, Zharkov MN, Gerasimov MV, Karshieva SS, Shirshova AD, Chrishtop VV, Kasyanov VA, Levin AA, Parfenov VA, Karalkin PA, Pereira FDAS, Petrov SV, Pyataev NA, Khesuani YD, Mironov VA, Sukhorukov GB (2021) Magnetic patterning of tissue spheroids using polymer microcapsules containing iron oxide nanoparticles. ACS Biomater Sci Eng 7:5206

23. Senthilkumar N, Sharma PK, Sood N, Bhalla N (2021) Designing magnetic nanoparticles for in vivo applications and understanding their fate inside human body. Coord Chem Rev 445:214082
24. Eivazzadeh-Keihan R, Chenab KK, Taheri-Ledari R, Mosafer J, Hashemi SM, Mokhtarzadeh A, Maleki A, Hamblin MR (2020) Recent advances in the application of mesoporous silica-based nanomaterials for bone tissue engineering. Mater Sci Eng C 107:110267
25. Fernandes MM, Correia DM, Ribeiro C, Castro N, Correia V, Lanceros-Mendez S (2019) Bioinspired three-dimensional magnetoactive scaffolds for bone tissue engineering. ACS Appl Mater Interfaces 11:45265–45275
26. Gao L, Tang Y, Wang C, Yao L, Zhang J, Gao R et al (2019) Highly efficient amphiphilic magnetic nanocomposites based on a simple sol-gel modification for adsorption of phthalate esters. J Colloid Interface Sci 552:142–152
27. Cojocaru FD, Balan V, Popa MI, Lobiuc A, Antoniac A, Antoniac IV, Verestiuc L (2019) Biopolymers—calcium phosphates composites with inclusions of magnetic nanoparticles for bone tissue engineering. Int J Biol Macromol 125:612–620
28. Bigham A, Aghajanian AH, Behzadzadeh S, Sokhani Z, Shojaei S, Kaviani Y, Hassanzadeh-Tabrizi SA (2019) Nanostructured magnetic Mg_2SiO_4-$CoFe_2O_4$ composite scaffold with multiple capabilities for bone tissue regeneration. Mater Sci Eng C 99:83–95
29. Li X, Wei Z, Lv H, Wu L, Cui Y, Yao H, Li J, Zhang H, Yang B, Jiang J (2019) Iron oxide nanoparticles promote the migration of mesenchymal stem cells to injury sites. Int J Nanomed 14:573–589
30. Alphandéry E (2020) Iron oxide nanoparticles for therapeutic applications. Drug Discov Today 25:141–149
31. Wang S, Hou Y (2024) New types of magnetic nanoparticles for stimuli-responsive theranostic nanoplatforms. Adv Sci (Weinh) 11:e2305459
32. Rastgoo Oskoui P, Rezvani M (2025) Revolution in cancer treatment methods: perspective review of factors affecting the final results of nanoparticles used in magnetic fluid hyperthermia. Health Sci Rev 14:100212
33. Souiade L, Domingo-Diez J, Alcaide C, Gámez B, Gámez L, Ramos M, Serrano Olmedo JJ (2023) Improving the efficacy of magnetic nanoparticle-mediated hyperthermia using trapezoidal pulsed electromagnetic fields as an in vitro anticancer treatment in melanoma and glioblastoma multiforme cell lines. Int. J. Mol. Sci. 24:15933
34. Szwed M, Marczak A (2024) Application of nanoparticles for magnetic hyperthermia for cancer treatment—the current state of knowledge. Cancers (Basel) 16:1156
35. Shestovskaya MV, Luss AL, Bezborodova OA, Makarov VV, Keskinov AA (2023) Iron oxide nanoparticles in cancer treatment: cell responses and the potency to improve radiosensitivity. Pharmaceutics 15:2406
36. Abedini-Nassab R, Miandoab MP, Sa¸smaz M (2021) Microfluidic synthesis, control, and sensing of magnetic nanoparticles: a review. Micromachines 12:768
37. Peng L, Luo Y, Xiong H, Yao S, Zhu M, Song H (2020) A novel amperometric glucose biosensor based on Fe_3O_4-chitosan-β-cyclodextrin/MWCNTs nanobiocomposite. Electroanalysis 33:723–732
38. Di Tocco A, Robledo SN, Osuna Y, Sandoval-Cortez J, Granero AM, Vettorazzi NR, Martínez JL, Segura EP, Iliná A, Zon MA, Arévalo FJ, Fernández H (2018) Development of an electrochemical biosensor for the determination of triglycerides in serum samples based on a lipase/magnetite-chitosan/copper oxide nanoparticles/multiwalled carbon nanotubes/pectin composite. Talanta 190:30–37
39. Wang Z, Feng X, Xiao F, Bai X, Xu Q, Xu H (2022) A novel PEG-mediated boric acid functionalized magnetic nanomaterials based fluorescence biosensor for the detection of Staphylococcus aureus. Microchem. J. 178:107379
40. Sahin S, Ozmen I, Bastemur GY, Ozkorucuklu SP (2019) Development of voltammetric glucose-6-phosphate biosensors based on the immobilization of glucose-6-phosphate dehydrogenase on polypyrrole- and chitosan-coated Fe(3)O(4) nanoparticles/polypyrrole nanocomposite films. Appl. Biochem. Biotechnol. 188:1145–1157

41. Lou C, Jing T, Zhou J, Tian J, Zheng Y, Wang C, Zhao Z, Lin J, Liu H, Zhao C, Guo Z (2020) Laccase immobilized polyaniline/magnetic graphene composite electrode for detecting hydroquinone. Int. J. Biol. Macromol. 149:1130–1138
42. Sardaremelli S, Hasanzadeh M, Seidi F (2021) Enzymatic recognition of hydrogen peroxide (H_2O_2) in human plasma samples using HRP immobilized on the surface of poly(arginine-toluidine blue)-Fe_3O_4 nanoparticles modified polydopamine: a novel biosensor. J. Mol. Recognit. 34:e2928
43. Achadu OJ, Nwaji N, Lee D, Lee J, Akinoglu EM, Giersig M, Park EY (2022) 3D hierarchically porous magnetic molybdenum trioxide@gold nanospheres as a nanogap-enhanced Raman scattering biosensor for SARS-CoV-2. Nanoscale Adv 4:871–883
44. Hu L, Cui J, Wang Y, Jia J (2022) An ultrasensitive electrochemical biosensor for bisphenol a based on aptamer-modified MrGO@AuNPs and ssDNA-functionalized AuNP@MBs synergistic amplification. Chemosphere 311:137154
45. Pandit C, Alajangi HK, Singh J, Khajuria A, Sharma A, Hassan MS, Parida M, Semwal AD, Gopalan N, Sharma RK, Suttee A, Soni U, Singh B, Sapra S, Barnwal RP, Singh G, Kaur IP (2022) Development of magnetic nanoparticle assisted aptamer-quantum dot based biosensor for the detection of Escherichia coli in water samples. Sci Total Environ 831:154857
46. Singh N, Jenkins GJ, Asadi R, Doak SH (2010) Potential toxicity of superparamagnetic iron oxide nanoparticles (SPION). Nano Rev 1:5358
47. Li X, Wang L, Fan Y, Feng Q, Cui FZ (2012) Biocompatibility and toxicity of nanoparticles and nanotubes. J Nanomater 2012:1–19
48. Singh A, Kukreti R, Saso L, Kukreti S (2019) Oxidative stress: a key modulator in neurodegenerative diseases. Molecules 24:1583
49. Nadar SS, Kelkar RK, Pise PV, Patil NP, Patil SP, Chaubal-Durve NS, Bhange VP, Tiwari MS, Patil PD (2021) The untapped potential of magnetic nanoparticles for forensic investigations: a comprehensive review. Talanta 230:122297
50. Perillo B, Di Donato M, Pezon A, Di Zazzo E, Giovannelli P, Galasso G, Castoria G, Migliaccio A (2020) ROS in cancer therapy: the bright side of the moon. Exp Mol Med 52:192–203

Chapter 3
Functionalized Magnetic Nanomaterials (FMNs) in Drug Delivery Systems

Jyoti Jain, Mahipal Bishnoi, and Deepak Sharma

Abstract Magnetic nanoparticles with surface modifications have attracted considerable attention as a versatile tool in drug delivery systems due to their exclusive combination of magnetic characteristics and the ability to be modified with specific functional groups. Typically made up of magnetic materials such as oxides of iron, these nanoparticles are tailored on their surface with various ligands, polymers, or antibodies to enhance their stability, bioavailability, and capacity for drug loading. The functionalization process not only improves biocompatibility but also allows for the binding of medical substances such as peptides, nucleic acids, or minor molecules, facilitating controlled and sustained drug liberation. A key benefit of using magnetic nanoparticles for targeted drug transport is their capability to be directed toward specific tissues or organs through an external magnetic field. This targeted approach significantly reduces systemic toxicity and minimizes off-target effects commonly related to conventional drug delivery methods. In addition, the applied magnetic field can improve the nanoparticles' ability to penetrate physiological barriers, such as the blood–brain barrier, making it possible to target diseases that were challenging to manage before. This chapter provides a detailed analysis of the role of functionalized magnetic nanomaterials (FMNs) in drug delivery, focusing on their capacity to transport and release therapeutic agents at targeted sites within the body. Key topics include controlled release mechanisms, biocompatibility considerations, and real-world case studies, highlighting the potential of FMNs in advancing precision medicine and improving therapeutic outcomes.

Keywords Functionalised magnetic nanomaterials · Drug delivery · Cancer treatment · Gene therapy · Biocompatibility · Blood–brain interface

J. Jain (✉) · M. Bishnoi · D. Sharma
Department of Chemistry, University of Rajasthan, Jaipur, Rajasthan, India
e-mail: jyoti.jain123@gmail.com

S. Gulati (ed.), *Functionalized Magnetic Nanomaterials*, Nanostructure Science and Technology, https://doi.org/10.1007/978-3-031-97199-0_3

3.1 Introduction

The notion of utilizing functionalized magnetic nanomaterials (FMNs) in targeted therapy was first explored in 1970 by Senyei and Widder. This approach involves attaching therapeutic agents, like chemotherapy medicine or DNA for gene therapy, to the surface of magnetic support or encapsulating them within a polymeric framework containing magnetic materials. Upon administration, typically via oral or intravenous injection, an applied strong-gradient magnetic field directs and localizes these particle-drug complexes to target-specific regions within the body [1]. Magnetic nanomaterials have significant promise in modern medicine. Despite ongoing concerns about the prolonged safety of metal nanoparticles, their utility across various biomedical fields, including site-specific imaging, cancer detection, neurodegenerative disease treatment, AIDS interventions, and therapies for ocular and respiratory conditions, is well-established [2]. A key application of magnetic nanoparticles is drug delivery, where their large surface area allows them to carry substantial amounts of drugs or other therapeutic agents. Additionally, incorporating a core with unique properties, such as a magnetic core, adds to their versatility. However, creating core–shell nanoparticles presents challenges, and while various methods exist, difficulties arise in developing core–shell structures [3]. Materials like refined metals (e.g., iron and nickel), metal blends, and ferrites (viz. $Mn_{0.6}Zn_{0.4}Fe_2O_4$, cobalt ferrite) are commonly used in the production of magnetic nanoparticles for clinical applications, including magnetothermal therapy and magnetic resonance imaging contrast agents. In spite of their elevated saturation magnetization, pure metals are unsuitable for applications within living organisms due to their susceptibility to oxidation and toxicity. As a result, iron oxides such as Fe_3O_4 and γ-Fe_2O_3 have become the dominant materials owing to their excellent stability towards chemicals, colloidal stability, and biocompatibility [4].

The utilization of MNPs in healthcare applications has expanded significantly in recent times, as shown in Fig. 3.1. Notable innovations have been made in addressing key challenges in clinical research. These include toxicity evaluation in both living systems and controlled conditions, ensuring the stability of surface-modified nanoparticles before clinical use, bridging in vitro experiments with real-world biology, and using theoretical models to understand complex biomedical interactions. However, new challenges have emerged, including developing MNP-based diagnostic tools for SARS-CoV-2 and integrating magnetic microrobots into biomedical applications [5].

Once MNPs are accumulated at the targeted location, therapeutic compounds are discharged via various mechanisms, including enzymatic cleavage, charge-based interactions, pH-induced changes, polymer matrix degradation, or thermal activation of the magnetic particles. This targeted delivery significantly reduces the adverse effects associated with conventional chemotherapy, which can affect both healthy and malignant cells. The exact localization and managed release of pharmaceutical agents enhance treatment efficacy by minimizing systemic exposure to toxic compounds [6].

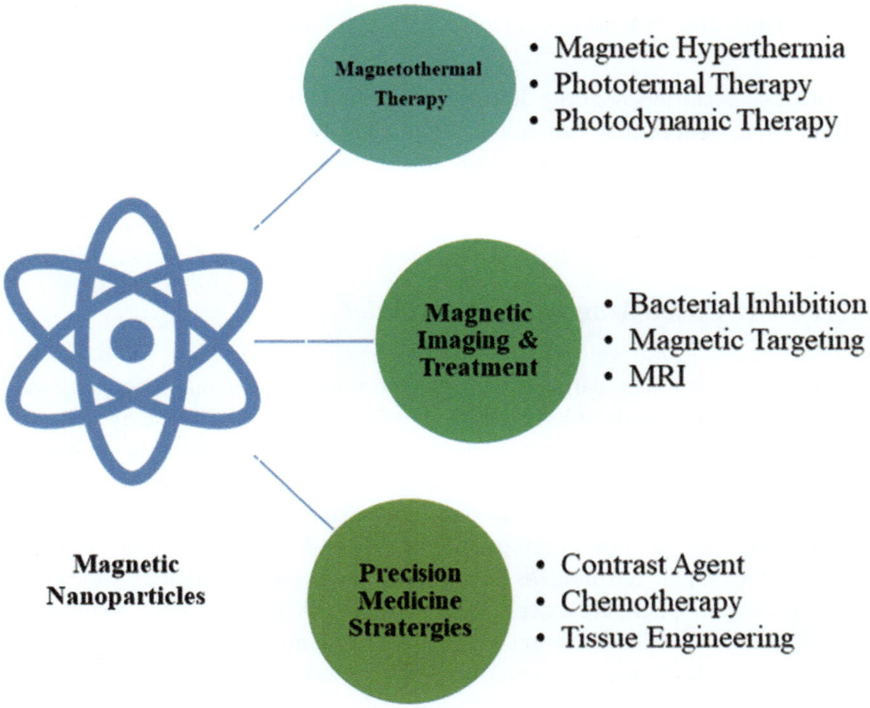

Fig. 3.1 Magnetic nanoparticles and their diverse therapeutic applications

MNPs are commonly covered with organic matrices like complex sugars, fatty lipid molecules, or different macromolecules to enhance suspension stability. A key feature of MNPs is their superparamagnetic behaviour, which means they display zero remanent magnetic charges when the applied field of magnetism is removed. This property prevents aggregation and facilitates their eventual excretion from the body, making them ideal for biomedical applications [4].

However, a major challenge for MNPs is their limited dispersibility in solvents, which can hinder cargo loading and delivery efficiency. Ensuring colloidal stability is critical for effective drug delivery systems [7]. Polymeric encapsulation offers a promising solution by enhancing stability, protecting against protein degradation, and improving drug bioavailability.

Magnetic targeting works by applying a gradient of magnetic flux, generating a directional force acting on the particle-drug system, guiding and localizing it to the desired target location. Effective trapping and retention of the complex depend on several factors, including the rate of blood flow, the structural properties of the carrier molecules, the magnetic characteristics of the nanoparticles, the intensity of the magnetic flux, and the steepness of the magnetic field gradient. Proper control of these parameters is essential for achieving efficient and selective drug delivery [4].

3.2 Magnetic Nanocarriers for Drug Transport

Magnetic nanocarriers are drug delivery systems that use magnetic nanoparticles to directly aim at a specific location in the biological system. By applying an extrinsic magnetic flux, they enable precise drug release and minimize side effects. Surface functionalization enhances stability, bioavailability, and controlled release.

3.2.1 Gel-Based Magnetic Nanocarriers

Superparamagnetic nanoparticles incorporated into polymer matrices, particularly microgels, show considerable potential as magnetic carriers. These systems offer significant advantages, including efficient drug encapsulation, multiple binding sites for bioconjugation, and enhanced mechanical and chemical stability in targeted biological environments. Vekas et al. developed magneto-responsive microgels with high saturation magnetization using a mini-emulsion-based approach. They incorporated a stable organic ferrofluid as the core, with Fe_3O_4 nanoparticles coated in oleic acid, packed into spherical clusters, and encapsulated in polymer shells ranging from 40 to 350 nm. When loaded with the anticancer drug mitoxantrone, these microgels were tested on the Jurkat cell line. The results demonstrated effective drug penetration and initiation of apoptosis based on dose and exposure time. Unloaded microgels showed no cytotoxicity, emphasizing the potential of mitoxantrone-loaded microgels for targeted magnetic drug delivery [7].

3.2.2 Magnetic Pure Metal Nanoparticles

Magnetic metal nanoparticles have achieved notable interest, particularly in medicine, due to their effectiveness in increasing drug delivery and overcoming challenges like multidrug resistance. In addition to drug delivery, MNPs are used in diagnostics, biocompatible materials, and nutraceuticals. Metals such as silver, Au, Pd, titanium, zinc, etc., offer adjustable light-related properties and can be engineered to attach targeting agents or bioactive molecules via hydrogen bonding, covalent bonding, or electrostatic forces. MNPs can carry multiple drugs, improving therapeutic effectiveness by providing enhanced stability, extended circulation, and precise targeting [8].

3.2.3 Metal-Oxide Nanoparticles

Nanoparticles, such as oxides of zinc and titanium, show great potential as strategies for therapeutic delivery in treating cancer, diabetes, and inflammatory conditions. Zhang et al. proposed a method to synthesize ZnO nanoparticles as carriers for the anticancer agent daunorubicin through a straightforward single-step process conducted under laboratory conditions. Their study found that ZnO NPs reduced the cytotoxicity of the drug while enhancing the targeting of cancer cells through reactive oxygen species in human hepatocarcinoma cells [9]. Various TiO_2 nanostructures, including nanoparticles, nanotubes, matrices, capsules, and whiskers, are also gaining attention for delivering anti-cancer drugs like gambogic acid, cisplatin, and doxorubicin. TiO_2 whiskers, for example, significantly enhanced daunorubicin cytotoxicity by increasing its intracellular concentration in SMMC-7721 cells [10].

3.2.4 Metal Sulphide Nanoparticles

Poly(dopamine)-coated (PDA) hollow mesoporous nickel sulphide nanomaterial has been developed as a delivery vehicle for doxorubicin drug, achieving 66.9% encapsulation yield and 7.1% retention capacity of the drug. The high delivery performance is due to the hollow mesoporous structure and strong doxorubicin drug—PDA coating interactions [11]. Hou et al. presented a drug delivery platform formed from copper sulphide nanoparticles for cancer treatment and theranostic applications. Their novel tumor-targeting strategy, diffusion molecular retention, uses transferrin-modified hollow mesoporous copper sulphide nanoparticles, allowing extensive diffusion through the tumor and retention after peritumoral injection [12].

3.2.5 Silica-Coated Magnetic Nanoparticles

Silica coatings on magnetic nanoparticles form core–shell structures that enhance stability, photostability, and water solubility while preventing aggregation and reducing toxicity. Two methods stabilize MNPs with silica: one shields magnetic interactions, and the other imparts a negative charge for improved repulsion. Chen et al. developed a PEGylated mesoporous nano-theragnostic combining MRI and ultrasound for cancer imaging and chemotherapy. They incorporated pH-sensitive manganese oxide nanoparticles into silica nanocapsules, enhancing MRI performance by 11-fold and overcoming cancer cell multidrug resistance [13]. Sun et al. [14] coated superparamagnetic magnetite nanoparticles with amorphous silica using the Stöber process, while Xia et al. [15] developed silica-coated superparamagnetic iron oxide nanomaterials by utilizing ferrofluids coupled with the Stöber method. Kostiv and coworkers demonstrated that adding polyethylene glycol to the surface of iron oxide and silica nanoparticles improved their biocompatibility. The polyethylene glycolated nanoparticles exhibited improved cell viability at higher concentrations when tested with murine neural stem cells, in contrast to uncoated Fe_3O_4 and SiO_2 nanoparticles, which led to a 50% decrease in cell viability even at a lower dose. This result emphasizes the significant contribution of polyethylene glycol(PEG) in improving the biocompatibility of nanoparticles. Additionally, PEG functionalization has been shown to improve hemocompatibility, as evidenced by studies on chromium-doped zinc gallate and diatomite-based NPs [16].

3.3 Controlled Release Mechanisms in Drug Delivery

3.3.1 EPR-Driven Passive Targeting

The improved permeability and retention (EPR) mechanism describes the accumulation of nanoparticles and macromolecules within cancer tissue, resulting from the irregular permeability of blood vessels and compromised lymphatic drainage. This passive targeting strategy is commonly used in nanomedicine to improve drug transport, enhancing treatment effectiveness while reducing overall toxicity in cancer therapy. First identified by Maeda et al. in 1986, the EPR effect occurs because defective lymphatic drainage prevents the clearance of magnetic nanoparticles (MNPs), allowing them to accumulate in tumor sites as illustrated in Fig. 3.2. This enables passive targeting without the need for specific ligands [17].

For effective EPR-based drug delivery, the drug-MNP complex must be smaller than the porous structure of tumor vessels to facilitate extravasation. As soon as in the

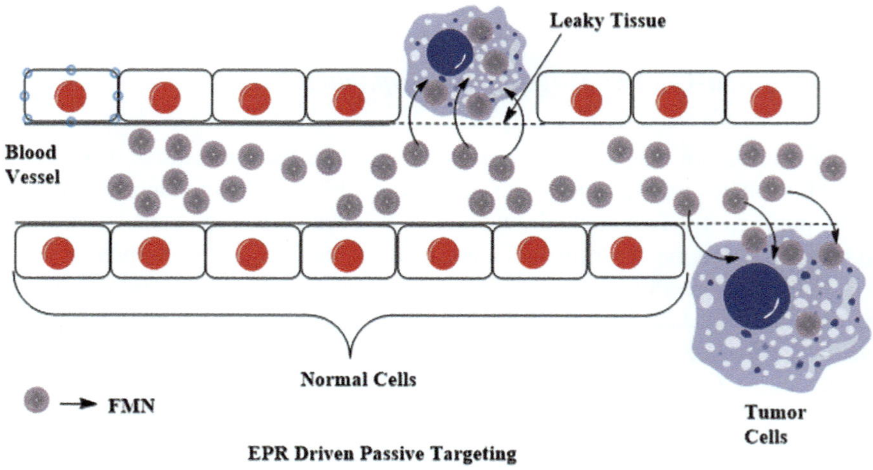

Fig. 3.2 EPR-driven passive targeting

tumor microenvironment, nanoparticle penetration is governed by diffusion, which is inversely related to particle size. Therefore, optimizing nanoparticle size has been a key focus in maximizing EPR-mediated drug delivery [17].

3.3.2 Ligand-Mediated Active Targeting

Ligand-mediated active targeting enhances drug delivery specificity by conjugating ligands, like peptides, antibodies, or micro molecules, to nanoparticles or drug transport systems. These ligands specifically bind to overexpressed receptors on diseased cells, such as cancer cells, enabling targeted treatment. This method improves receptor-facilitated endocytic drug delivery into cells, as shown in Fig. 3.3, which increases treatment efficacy and reduces off-target effects.

Mao and his team created a polysiloxane-based PEO-b-Pγ MPS copolymer-coated iron oxide nanoparticles, which imparted anti-biofouling "stealth" characteristics. This coating prolonged the circulation time in the blood flow and reduced nonspecific incorporation by macrophages and the reticuloendothelial system. When HER2- or EGF receptor-specific antibodies were covalently linked to the coated iron oxide nanoparticles, the nanoparticles effectively targeted breast cancer cells overexpressing HER2 or EGF receptor. In comparison, non-targeted iron oxide nanoparticles showed minimal cellular absorption. Additionally, receptor-specific binding and internalization were blocked when excess free antibodies were introduced. The stealth properties of these nanoparticles enabled effective cancer cell targeting, making them promising candidates for molecular-level detection and image-assisted drug transport [18].

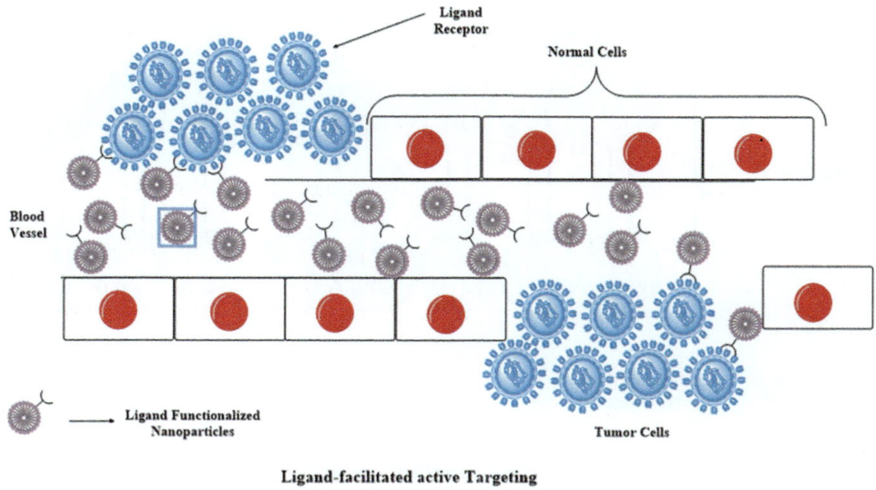

Fig. 3.3 Ligand-mediated active targeting

3.3.3 Surface Modification with Targeting Ligands

Enhancing the specificity of functional magnetic nanoparticles (FMNs) for biomedical applications involves surface modification with targeting ligands, as demonstrated in Fig. 3.4. By attaching specific ligands, FMNs can be directed to cancer cells overexpressing receptors like HER2 or EGFR, improving targeted therapy. Ligands are attached to magnetic nanoparticles (MNPs) through covalent or non-covalent interactions using surface groups like amine, carboxyl, hydroxyl, and aldehyde, facilitating direct attachment or linkage via cross-linkers for better drug delivery and imaging [18].

Yang et al. developed biodegradable, amphiphilic polymer-coated magnetic iron oxide nanoparticles targeting the urokinase plasminogen activator receptor. These nanoparticles were conjugated with the urokinase plasminogen activator N-terminal fragment and encapsulated hydrophobic chemotherapeutic drugs like doxorubicin. The system demonstrated efficient drug release at pH 4.0–5.0 and higher drug concentrations in breast cancer cells, showing promise for selective cancer treatment [19].

Wang and colleagues created a multifunctional thermally cured superparamagnetic iron oxide nanomaterial for prostate tumor detection and targeted drug transport. By employing the A10 RNA aptamer designed for specific binding to prostate-specific membrane pathogen markers, they designed a system that integrated

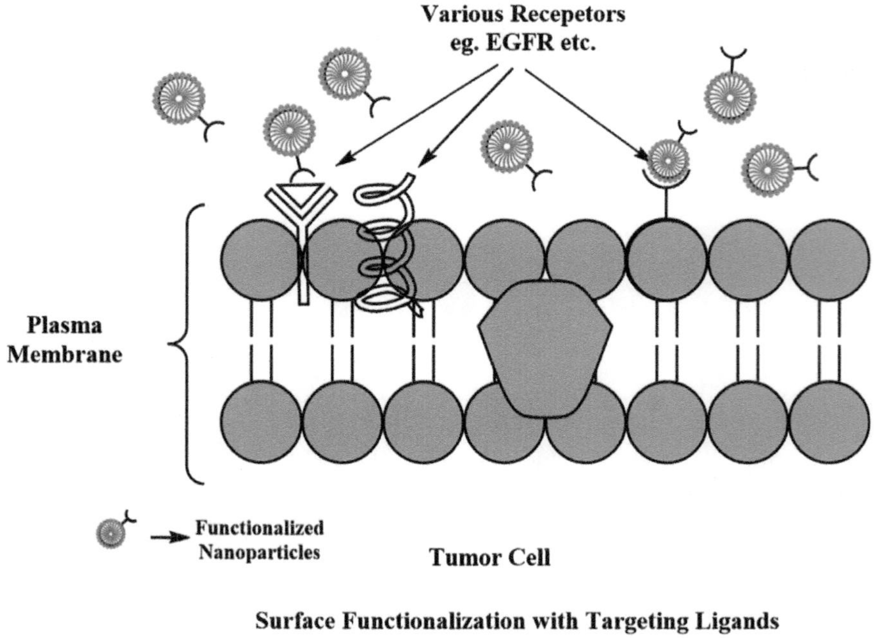

Fig. 3.4 Surface modification with targeting ligands

imaging and treatment for prostate tumors with reduced toxicity [20]. Cristofolini and colleagues created a magnetic hybrid nanostructure made of Surface-engineered iron oxide nanoparticles featuring a caffeic acid layer, calcium phosphate support, and PEG-polyanion modification for siRNA transport. These nanoparticles effectively delivered siRNA into HER2-positive breast cancer cells under magnetic guidance, achieving successful gene silencing without cytotoxicity [21].

3.4 Biocompatibility

The biological compatibility of functionalized magnetic nanoparticles (MNPs) is essential for their safe application in medical fields like drug delivery, imaging, and diagnostics. Functionalization involves modifying the nanoparticle surface with specific molecules to enhance stability, targeting capabilities, and interaction with biological systems. However, these modifications must not cause toxicity or adverse reactions when interacting with the human body [22].

Shaobing Zhou and colleagues investigated the biocompatibility of superparamagnetic nanoparticles of Fe_3O_4, coated with biodegradable polymers such as PEG and poly (ethylene glycol)-co-poly(d,l-lactide) (PELA). Their study, using the MTT assay, demonstrated that these nanoparticles displayed outstanding biocompatibility. The functionalized nanoparticles are excellently dispersible in both water and organic liquid solvents and retain high potential as magnetic carriers in medical technologies [23].

Dextran is often utilized to minimize the toxicity of nanoparticles, especially in iron-oxide NPs. A study by de Oliveira and coworkers found that dextran-coated iron-oxide nanoparticles exhibited minimal toxicity and induced no significant changes in zebrafish larvae. Additional research by Balas and colleagues on Jurkat cells, along with Wu and their team's work on human monocytes, demonstrated minimal toxicity and no substantial impact on cell viability or apoptosis, confirming the biocompatibility of dextran-coated nanoparticles [24].

Studies have shown that well-designed nanoparticles with appropriate surface modifications exhibit minimal cytotoxicity and do not adversely affect cell viability or induce apoptosis, making them promising candidates for clinical use. However, additional studies are required to fully understand their prolonged effects and confirm their reliability in diverse biological environments.

3.5 Real-World Case Studies

3.5.1 Cancer Therapy

FMNs can deliver chemotherapy drugs instantly to the tumor location, enhancing efficacy and minimizing harmful effects through magnetic precision targeting with regulated release. This section explores various chemotherapy drugs used in cancer treatment via FMNs.

a. **Doxorubicin**: Doxorubicin (DOX) is a widely used chemotherapy drug, but its toxicity remains a challenge. To minimize this, magnetic nanoparticles modified with different surface functionalities have been extensively studied as drug transport systems for DOX. Research has focused on improving biocompatibility and targeting efficiency to enhance drug delivery effectiveness. Sailaraja et al. developed chitosan-based magneto-responsive ferrogels for controlled drug release and heat-based cancer treatment. These ferrogels exhibit high drug loading efficiency (~40%) for DOX and demonstrate tumor-targeted, pH-sensitive expansion and biodegradability. The ferrogels show significant magnetic hyperthermia potential, releasing DOX efficiently under an external magnetic field and displaying excellent biocompatibility [25].

   Jeong et al. developed GSH-responsive, magnetic-sensitive nanocomposites for dual-targeting therapy. These nanocomposites were made by grafting methoxy poly(ethylene glycol) (MePEG) onto chitosan and linking DOX via a disulfide bond. Iron oxide nanoparticles were conjugated for magnetic sensitivity. The presence of glutathione accelerated drug release, and external magnetic fields improved drug delivery to cancer cells, showing significant potential for targeted in vivo cancer treatment [26].
b. **Methotrexate**: Methotrexate is a chemotherapy medicine utilized to treat cancers and autoimmune disorders by inhibiting DNA synthesis. To improve its delivery, magnetic drug delivery systems have been explored. One such system deployed amino acids like L-lysine functionalized magnetite nanoparticles. Manjili et al. synthesized arginine-coated iron oxide nanoparticles capped via a single-pot co-precipitation method, where methotrexate was linked to the nanoparticles. These nanoparticles were designed for controlled drug release, directed delivery to cancerous cells, and MRI contrast. The nanoparticles showed strong magnetic properties, no cytotoxicity in various cell lines, and excellent biocompatibility, supporting their use in cancer therapy and diagnosis [27].
c. **Curcumin**: Curcumin, a bioactive compound with anticancer potential, has been incorporated into magnetic nanostructures for enhanced drug delivery. One approach integrated Fe_3O_4 nanoparticles with layered double hydroxide (LDH), creating Fe_3O_4@Cur-LDH, which was further coated with polydopamine. This nanostructure demonstrated strong magnetic properties, high drug-loading capacity (38%), and pH-responsive regulated release. Laboratory tests revealed biocompatibility, cellular uptake, and effective growth inhibition of HepG2 cells,

suggesting its potential for cancer therapy. Similarly, Salehi et al. developed a magnetic nanocarrier using graphene quantum dots (GQDs) combined with Fe_3O_4 and folic acid for targeted delivery. The system exhibited superparamagnetic behaviour and tumor-specific targeting. Drug release studies showed pH-dependent release, with higher cytotoxicity in tumor cells, confirming the efficacy of folate (FA) for targeted drug transport. The model was non-toxic to normal cells, making it a promising candidate for therapeutic applications [28].

d. **Sorafenib**: **Sorafenib** is an oral kinase inhibitor used to treat advanced kidney, liver, and thyroid cancers by blocking proteins that promote tumor growth and angiogenesis. While it may cause fatigue and nausea, it improves progression-free survival. Hussein et al. explored iron oxide nanoparticles surface-modified with PEG (Polyethylene Glycol) and PVA (Polyvinyl Alcohol) as carriers for Sorafenib. The drug was encapsulated using Mg–Al-layered double hydroxide (MLDH), with FTIR and UV–Vis spectroscopy confirming both drug loading and release. About 85% of Sorafenib was released within 72 h, with 74% released at pH 4.8, following pseudo-second-order kinetics [29].

   Sorafenib-loaded magnetic microspheres for hepatocellular carcinoma treatment were developed by Denkbas et al. Alginate microspheres encapsulating superparamagnetic iron oxide nanoparticles demonstrated sustained drug release over 8 hours, showing strong anticancer effects and potential for hyperthermia treatment [25]. In this section, Table 3.1 highlights various nanomaterials used for targeted cancer drug delivery, including drugs like doxorubicin, methotrexate, taxol, etc. It highlights nanoparticle types, sizes, and their roles in enhancing drug delivery.

An overview of drugs and functionalized magnetic nanomaterials (FMNs) for cancer therapy is presented in Table 3.1.

In summary, functionalized magnetic nanoparticles (FMNs) offer promising potential in cancer treatment by enabling targeted drug transport, regulated release, and minimizing harmful effects, resulting in more effective and personalized treatments.

3.5.2 Neurological Disorders

Functionalized magnetic nanoparticles (FMNs) present a promising approach for targeting neurological diseases such as Parkinson's and Alzheimer's, owing to their ability to penetrate the blood–brain barrier. This capability enables FMNs to deliver neuroprotective agents with enhanced precision and effectiveness.

Neurodegenerative diseases such as Alzheimer's and Parkinson's are significant global health challenges. Other NDDs, including Amyotrophic Lateral Sclerosis, Huntington's disease, and Multiple Sclerosis, are also growing in prevalence, with numbers expected to triple by 2050 due to the aging population [37].

Table 3.1 Overview of drugs and functionalized magnetic nanomaterials (FMNs) used in cancer therapy

S. No	Drug	Nanomaterial	Function	Size (nm)	RefS.
1	Doxorubicin	PLGA-Fe_3O_4 NPs	Intratumoral drug delivery	200–300	[30]
2	Doxorubicin	Zn-doped Fe_3O_4 nano-octahedral core	Drug delivery	10–20	[31]
3	Doxorubicin	Iron oxide nanoparticles conjugated with folic acid	Drug delivery	23,40,67	[32]
4	Doxorubicin	Bacterial magnetosomes (BMs)	Drug delivery	71.02 ± 6.7	[33]
7	Taxol	Poly(acrylic acid) coated oxides of iron	Drug transport and MRI	90	[34]
8	Gemcitabine	poly ε-caprolactone (PCL) coated (Fe_3O_4)	Drug delivery	~164	[35]
9	Doxorubicin	Polyethylene glycol (PEG)-encapsulated magnetite nanoparticles	Drug delivery	7	[36]
10	Paclitaxel	Superparamagnetic oxides of Iron	Drug delivery and MRI	168.30 ± 2.80	[25]
11	Cisplatin	superparamagnetic nanoclusters	Drug delivery	90–100	[25]

Treating these conditions is difficult, primarily because of the obstacle of drug delivery across the blood–brain barrier. Nanoparticles (NPs) have emerged as effective drug carriers, facilitating the development of nanoparticle-focused drug transport systems that enhance drug durability, efficacy, and BBB penetration. Traditional methods for transporting drugs to the nervous system often yield limited clinical results. However, the distinct features of nanomaterials—such as their compact size and multifunctionality—allow for more precise targeting of disease sites [38].

FMNs, engineered to penetrate the blood–brain barrier, provide a targeted and effective strategy for the cure of diseases like Alzheimer's, Parkinson's, and Huntington's that affect the nervous system. This section examines various types of nanoparticles, with an emphasis on FMNs, and their potential in addressing these conditions.

a. **Gold Nanoparticles**

Nanoparticles can be engineered to bypass physiological barriers and provide controlled, sustained drug release, with targeted delivery to specific areas. For example, transferrin-linked gold nanoparticles improve brain uptake. OX26, 8D3, and R17217 antibodies reduce receptor saturation to avoid a contest with endogenous transferrin, though they may accumulate in other organs [39].

Magnetic actuation has enabled the creation of micro/nanorobots that replicate biological movements. Brain cell interactions with magnetic nanoparticles are influenced by shape, size, surface chemistry, and delivery method. To minimize immune responses during drug delivery, biocompatible polymers and PEG coatings are used.

Gold nanoparticles have diverse applications, including enhancing blood–brain barrier integrity and Brain-directed drug delivery related conditions like Alzheimer's and Parkinson's disease. They also aid in detecting Aβ aggregation in AD. Zhou and his coworker synthesized a color-based sensor using gold NPs to detect Aβ accumulation, demonstrating the potential for diagnostic applications [40].

b. **Silica Nanoparticles**

Silica nanoparticles (SiO_2-NPs) have been associated with neurotoxicity and neurodegeneration in mice. However, mesoporous silica nanoparticles are gaining attention due to their large surface area, high drug-loading capacity, and potential for functionalization. MSNPs functionalized with PEG-PEI can cross the blood–brain barrier without causing adverse effects. Additionally, silica nanoparticles conjugated with PEGylated quercetin show neuroprotective properties, suggesting potential for neurodegenerative disease treatment [41]. Silica nanoparticles conjugated with ligands like lactoferrin and succinic anhydride can improve drug durability and bioavailability. Furthermore, MSNPs coated with lipid-containing berberine have shown promise in the cure of Alzheimer's by inhibiting acetylcholinesterase, minimizing amyloid fibrillation, and lowering malondialdehyde levels.

c. **Cerium Oxide Nanoparticles**

Cerium oxide nanoparticles show significant clinical potential for neurodegenerative diseases due to their strong antioxidant and free radical-neutralizing properties. Nanoceria can cross the blood–brain barrier, protecting neurons from Aβ-induced mitochondrial damage and Cellular demise [42]. Furthermore, PEG-encapsulated CeONPs conjugated incorporating antibodies aimed at Aβ aggregates have been found to activate the pathway mediated by brain-derived neurotrophic factor in Alzheimer's disease models [43].

In summary, nanoparticles, including functionalized magnetic nanoparticles (FMNs), silica nanoparticles (SiO_2-NPs), cerium oxide nanoparticles (CeONPs), and gold nanoparticles (AuNPs), represent a transformative approach for treating neurodegenerative diseases. These particles can be tailored to traverse the blood–brain barrier and provide precision drug delivery with regulated release, improving treatment precision for conditions like Alzheimer's, Parkinson's, and Huntington's diseases.

3.5.3 Gene Therapy

Efficient delivery of nucleic acids using nanocarriers has become a key area of research in gene therapy. For successful gene delivery, the exogenous genetic material must reach the target cell nucleus, where it can produce the desired proteins. An ideal vector should transport the right amount of genetic material to specific cells, ensuring proper transgene expression to correct genetic defects, while being non-immunogenic and safe to avoid toxicity [44]. Three widely used approaches for modifying inorganic nanoparticles for gene delivery are: (i) employing positively charged nanoparticles to form complexes with negatively charged genetic material, (ii) directly binding genetic material to nanoparticles through an active linking agent, and (iii) utilizing a cationic surfactant polymer on the nanomaterials to improve their interaction with genetic material. Here, we describe some nanoparticles used for gene therapy.

a. **Gold nanoparticles**: Peng and his coworkers studied that gold nanoparticles were used for dual gene and antimicrobial treatment by linking antimicrobial peptides to cationic gold NPs, which enabled gene introduction into mesenchymal stem cells [45]. A common strategy for gold NP-driven gene delivery is to functionalize the surface of gold nanoparticles with cationic molecules like amino acids, cationic peptides, or compounds with tertiary amines. Gold NPs conjugated with oligonucleotides have proven effective in gene therapy applications. For instance, gold NPs attached to an antisense oligonucleotide targeting the BCR-ABL messenger RNA, which encodes an active tyrosine kinase, successfully silenced the gene and induced increased cell death in K562 cells, a model for leukemia [44].
b. **Iron oxide nanoparticles**: oxide nanoparticles of iron, including superparamagnetic ones, are favoured for gene delivery because they are cost-effective, exhibit low toxicity, biodegrade efficiently, and can be easily modified at the surface. Additionally, magnetic nanoparticles improve gene transport in both viral and recombinant vectors. One approach involves creating a gene delivery vehicle with an iron oxide NPs core with the shell of an alkylated polyethyleneimine, and siRNA attached to the NPs. Targeted delivery is achieved using high-field/high-gradient magnets. This method demonstrated high efficiency in downregulating luciferase in the fluc-4T1 cell line. Moreover, Shakil and his collaborator showed that iron oxide nanoparticles could function as therapeutic and diagnostic agents in breast tumor gene therapy, where applying a magnetic field enhanced DNA transfer and facilitated cellular compartment delivery [45]. Similarly, Jin and coworkers observed that iron oxide NPs improved gene silencing, directing B-cell lymphoma-2 protein in Ca9-22 oral carcinoma cells. Additionally, iron oxide with superparamagnetic properties NPs were effective in delivering anti-nef siRNA targeting HIV-1 nef into two different cell lines [46].

   In brief, FMNs can be used to carry DNA, RNA, or other therapeutic molecules, enhancing gene expression and silencing specific genes. They hold considerable potential for enhancing the effectiveness of gene therapy by enabling targeted, controlled, and safe delivery of genetic material to specific tissues or organs.

3.5.4 Antibiotic Delivery

The chemistry of nanoparticles (NPs) can be precisely tailored to incorporate various chemical functional groups, charges, and other characteristics. This flexibility enables nanoparticles to carry a significant number of antibiotic molecules, enhancing their targeted delivery to infectious bacterial cells. By attaching antimicrobial agents to nanoparticles, a more efficient strategy for tackling antibiotic-resistant infections can be developed. Unlike conventional antibiotic delivery methods, where drugs diffuse throughout a solution, concentrating antibiotics on the surface of NPs allows for a higher, more targeted dose to bacterial cells. This method can potentially improve the efficacy of less potent antibiotics by directing them straight to resistant bacteria [47].

Nanoparticles of iron oxide have shown antibacterial influence on *Staphylococcus aureus* when using Fe_3O_4/PVA nanoparticles. At a concentration of 3 mg/mL, Fe_3O_4/PVA NPs inhibited bacterial growth, resulting in a 20% live/dead ratio after 24 h. Furthermore, polymer vesicles containing iron oxide nanoparticles efficiently penetrated thick (20 μm) biofilms of *Staphylococcus epidermidis* under an applied magnetic field. These findings suggest that magnetic nanoparticles (MNPs) co-encapsulated with antibiotics like methicillin could effectively target and eliminate antibiotic-resistant infections in biofilms [48].

Cationic poly(2-(dimethylamino)-ethyl methacrylate)/Fe_3O_4 composite nanogels with multifunctional capabilities demonstrated strong antimicrobial properties. These nanogels exhibited effective antibacterial and antifungal activity at different levels (0–200 μg/mL) at 37 °C and 25 °C, respectively. At concentrations above 200 μg/mL [49], they fully suppressed E. coli growth, with minimal growth of *S. aureus* observed after 300 min. Some *C. albicans* colonies persisted at a concentration of 200 μg/mL. Additionally, a novel bio-nanosensor utilizing silver-coated magnetic nanoparticles functionalized with lectin was used for bacterial magnetic separation and surface-enhanced Raman scattering for detection and extraction.

Fe_3O_4 nanosized beads, coated with specific antibodies aimed at the SARS-CoV-2 spike protein, were used to detect antibodies against the virus. The test, conducted on 20 samples from COVID-19-recovered patients with anti-SARS-CoV-2 IgM/IgG antibodies, demonstrated a 100% concordance rate in just 15 min. This study showcases the capabilities of magnetic nanoparticles in diverse detection uses, further emphasizing their importance in virus detection [50].

Table 3.2 summarizes several nanoparticle-based approaches, including antimicrobial properties, targeted drug delivery, and diagnostic applications, for combating various diseases.

Nanomaterial-based antibiotic delivery improves targeting and effectiveness against bacterial infections, including drug-resistant strains. By functionalizing

Table 3.2 Overview of functionalized magnetic nanomaterials (FMNs) and their functions in bacterial detection and treatment

S. No	Nanomaterial	Function	Size (nm)	Drug/gene	Refs.
1	Silver core/ magnetic Fe_3O_4 shell nanoparticle	Bacterial disinfection	480–825	Photothermal ablation	[51]
2	Core–shell magnetic nanostructures	Antimicrobial activity		Chlorhexidine	[52]
3	Iron oxide nanoparticle	Antimicrobial activity	18–20		[53]
4	Ag NP and AuNP	Multiple-antibiotic-resistant bacteria	4 and 7 nm	Ampicillin	[54]
5	Magnetic silver nanoshells	Detection and inhibition of Bacteria	369.9 ± 22.700		[55]
6	Iron oxide MNPs	Detection of water-borne bacteria	380	Chains of magnetic n nanoparticle	[56]

nanoparticles like silver, iron oxide, and gold, antibiotics can be directly delivered to infection sites, enhancing stability and reducing side effects. Magnetic nanoparticles enable precise targeting, while controlled release boosts treatment efficiency, offering a promising solution for combating antibiotic resistance.

Conclusion: In conclusion, nanoparticles, particularly functionalized magnetic nanoparticles (FMNs), offer significant advancements in drug and gene delivery systems, particularly for challenging conditions such as cancer and neurological disorders. Their capacity to navigate biological barriers, improve targeted delivery, and enable controlled release makes them essential assets in contemporary medicine. Furthermore, their versatility in both therapeutic and diagnostic applications—ranging from antimicrobial resistance treatment to gene silencing—demonstrates their potential for transforming the scope of individualized medicine. As Studies continue to evolve, nanoparticles hold the promise of offering safer, more effective, and targeted treatment options, addressing the growing demand for novel solutions in drug delivery and disease management.

Acknowledgements J. Jain* acknowledges the funding support through the Start-Up Grant (Award No. 30-572/2021 BSR) provided by the UGC, MHRD, New Delhi.

Abbreviation

AD Alzheimer's disease

BBB	Blood–brain barrier
EGF	Epidermal-derived growth factor
FMNs	Functionalized magnetic nanoparticles
GQDs	Graphene quantum dots
HER2 receptor	Human epidermal growth factor 2 receptor
LDA	Layered double hydroxide
MePEG	Methoxy poly (ethylene glycol)
MLDH	Magnesium–aluminum-layered double hydroxide
MNPs	Magnetic nanoparticle
MRI	Magnetic resonance imaging
MSNPs	Mesoporous silica nanoparticles
MTT	3-(4,5-Dimethylthiazolyl-2)-2,5-diphenyltetrazolium bromide)
NDDs	Novel Drug Delivery System
PDA	Poly(dopamine)
PELA	Copolymer of poly(ethylene glycol) and poly(d,l-lactide)
PEO-b-Pγ MPS	Poly(ethylene oxide)-block-poly(γ-methacryloxypropyltrimethoxysilane)
PVA	Polyvinyl alcohol

References

1. Senyei A, Widder K, Czerlinski G (1978) Magnetic guidance of drug-carrying microspheres. J Appl Phys 49(6):3578–3583
2. Gao X, Cui Y, Levenson RM, Chung LW, Nie S (2004) In vivo cancer targeting and imaging with semiconductor quantum dots. Nat Biotechnol 22(8):969–976
3. Teste B, Malloggi F, Gassner A-L, Georgelin T, Siaugue J-M, Varenne A, Girault H, Descroix S (2011) Magnetic core shell nanoparticles trapping in a microdevice generating high magnetic gradient. Lab Chip 11(5):833–840
4. Mittal A, Roy I, Gandhi S (2022) Magnetic nanoparticles: an overview for biomedical applications. Magnetochemistry 8(9):107
5. Martins PM, Lima AC, Ribeiro S, Lanceros-Mendez S, Martins P (2021) Magnetic nanoparticles for biomedical applications: from the soul of the earth to the deep history of ourselves. ACS Appl Bio Mater 4(8):5839–5870
6. Unsoy G, Khodadust R, Yalcin S, Mutlu P, Gunduz U (2014) Synthesis of Doxorubicin loaded magnetic chitosan nanoparticles for pH responsive targeted drug delivery. Eur J Pharm Sci 62:243–250
7. Jiang K, Zhang L, Bao G (2021) Magnetic iron oxide nanoparticles for biomedical applications. Curr Opin Biomed Eng 20:100330
8. Thanki K, Gangwal RP, Sangamwar AT, Jain S (2013) Oral delivery of anticancer drugs: challenges and opportunities. J Control Release 170(1):15–40
9. Zhang H, Chen B, Jiang H, Wang C, Wang H, Wang X (2011) A strategy for ZnO nanorod mediated multi-mode cancer treatment. Biomaterials 32(7):1906–1914
10. Chen Y, Wan Y, Wang Y, Zhang H, Jiao Z (2011) Anticancer efficacy enhancement and attenuation of side effects of doxorubicin with titanium dioxide nanoparticles. Int J Nanomed 2321–2326
11. Yang R, Li R, Zhang L, Xu Z, Kang Y, Xue P (2020) Facile synthesis of hollow mesoporous nickel sulfide nanoparticles for highly efficient combinatorial photothermal–chemotherapy of cancer. J Mater Chem B 8(34):7766–7776

12. Hou L, Shan X, Hao L, Feng Q, Zhang Z (2017) Copper sulfide nanoparticle-based localized drug delivery system as an effective cancer synergistic treatment and theranostic platform. Acta Biomater 54:307–320
13. Chen Y, Yin Q, Ji X, Zhang S, Chen H, Zheng Y, Sun Y, Qu H, Wang Z, Li Y, Wang X, Zhang K, Zhang L, Shi J (2012) Manganese oxide-based multifunctionalized mesoporous silica nanoparticles for pH-responsive MRI, ultrasonography and circumvention of MDR in cancer cells. Biomaterials 33(29):7126–7137
14. Sun Y, Duan L, Guo Z, DuanMu Y, Ma M, Xu L, Zhang Y, Gu N (2005) An improved way to prepare superparamagnetic magnetite-silica core-shell nanoparticles for possible biological application. J Magn Magn Mater 285(1–2):65–70
15. Im SH, Herricks T, Lee YT, Xia Y (2005) Synthesis and characterization of monodisperse silica colloids loaded with superparamagnetic iron oxide nanoparticles. Chem Phys Lett 401(1–3):19–23
16. Kostiva U, Patsula V, Šlouf M, Pongrac IM, Škokić S, Radmilović MD, Pavičić I, Vinković I, Gajović S, Horák D (2017) Physico-chemical characteristics, biocompatibility, and MRI applicability of novel monodisperse PEG-modified magnetic Fe_3O4 & SiO_2 core–shell nanoparticles. RSC Adv 7(15):8786–8797
17. Matsumura Y, Maeda H (1986) A new concept for macromolecular therapeutics in cancer chemotherapy: mechanism of tumoritropic accumulation of proteins and the antitumor agent smancs. Cancer Res 46(12_Part_1):6387–6392
18. Huang J, Li Y, Orza A, Lu Q, Guo P, Wang L, Yang L, Mao H (2016) Magnetic nanoparticle facilitated drug delivery for cancer therapy with targeted and image-guided approaches. Adv Funct Mater 26(22):3818–3836
19. Chen H, Wang L, Yu Q, Qian W, Tiwari D, Yi H, Wang AY, Huang J, Yang L, Mao H (2013) Anti-HER2 antibody and ScFvEGFR-conjugated antifouling magnetic iron oxide nanoparticles for targeting and magnetic resonance imaging of breast cancer. Int J Nanomed 3781–3794
20. Wang AZ, Bagalkot V, Vasilliou CC, Gu F, Alexis F, Zhang L, Shaikh M, Yuet K, Cima MJ, Langer R, Kantoff PW, Bander NH, Jon S, Farokhzad OC (2008) Superparamagnetic iron oxide nanoparticle–aptamer bioconjugates for combined prostate cancer imaging and therapy. ChemMedChem 3(9):1311
21. Cristofolini T, Dalmina M, Sierra JA, Silva AH, Pasa AA, Pittella F, Creczynski-Pasa TB (2020) Multifunctional hybrid nanoparticles as magnetic delivery systems for siRNA targeting the HER2 gene in breast cancer cells. Mater Sci Eng, C 109:110555
22. Sanità G, Carrese B, Lamberti A (2020) Nanoparticle surface functionalization: how to improve biocompatibility and cellular internalization. Front Mol Biosci 7:587012
23. Sun L, Huang C, Gong T, Zhou S (2010) A biocompatible approach to surface modification: biodegradable polymer functionalized super-paramagnetic iron oxide nanoparticles. Mater Sci Eng, C 30(4):583–589
24. de Oliveira GMT, de Oliveira EMN, Pereira TCB, Papaléo RM, Bogo MR (2017) Implications of exposure to dextran-coated and uncoated iron oxide nanoparticles to developmental toxicity in zebrafish. J Nanopart Res 19:1–16
25. Pusta A, Tertis M, Crăciunescu I, Turcu R, Mirel S, Cristea C (2023) Recent advances in the development of drug delivery applications of magnetic nanomaterials. Pharmaceutics 15(7):1872
26. Yoon HM, Kang MS, Choi GE, Kim YJ, Bae CH, Yu YB, Jeong YI (2021) Stimuli-responsive drug delivery of doxorubicin using magnetic nanoparticle conjugated poly (ethylene glycol)-g-chitosan copolymer. Int J Mol Sci 22(23):13169
27. Attari E, Nosrati H, Danafar H, Kheiri H (2019) Methotrexate anticancer drug delivery to breast cancer cell lines by iron oxide magnetic based nanocarrier. J Biomed Mater Res, Part A 107(11):2492–2500
28. Zadeh ES, Ghanbari N, Salehi Z, Derakhti S, Amoabediny G, Akbari M, Tokmedash MA (2023) Smart pH-responsive magnetic graphene quantum dots nanocarriers for anticancer drug delivery of curcumin. Mater Chem Phys 297:127336

29. Ebadi M, Bullo S, Buskara K, Hussein MZ, Fakurazi S, Pastorin G (2020) Release of a liver anticancer drug, sorafenib from its PVA/LDH-and PEG/LDH-coated iron oxide nanoparticles for drug delivery applications. Sci Rep 10(1):21521
30. Jia Y, Yuan M, Yuan H, Huang X, Sui X, Cui X, Tang F, Peng J, Chen J, Lu S, Xu W, Zhang L, Guo Q (2012) Co-encapsulation of magnetic Fe3O4 nanoparticles and doxorubicin into biodegradable PLGA nanocarriers for intratumoral drug delivery. Int J Nanomed 1697–1708
31. Chen Z, Peng Y, Li Y, Xie X, Wei X, Yang G, Zhang H, Li N, Li T, Qin X, Li S, Wu C, You F, Yang H, Liu Y (2021) Aptamer-dendrimer functionalized magnetic nano-octahedrons: theranostic drug/gene delivery platform for near-infrared/magnetic resonance imaging-guided magneto chemotherapy. ACS Nano 15(10):16683–16696
32. Huang Y, Mao K, Zhang B, Zhao Y (2017) Superparamagnetic iron oxide nanoparticles conjugated with folic acid for dual target-specific drug delivery and MRI in cancer theranostics. Mater Sci Eng, C 70:763–771
33. Sun JB, Duan JH, Dai SL, Ren J, Zhang YD, Tian JS, Li Y (2007) In vitro and in vivo antitumor effects of doxorubicin loaded with bacterial magnetosomes (DBMs) on H22 cells: the magnetic bio-nanoparticles as drug carriers. Cancer Lett 258(1):109–117
34. Santra S, Kaittanis C, Grimm J, Perez JM (2009) Drug/dye-loaded, multifunctional iron oxide nanoparticles for combined targeted cancer therapy and dual optical/magnetic resonance imaging. Small 5(16):1862–1868
35. Gang J, Park SB, Hyung W, Choi EH, Wen J, Kim HS, Shul YG, Haam S, Song SY (2007) Magnetic poly ε-caprolactone nanoparticles containing Fe3O4 and gemcitabine enhance antitumor effect in pancreatic cancer xenograft mouse model. J Drug Target 15(6):445–453
36. Zhang J, Rana S, Srivastava RS, Misra RDK (2008) On the chemical synthesis and drug delivery response of folate receptor-activated, polyethylene glycol-functionalized magnetite nanoparticles. Acta Biomater 4(1):40–48
37. Ogonowski NS, García-Marín LM, Fernando AS, Flores-Ocampo V, Rentería ME (2024) Impact of genetic predisposition to late-onset neurodegenerative diseases on early life outcomes and brain structure. Transl Psychiatry 14(1):185
38. Duan L, Li X, Ji R, Hao Z, Kong M, Wen X, Guan F, Ma S (2023) Nanoparticle-based drug delivery systems: an inspiring therapeutic strategy for neurodegenerative diseases. Polymers 15(9):2196
39. Sharma G, Lakkadwala S, Modgil A, Singh J (2016) The role of cell-penetrating peptide and transferrin on enhanced delivery of drug to brain. Int J Mol Sci 17(6):806
40. Huang H, Liu R, Yang J, Dai J, Fan S, Pi J, Wei Y, Guo X (2023) Gold nanoparticles: construction for drug delivery and application in cancer immunotherapy. Pharmaceutics 15(7):1868
41. Bharti C, Nagaich U, Pal AK, Gulati N (2015) Mesoporous silica nanoparticles in target drug delivery system: a review. Int J Pharm Invest 5(3):124
42. Dowding JM, Song W, Bossy K, Karakoti A, Kumar A, Kim A, Bossy B, Seal S, Ellisman MH, Perkins G, Self WT, Bossy-Wetzel E (2014) Cerium oxide nanoparticles protect against Aβ-induced mitochondrial fragmentation and neuronal cell death. Cell Death Differ 21(10):1622–1632
43. Cimini A, D'Angelo B, Das S, Gentile R, Benedetti E, Singh V, Monaco AM, Santucci S, Seal S (2012) Antibody-conjugated PEGylated cerium oxide nanoparticles for specific targeting of Aβ aggregates modulate neuronal survival pathways. Acta Biomater 8(6):2056–2067
44. Goswami S, Isaev NK, Salmina AB, Rana SVS, Illarioshkin SN, Verma Y Brain Disorders. Neurodegenerative Dis 4:5
45. Hamimed S, Jabberi M, Chatti A (2022) Nanotechnology in drug and gene delivery. Naunyn Schmiedebergs Arch Pharmacol 395(7):769–787
46. Kamalzare S, Noormohammadi Z, Rahimi P, Atyabi F, Irani S, Tekie FSM, Mottaghitalab F (2019) Carboxymethyl dextran-trimethyl chitosan coated superparamagnetic iron oxide nanoparticles: An effective siRNA delivery system for HIV-1 Nef. J Cell Physiol 234(11):20554–20565
47. Lee NY, Ko WC, Hsueh PR (2019) Nanoparticles in the treatment of infections caused by multidrug-resistant organisms. Front Pharmacol 10:1153

48. Geilich BM, Gelfat I, Sridhar S, van de Ven AL, Webster TJ (2017) Superparamagnetic iron oxide-encapsulating polymersome nanocarriers for biofilm eradication. Biomaterials 119:78–85
49. Gao F, Qi Q, Wu X, Yu J, Yao J, Cao Z, Cui Q (2021) Multifunctional poly (quaternary ammonium)/Fe3O4 composite nanogels for integration of antibacterial and degradable magnetic redox-responsive properties. Colloids Surf, A 615:126235
50. Shen Q, Liang H, Tian J, Zhou C, Gao A, Wang D, Ni J, Cui D (2020) A magnetic nanoparticle labeled immunochromatography kit for SARS-CoV-2 infection diagnosis. Nano Biomed Eng 12(4):325–330
51. Fang W, Zhang H, Wang X, Wei W, Shen Y, Yu J, Liang J, Zheng J, Shen Y (2017) Facile synthesis of tunable plasmonic silver core/magnetic Fe_3O_4 shell nanoparticles for rapid capture and effective photothermal ablation of bacterial pathogens. New J Chem 41(18):10155–10164
52. Tokajuk G, Niemirowicz K, Deptuła P, Piktel E, Cieśluk M, Wilczewska AZ, Bucki R (2017) Use of magnetic nanoparticles as a drug delivery system to improve chlorhexidine antimicrobial activity. Int J Nanomed 7833–7846
53. Arakha M, Pal S, Samantarrai D, Panigrahi TK, Mallick BC, Pramanik K, Mallick B, Jha S (2015) Antimicrobial activity of iron oxide nanoparticle upon modulation of nanoparticle-bacteria interface. Sci Rep 5(1):14813
54. Brown AN, Smith K, Samuels TA, Lu J, Obare SO, Scott ME (2012) Nanoparticles functionalized with ampicillin destroy multiple-antibiotic-resistant isolates of Pseudomonas aeruginosa and Enterobacter aerogenes and methicillin-resistant Staphylococcus aureus. Appl Environ Microbiol 78(8):2768–2774
55. Vedarethinam V, Huang L, Xu W, Zhang R, Gurav DD, Sun X, Qian K (2019) Detection and Inhibition of Bacteria on a Dual-Functional Silver Platform. Small 15(3):1803051
56. Lee H, Han H, Jeon S (2020) Baleen-mimicking virtual filters for rapid detection of pathogenic bacteria in water using magnetic nanoparticle chains and a Halbach ring. ACS Sens 5(11):3432–3437

Chapter 4
Magnetic Hyperthermia Using Magnetic Nanoparticles: A Promising Strategy in Cancer Treatment

Shefali Shukla, Sreya Jayan, P. K. Hasna, and Shikha Gulati

Abstract Magnetic hyperthermia has emerged as a promising cancer treatment strategy wherein magnetic nanoparticles (MNPs) are employed to generate localized heat under the influence of an alternating magnetic field (AMF). These nanoparticles are responsible for inducing cell death selectively at the tumor site without affecting healthy cells. This chapter explores various modifications of magnetic nanoparticles in inducing hyperthermia, including their combination with chemotherapy and radiotherapy, to enhance therapeutic efficacy. Special emphasis is placed on surface functionalization strategies, such as coating of polymer/biopolymers, silica, graphene oxide, and dendrimers; ligand exchange; surfactant modifications; and host–guest complexation, which improve biocompatibility, targeting efficiency, and drug-loading capacity of the nanostructures. Additionally, the role of iron oxide-based MNPs in ferroptosis, an emerging cancer cell death pathway, is discussed. While magnetic nanocarriers offer immense potential in targeted drug delivery, various challenges such as toxicity, stability, target specificity, biocompatibility, and controlled drug release remain key hurdles for clinical translation. Addressing these issues through innovative nanomaterial engineering will pave the way for more effective and safer cancer therapies.

Keywords Magnetic hyperthermia · Nanoparticles · Cancer therapy · Surface functionalization · Drug delivery

4.1 Introduction

Cancer is one of the leading causes of death in the world. As per WHO data, in 2022, there were around 20 million new cases of cancer and around 10 million deaths reported worldwide due to cancer. Many methods, such as chemotherapy, radiotherapy, immunotherapy, hormone therapy, stem cell transplant, and surgery,

S. Shukla (✉) · S. Jayan · P. K. Hasna · S. Gulati
Department of Chemistry, Sri Venkateswara College, University of Delhi, Delhi, India
e-mail: shefali.shukla@yahoo.co.in

S. Gulati (ed.), *Functionalized Magnetic Nanomaterials*, Nanostructure Science and Technology, https://doi.org/10.1007/978-3-031-97199-0_4

have been used conventionally to treat cancer. However, these methods are often associated with numerous limitations, including non-specific targeting, severe side effects, development of multi-drug resistance over a period of time, etc. Also, some of these methods cause nausea, vomiting, bleeding, anemia, loss of appetite, constipation, etc. [1]. Since most conventional methods often cause significant damage to healthy tissues, they lead to multiple side effects that further complicate treatment outcomes. These methods reduce the quality and life expectancy of cancer patients, which further makes it imperative to explore alternative strategies.

To address these challenges, researchers are actively involved worldwide in developing novel therapeutic approaches that enhance treatment efficiency while minimizing adverse effects. Recent advances in the field of nanotechnology have paved the way for more precise and effective cancer therapy, where nanomaterials play a significant role in diagnosis, drug delivery, and treatment [2]. Recent cancer treatments are based on methodologies that offer precision, improved bioavailability, and controlled therapeutic release of the drug molecule. Among these, magnetic nanoparticles (MNPs) have emerged as a highly promising tool, providing a multifaceted approach to cancer therapy through targeted drug delivery, hyperthermia treatment, and real-time imaging. These MNPs are small-sized molecules with supermagnetic properties, and these unique features can be used in the diagnosis and early treatment of various cancers [3].

Increase in body temperature artificially by delivering the heat drawn from external sources for the purpose of inhibiting the growth, shrinking the tumor size, or completely destroying the cancerous cells is hyperthermia treatment (HT). Magnetic hyperthermia has emerged as an advanced oncological technique that utilizes MNPs to generate heat when exposed to AMF. This process raises the temperature of tumor tissues to 42–45 °C, and destroys the malignant cells with fewer side effects. The MNPs, typically composed of iron, nickel [4], cobalt, or their oxides, or a mixture of metal and some polymeric coatings that exhibit a strong response to external magnetic fields. Their unique magnetic properties make them highly valuable in biomedical applications like molecular imaging, biosensing, and targeted and controlled drug delivery [5]. Iron oxide-based MNPs play a significant role in cancer therapy by inducing ferroptosis, which is a regulated form of cell death. Ferroptosis involves the release of Fe^{2+}/Fe^{3+} ions in the TME that produce reactive oxygen species (ROS) and hence trigger lipid peroxidation and oxidative stress, leading to cell death [6].

One of the key challenges in advancing MNP-based cancer therapies is the development of multifunctional nanoparticles capable of selectively binding to diseased tissues or cancer cells. Achieving this requires sophisticated surface modifications, such as functionalizing MNPs with targeting ligands, antibodies, or peptides that recognize and attach specifically to cancerous cells [7]. These modifications not only enhance the selectivity of MNPs but also enable their integration with various therapeutic strategies, such as hyperthermia, controlled drug release, and real-time imaging. The ongoing research in this field aims to optimize the synthesis and functionalization of MNPs to improve their stability, biocompatibility, and efficiency in cancer diagnostics and treatment. The chemical composition, size, shape, morphology, and magnetic properties are crucial factors in determining their suitability for biomedical applications [8]. Furthermore, their magnetic behavior and in vivo effectiveness can be optimized by applying a biocompatible coating, enhancing their stability and specificity for targeted applications within the human body.

MNPs are typically composed of materials like iron oxide (Fe_3O_4) that are capable of exhibiting superparamagnetic behavior, a property that allows them to respond to external magnetic fields without retaining residual magnetism. This property helps in the targeted action of killing cancer cells without affecting surrounding healthy tissues. The present chapter provides details of research progress in oncological treatment, with a special emphasis on surface alterations of these architectures to overcome various limitations associated. These modifications not only help in more promising use of MNPs in cancer treatment but also in diagnosis and real-time monitoring of treatment, which represents a significant step towards the development of safer and more effective cancer therapies.

4.2 Hyperthermia-Induced Cell Death: Mechanisms and Therapeutic Potential

The term "hyperthermia" refers to the local, regional, or generalized increase in body temperature. It is an artificial way of increasing the body tissue temperature by delivering heat from external sources. HT elevates the temperature of cancerous tissues to 42–45 °C and can induce cancer cell death by enhancing the effectiveness of chemotherapy and radiotherapy. Hyperthermia induces several physiological changes in cells that impact their structure and function. The cancerous cells may experience denaturation of cellular proteins, leading to disruption of the cytoskeleton and membrane instability, which could result in increased permeability. Additionally, hyperthermia may disrupt the DNA repair mechanism, making the cell more susceptible to apoptosis via chemotherapy or radiation therapy. This thermal increase also alters blood flow towards tumor cells, promotes oxidative stress by increasing ROS production, leading to mitochondrial dysfunction and apoptosis. A general mechanism of action of magnetic-nanoparticles-induced hyperthermia and eventual tumor reduction/shrinkage is depicted in Fig. 4.1.

There are three primary clinical techniques for applying high temperatures to treat cancer: whole body hyperthermia (WBH), regional hyperthermia (RHT), and local hyperthermia (LHT) [9]. Various energy sources are used to heat tumors, including microwaves, radiofrequency, ultrasound, and nanoparticles. Whole body hyperthermia is used to treat metastatic cancer, where the entire body is heated to a high temperature. Regional hyperthermia involves heating a specific region of the body, such as an organ or limb, to a temperature that is not hot enough to destroy cancer cells completely. This technique is often combined with radiation therapy or chemotherapy. Local hyperthermia is used to heat a small region, typically a tumor, to a high temperature to destroy cancer cells. This technique can be used to treat

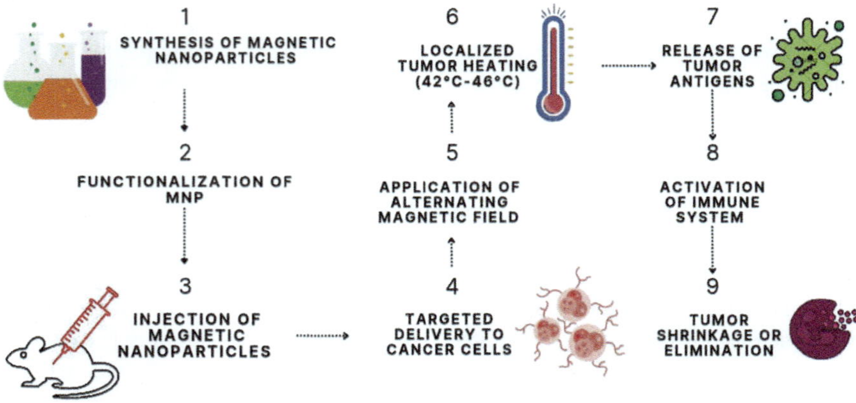

Fig. 4.1 General mechanism of magnetic-hyperthermia-induced cancer cell apoptosis

surface tumors using applicators or antennas that emit radio waves or microwaves. There are various techniques to induce hyperthermia, such as isolated perfusion, where the blood supply to a specific region (such as a limb or an organ) is temporarily separated from the systemic circulation. During this process, heated chemotherapy or heated blood is perfused (circulated) through the isolated area to achieve localized hyperthermia. Ultrasonic hyperthermia and radiofrequency hyperthermia use ultrasonic sound waves or radiofrequency waves to heat the tumor cells.

Despite its effectiveness, cancer treatment using hyperthermia faces several limitations that include uneven heat distribution because of variations in tissue composition, blood flow and size; development of thermal resistance because of upregulation of heat shock proteins; inefficiency to reach out to deep rooted tumors; increased blood flow at tumor site that results in lowering of temperature; and potential damage to healthy tissues.

4.2.1 Magnetic Hyperthermia—Combined with Radiation Therapy

Radiation therapy is a widely used cancer treatment that employs ionizing radiation to damage the DNA of malignant cells, ultimately leading to cell death. The most common type of radiation therapy is External Beam Radiation Therapy (EBRT), in which radiation is directed to the tumour site by using a machine outside the body. It is of various types like Intensity-Modulated Radiation Therapy (IMRT), Stereotactic Body Radiotherapy (SBRT), Stereotactic Radiosurgery (SRS), and Proton Therapy. Another type of radiation therapy includes Internal Radiation Therapy (Brachytherapy), where radioactive sources are placed inside or near the tumor. In this method, the radiations are given either continuously over a long period of time or in short, high-intensity sessions. The third radiotherapy type includes Systemic Radiation Therapy, where radioactive pharmaceuticals are injected, swallowed, or infused to target cancer cells throughout the body. Though radiation therapy is very effective in cancer treatment but owing to its limitations like radiation resistance, damage to surrounding healthy tissues, and insufficient tumor eradication, especially in hypoxic tumor regions, its therapeutic success is low. That is the reason researchers are exploring various combination therapies to enhance the effectiveness of the therapy while minimizing its side effects. One such promising approach is the integration of magnetic hyperthermia with radiation therapy.

Magnetic hyperthermia (MH) is a promising cancer treatment that involves the use of MNPs to generate localized heat under an AMF. These MNPs exhibit superparamagnetic behavior, allowing them to respond to external magnetic fields without retaining residual magnetism. This property makes them ideal for applications in cancer therapy, where precise targeting and controlled therapeutic delivery are critical. When combined with radiation or chemotherapy, it can enhance therapeutic efficacy through various mechanisms [10, 11]. The combination of MH and RT presents a promising approach in overcoming the limitations observed in each therapy when used individually. As discussed, the radiation therapy relies on ionizing radiation to damage cancer cell DNA, and the magnetic hyperthermia utilizes MNPs to generate localized heat when exposed to AMF. This combination of these two therapeutic methodologies offers greater efficacy with fewer side effects through several key mechanisms, like.

- Hyperthermia increases the sensitivity to radiation by weakening the DNA repair systems of tumour cells.
- As many tumors are resistant to radiation due to low oxygen levels, hyperthermia improves oxygenation by increasing blood flow to the tumor, thereby enhancing radiation effectiveness.
- Enhanced oxygenation leads to greater production of ROS upon irradiation, increasing DNA damage.
- Hyperthermia-induced stress enhances radiation-induced apoptosis and promotes immune system activation against tumors.
- Heat stress can cause mitochondrial dysfunction or arrest the cell cycle in G2/M phase and thus activate apoptosis.

4.2.2 Magnetic Hyperthermia—Combined with Chemotherapy

Functionalized MH can significantly enhance chemotherapy effectiveness through multiple pathways.

- Heating enhances the drug absorption into cancer cells as MH improves the cell permeability and also disrupts the membrane integrity of the cancer cell.
- MNPs can be functionalized for the targeted delivery of chemotherapeutic agents.

4.3 Characteristics of Magnetic Nanoparticles

MNPs are the core materials used in MH due to their ability to generate heat under AMF. The amplitude and the applied magnetic field, as well as the structural or magnetic characteristics of MNPs, are some of the variables that affect their heating efficiency. Their characteristics play a crucial role in determining the efficacy, biocompatibility, and targeting efficiency of hyperthermia-based cancer therapy. Magnetic properties of nanoparticles (NPs) are dominated by two main features: finite-size effects and surface effects. The surface effects arise due to the symmetry breaking of the crystal structure at the nanoparticle surface, oxidation, dangling bonds, surface strain, and other structural distortions. These effects play a crucial role in determining their superparamagnetic behavior, magnetic anisotropy, and heating efficiency in magnetic hyperthermia applications.

- Finite-size effects- In most medical applications, the desired size of the nanoparticles is typically approximately 10–50 nm. In this range of sizes, usually a nanoparticle shows supermagnetic behaviour. In this size range, the NPs exhibit magnetism only under an external field, which prevents aggregation of nanoparticles, making them ideal for biomedical applications.
- Surface effects- Surface effects become significant as the particle size decreases because the ratio of the number of surface atoms to the core atoms increases. In nanoparticles, surface atoms experience reduced coordination and broken exchange interactions, leading to misalignment of magnetic moments.

In the superparamagnetic state, a nanoparticle possesses a large magnetic moment, responds rapidly to applied magnetic fields, and exhibits negligible coercivity and remanence, similar to a giant paramagnetic atom. For magnetic hyperthermia applications, MNPs must exhibit high saturation magnetization (Ms) to ensure efficient heat generation under AMF. A high Ms value enables stronger heating effects in tumor cells and allows better control over MNP movement in blood circulation via an external magnetic field.

To be effective in hyperthermia therapy, MNPs must meet two key criteria: high heating power and good stability. Superparamagnetic MNPs are favored for their excellent stability and property to lose magnetization in the absence of an external magnetic field. This behavior occurs at temperatures above the blocking temperature (Tb), where thermal energy overcomes magnetic anisotropy, leading to free magnetic moment fluctuations. However, as MNP size decreases, the dipolar interactions between particles also weaken, which influences heating efficiency. Studies have shown that high concentrations of MNPs can suppress relaxation times (Néel and Brownian relaxation) due to increased interparticle interactions, affecting the overall heat dissipation in hyperthermia applications.

4.4 Types of Nanoparticles Used in Hyperthermia

Various magnetic nanoparticles, when exposed to AMF, generate localized heat or Magnetic hyperthermia. These nanoparticles differ in composition, structural properties, and modifications for their heating efficiency, biocompatibility, and targeting capabilities. The major types of nanoparticles used in hyperthermia include (Fig. 4.2).

Metal Oxide Nanoparticles (MONPs): Metal oxide nanoparticles are nanostructured materials that are widely used in biomedical applications due to their stability, biocompatibility, and tunable magnetic properties. Amongst the various MONPs, Iron oxide NPs (e.g., Fe_3O_4, $MnFe_2O_4$) are the most studied due to their excellent heating efficiency. Various groups of researchers have functionalized these nanoparticles using PEG, dextran, silica, etc., to improve their biocompatibility and stability in biological systems.

Metal Alloy Nanoparticles: Two or more metallic elements (e.g., FeCo, FePt, FeNi) are used to form an alloy to enhance the magnetic anisotropy, saturation magnetization, and thermal stability as compared to single-metal nanoparticles. These metal alloys exhibit superior hyperthermia and are also functionalized using silica or other polymeric coatings to enhance biocompatibility and stability.

Noble Metal-CoatedMagnetic Nanoparticles: The coating of noble metals like Au, Ag, Pt (e.g., Fe_3O_4@Au) enhances colloidal stability and prevents oxidation. These coatings improve the biocompatibility and targeted drug delivery.

Core–Shell Nanoparticles: Core–shell MNPs consist of a magnetic core (e.g., Fe_3O_4, $MnFe_2O_4$) surrounded by a functional shell (e.g., silica, gold, polymers, carbon), which enhances their stability, biocompatibility, and multifunctionality. The core provides magnetic properties for hyperthermia, imaging, and the functional shell is responsible for the prevention of oxidation, reduction in toxicity, drug targeting, and also provides a surface for further modifications. Examples include

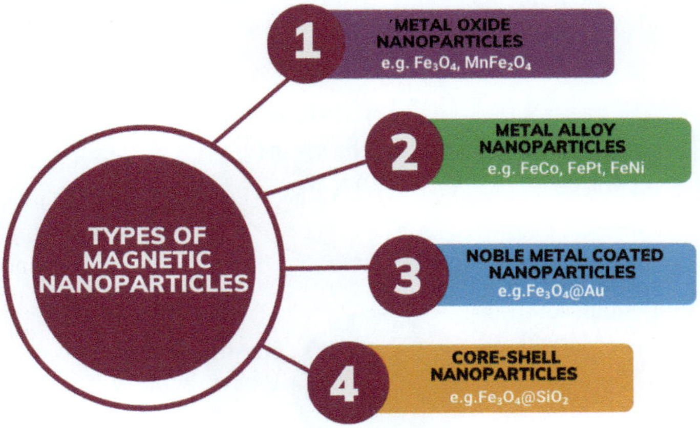

Fig. 4.2 Types of magnetic nanoparticles used in hyperthermia

silica-coated magnetic nanoparticles ($Fe_3O_4@SiO_2$), polymer (chitosan, PEG, or liposomes)-coated magnetic nanoparticles [12].

4.5 Strategies of Synthesis of MNPs

Synthesis of MNPs is the first step towards designing their properties in various applications. In the literature, several chemical and physical methods have been reported for the preparation of MNPs of specific size, shape, crystalline structure, and magnetic properties Fig. 4.3. The therapeutic potential of MNPs in cancer treatment is significantly influenced by these factors as these factors determine their biocompatibility, cellular uptake, magnetic response, and heating efficiency in hyperthermia therapy. For instance, smaller nanoparticles (5–50 nm) exhibit enhanced tumor penetration and cellular uptake, while larger ones (>100 nm) may be rapidly cleared by the immune system. Spherical MNPs offer better circulation stability, whereas rod- or cube-shaped MNPs show higher magnetic anisotropy, improving heating efficiency in hyperthermia. Highly crystalline MNPs possess greater magnetic saturation, leading to more efficient heat generation in magnetic hyperthermia therapy. Each method of preparation of MNPs has distinct advantages and limitations, influencing factors such as size uniformity, magnetic behavior, and biomedical applicability as summarized in Table 4.1.

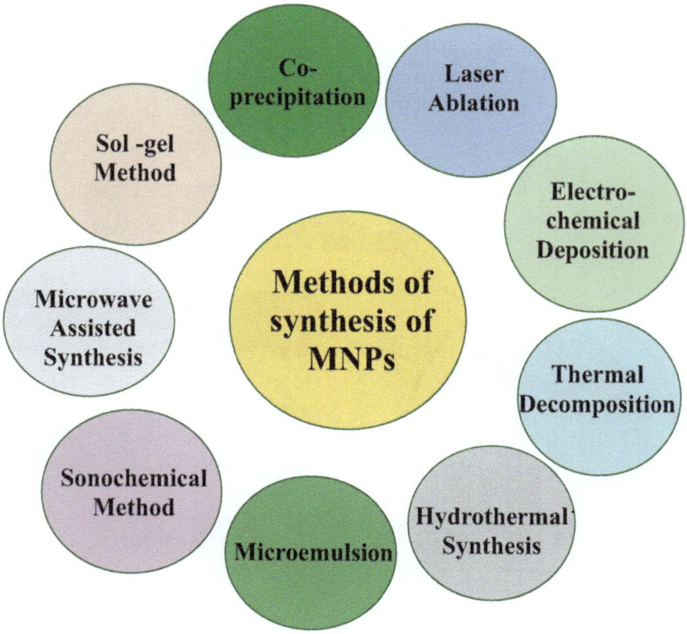

Fig. 4.3 Some common methods of synthesis of magnetic nanoparticles

Table 4.1 Some common methods of synthesis of magnetic nanoparticles

Preparation method	Synthetic procedure	Properties of MNPs	Advantages	Limitations	References
Co-precipitation	Chemical reaction of Fe^{2+} and Fe^{3+} in alkaline medium	Superparamagnetic, high saturation magnetization, moderate crystallinity	Simple, cost-effective, high yield	Polydispersity, aggregation	[13–15]
Thermal decomposition	Decomposition of metal precursors (e.g., iron penta carbonyl) at high temperatures (200–300 °C) in organic solvents	High crystallinity, monodisperse, tunable size	Monodisperse, high crystallinity	Requires high temperature, organic solvents	[16]
Hydrothermal synthesis	Iron salts are treated at high pressure (10–100 atm) and temperature (150–300 °C) in solvents like water, glycol	High saturation magnetization, controlled size, good dispersion	Good size control, high crystallinity	Requires autoclave, long synthesis time	[17]
Sol–Gel method	Formation of a gel network from metal salt precursors	Good magnetic response, stable dispersion, high purity	High purity, good control over composition	Time-consuming, requires high-temperature annealing	[18]
Microemulsion	Formation of nanodroplets in surfactant-stabilized emulsions. Surfactants like CTAB, along with co-surfactants (to control MNP growth and nucleation), are used	Small particle size, uniform morphology, high surface area	Good size control, tunable shape	Low yield, requires surfactants	[19, 20]
Electrochemical deposition	Electrochemical reduction of metal salts on electrodes	High magnetic moment, tunable composition, excellent surface functionalization	High purity, controlled particle growth	Limited to conductive substrates	[21]
Laser ablation	High-energy laser irradiation of metal targets in liquid	High purity, controlled structure, non-toxic	No chemical contamination, high purity	Requires specialized equipment, low yield	[22, 23]

(continued)

Table 4.1 (continued)

Preparation method	Synthetic procedure	Properties of MNPs	Advantages	Limitations	References
Sonochemical method	High-intensity ultrasound induces cavitation, leading to nanoparticle formation	High surface area, narrow size distribution, good magnetic properties	Fast reaction time, uniform size, good crystallinity	Requires special equipment, limited scalability	[24, 25]
Microwave-assisted synthesis	Microwave radiation provides uniform heating for rapid nanoparticle formation	Monodisperse, high saturation magnetization, well-dispersed particles	Short reaction time, energy efficient, high crystallinity	Expensive equipment, limited scalability	[26, 27]

4.6 Strategies of Functionalization of MNPs in Biomedical Applications

Functionalization of MNPs is a crucial step that enables these nanostructures to be used in diverse biomedical applications by enhancing their therapeutic potential. For instance, iron oxide-based nanoparticles (Fe_3O_4, γ-Fe_2O_3) are the most widely used MONPs and exhibit excellent magnetic properties, making them ideal for applications such as magnetic hyperthermia, targeted drug delivery, biosensing, and MRI contrast enhancement. However, non-functionalized MNPs tend to aggregate, oxidize, and exhibit limited bioavailability, thus surface modifications are necessary to improve their performance as therapeutic agents [28]. Various groups of researchers have explored different functionalization strategies to address these challenges. Polymer coatings (e.g., PEG, chitosan) prevent immune clearance and enhance circulation time, and silica coatings improve chemical stability and provide a means for further functionalization. Noble metal coatings (gold, silver) allow for dual-modal therapy, combining magnetic hyperthermia and photothermal therapy. Additionally, ligand exchange strategies (e.g., oleic acid, citric acid) improve dispersibility and hydrophilicity, whereas biopolymer coatings (e.g., hyaluronic acid, collagen) enhance cellular interactions and enable targeted therapy. These surface modifications also help in reducing the toxicity and facilitate a multi-modal approach and combination therapy, ensuring greater efficiency in cancer treatment. A comparative overview of various functionalization strategies, highlighting their functional groups and advantages in biomedical applications, is given in Table 4.2 which can be helpful in the development of the next-generation of multifunctional MNPs.

4.7 Surface Functionalization for Targeted Drug Delivery

Targeted drug delivery is the most important gift of modern advancements in the field of medicine, particularly in cancer therapy. Traditional drug delivery methods are often associated with multiple side effects like poor bioavailability, systemic toxicity, and non-specific distribution that adversely affect the healthy cells, leading to severe side effects and reduced therapeutic efficacy. In targeted drug delivery, various strategies are adopted that ensure the delivery of therapeutic agents at the diseased site specifically, that minimize the adverse effects, reduce the required dosage, drug resistance, and hence the overall treatment outcomes. As the uncovered MNPs are associated with various shortcomings, functionalization not only helps in stability of these MNPs but the coating done with antibodies, peptides, aptamers, or ligands (e.g., folic acid, transferrin), allows MNPs to bind selectively to target cells, which enhances specificity in drug delivery as depicted in Table 4.3. These coatings can be made smart by introducing stimuli responsiveness, through pH, temperature, or redox-responsive materials, which allows the drug to be released at the desired site

Table 4.2 Comparison of functionalization strategies, their advantages, and biomedical applications

Functionalization strategy	Functional group	Advantages in biomedical applications
Polymer/bio-polymer Coating	Polyethylene glycol (peg), polyvinylpyrrolidone (PVP), chitosan, dextran, hyaluronic acid (HA), alginate, collagen, polyaspartamide	Improves biocompatibility, dispersibility, enhances physiological stability, prevents aggregation, targeted drug delivery, MRI contrast agents
Silica coating	SiO_2	Provides chemical stability, improves dispersibility in aqueous medium, prevents oxidation, allows further functionalization, targeted drug delivery, MRI contrast agents
Gold coating	Au	Stabilizes core against oxidation, enhances biocompatibility, bioimaging, further functionalization, photothermal therapy
Graphene oxide	Graphene oxide	High surface area, excellent stability, enables drug loading, targeted drug delivery, biosensing
Dendrimer	Polyamidoamine (PAMAM), poly(propyleneimine) (PPI)	Provides multivalent interactions, increases drug loading efficiency
Ligand exchange	Oleic acid, citric acid, dopamine	Provides hydrophilicity, stabilizes MNPs, allows bioconjugation, targeted drug delivery, biosensing
Surfactant	Oleic acid, SDS (sodium dodecyl sulfate), CTAB (cetyltrimethylammonium bromide)	Controls size and dispersibility, improves colloidal, targeted drug delivery, MRI contrast agent
Host–guest	Host (Cyclodextrins (CDs), Cucurbiturils (CBs), Calixarenes); Guest (anti-cancer drugs, peptides, metal ions)	Enhances drug solubility, biocompatibility, controlled release, stability, selective binding, targeted drug delivery, chemotherapy, theranostics

only under specific conditions. The extent and rate of release of the drug molecule can be quantified by various assays [28].

Table 4.3 Surface functionalization for magnetic nanoparticles in targeted drug delivery

MNP core	Coating	Therapeutic agent	Functionalized MNP	Experimental findings	Characterization	References
Fe_3O_4	PEI-FA grafted on SBA-15 (mesoporous silica)	DOX	Fe_3O_4@SiO_2@SBA-15 labelled FA	Targeted, pH-controlled release of DOX, high internalization, biocompatible, IC50 of MCF-7 was reported	UV–Vis, FT-IR, TGA, XRD, SEM, VSM, EDX	[34]
Fe_3O_4	Chitosan modified, with Silica/without silica	Silymarin	SIL-Cs-f-SiO_2@Fe_3O_4 SIL-Cs-f-@Fe_3O_4	With silica the MTT and IC50 values on MCF-7 were 110.03 and 73.56 $\mu g ml^{-1}$	MTT, IC50	[35]
SPION	Citrate	–	SPION@citrate	High colloidal stability, anticancer property, non-spilling loading into primary human CD3+ T cells with 1.4 pg iron loading/cell	TEM, Magnetic susceptibility, hydrodynamic z-average size, zeta potential	[36]
^{166}Ho-doped SPION	Au	Monoclonal antibody trastuzum (Tmab)	Tmab conjugated [^{166}Ho]SPION@Au	Multi-modal approach combining MH and immunotherapy, Cytotoxicity towards SK-OV-3 ovarian cancer cell line with overexpressed HER 2+ receptors	TEM, TGA	[37]

(continued)

Table 4.3 (continued)

MNP core	Coating	Therapeutic agent	Functionalized MNP	Experimental findings	Characterization	References
SPION	a. BSA-caffeic acid b. citrate	Mitoxantrone	$SPION^{MTO}$	Targeted drug delivery to A375M melanoma cells and fibroblasts	AES, HPLC–UV, fluorescence, zeta potential at physiological pH	[38]
SPIONs, Cubosomes	Hyaluronic acid (HA for CD44) and antibodies for CD221	Helenalin	Cub^{fun} (HA + Abs + SPION)	Potential cytotoxic agent against rhabdomyosarcoma, arrested cells at G2/M phase of cell cycle	SAXS, cryo-EM	[39]
Fe_3O_4	Poly-L-Lysine followed by Au followed by SiO2, later functionalized with Oligonucleotides to block the pores of SiO_2	–	NIR-responsive DNA gated Fe_3O_4@Au@mSiO_2	Controlled delivery of MNPs, synergistic therapy of chemotherapy and phototherapy along with magnetic therapy, drug release is controlled remotely by NIR, MRI diagnosis	TEM, HR-TEM, SEM, UV–Vis, cytotoxicity assays,	[40]
MNPs	Functionalized Graphene Oxide forms the core with 2 more layers, viz. PVA, PEG, and nilotinib (2nd layer), Sodium alginate, PEG, poly(lactic-co-glycolic acid), PLA, and nilotinib gel (3rd layer)	Nilotinib	–	pH-dependent release of drug from three layers, with faster release at acidic pH (TME)	UV–Vis, FT-IR, SEM, TEM, TGA, VSM	[41]

(continued)

Table 4.3 (continued)

MNP core	Coating	Therapeutic agent	Functionalized MNP	Experimental findings	Characterization	References
MNPs	Reduced graphene oxide forms the core, functionalized with MNP, then 4-Hydroxycoumarin (4HC, a photosensitizer) was attached via allylamine (AA)	Camptothecin (CPT)	MrGO-AA-g-4HC	CPT exhibited higher toxicity against MCF-f as compared to the normal fibroblast cell line, demonstrated great results in vivo in synergistic cancer treatment through chemophodynamic therapy	SEM, HR-TEM, FT-IR, AFM, XRD, DLS	[42]
IONP	Fourth (G4) generation of polyamidoamine (PAMAM) dendrimer functionalized	–	G4@IONP	Great biocompatibility, stability, in vivo apoptosis of breast cancer cells as induced by elevated Bax/Bcl-2 ratio and reduction in tumor angiogenesis	TEM, FTIR, zeta potential	[43]
SPION	Triblock copolymer of PAMAM-b-PDMAEMA-b-PPEGMA	DOX	SPION@triblockdendrimer@DOX	Water-soluble functionalized MNPs with uniform particle size of <30 nm, pH-responsive and controlled drug release, biocompatible	SEM, TEM, FT-IR, TGA, XRD, DLS, MTT Assay	[44]

(continued)

Table 4.3 (continued)

MNP core	Coating	Therapeutic agent	Functionalized MNP	Experimental findings	Characterization	References
Fe_3O_4	Chitosan	Telmisartan	MNP-CS-TEL	High drug loading, pH-responsive, controlled, dose-dependent cytotoxic behaviour in PC-3 human prostate cancer cell line	TEM, EDS, XRD, FTIR, TGA	[45]
Fe_3O_4	Oleic Acid (OA)	–	Fe_3O_4 /OA	OA-coated monodisperse MNPs with a size of 7.8 ± 1.9 nm, MNPs exhibit good dispersibility and supermagnetism, saturation magnetisation, and particle size were proportional	SEM, TEM, FT-IR, TGA, XRD, EDXRF, VSM	[46]
SPION	Au followed by EGF-α-lipoic acid and chlorin e6(Ce6)-cysteamine	–	SPIONs@Au	PTT results showed potential anticancer properties, EGF-α-lipoic acid and chlorin e6(Ce6)-cysteamine were observed in cytoplasm as well in nuclei in TNBC (MDA-MB-468) cells	TEM, DLS, EDS, PTT	[15]

(continued)

Table 4.3 (continued)

MNP core	Coating	Therapeutic agent	Functionalized MNP	Experimental findings	Characterization	References
SPION	Vinylated β-CD integrated with $Fe_3O_4@SiO_2$	DOX	$Fe_3O_4@SiO_2$	pH-dependent DOX release,	FTIR, SEM, TEM, XRD, VSM, and UV–Vis	[47]
MNPs	Polydopamine	DOX	magneticDOX-IP	Targeted drug delivery, improved distribution of the drug in breast adenocarcinoma in BALB/c mice	FTIR, SEM, FAAS, HPLC	[48]

4.7.1 Superparamagnetic Iron Oxide Nanoparticles (SPIONs)

SPIONS are a type of magnetic nanoparticle composed of iron oxide (Fe_3O_4 or γ-Fe_2O_3) that are specifically small enough (<20 nm) to exhibit superparamagnetism at the nanoscale. This means they display strong magnetization under an external magnetic field but have no residual magnetism once the field is removed, preventing aggregation. Due to their biocompatibility, stability, and high magnetic responsiveness, SPIONs are widely used in biomedical applications, including magnetic hyperthermia, targeted drug delivery, MRI contrast enhancement, and biosensing. Their surface can be functionalized with polymers, biomolecules, or silica to enhance their dispersibility and targeting efficiency in biological systems.

4.7.2 Polymer/Biopolymer-Coated Magnetic Nanoparticles

Polymer coating on MNPs is a crucial surface modification strategy for enhancing the stability, biocompatibility, and functional properties. The polymer coatings not only help to improve the dispersibility and circulation time in the biological system but also prevent the nonspecific interactions with the surroundings, like oxidation and aggregation of iron MNPs [29, 30]. In literature, there are many coatings introduced on iron MNPs using natural (dextran, chitosan, alginate) and synthetic polymers (PEG, PVA, polylactic-co-glycolic Acid (PLGA) [31]. Amphiphilic polymer, e.g., modified poly (maleic anhydride), has been shown to work as a linker where the hydrophobic fraction interacts with the MNP layer and the hydrophilic fraction helps in conjugating other molecules for better compatibility and stability [32]. The polymer layer helped in the dispersibility of the nanoarchitecture in aqueous conditions and provides a surface for further functionalization of these MNPs' backbone. Amongst various polymers, PEG is one of the most commonly used polymer coating [33], which not only improve the biocompatibility but also improves the cellular internalization, leading to better apoptotic effect of the iron MNPs.

4.7.3 Silica-Coated Magnetic Nanoparticles

Silica-coated MNPs are widely used in various biomedical applications, especially in cancer treatment, as these have high chemical stability and low toxicity in physiological conditions. The silica shell not only prevents the aggregation and oxidation, which is a prominent problem associated with bare MNPs, but also provides a hydrophilic surface that improves their dispersibility in physiological fluids. The silica coating provides an opportunity for covalent attachment of various molecules (targeting ligands/ therapeutic agents), making them highly effective for targeted drug delivery, bioimaging, and magnetic hyperthermia. The porous structure of mesoporous silica-coated MNPs further enables controlled drug release, enhancing therapeutic efficiency in cancer treatment.

4.7.4 Graphene Oxide Coated Magnetic Nanoparticles (GO-MNPs)

GO-MNPs are hybrid nanostructures where the characteristic features of GO, like high surface area, mechanical strength, and easy functionalization, are combined with the magnetic properties of iron oxide nanoparticles (Fe_3O_4 or γ-Fe_2O_3). The GO generally forms the inner core, and MNPs are coated with oxygen-containing functional groups (–OH, –COOH, –C=O) in GO that facilitate strong binding with MNPs. This binding improves the colloidal stability of the hybrid and also provides a platform for further surface functionalization for biomedical applications. Functionalized GO-MNPs exhibit improved dispersibility in aqueous media, enhanced drug-loading capacity, along with targeted delivery of drug molecules. These characteristics make them potential and effective candidates for cancer therapy, magnetic hyperthermia, and multimodal imaging. Their high thermal and electrical conductivity, because of the GO core, also allows for synergistic photothermal and magnetic hyperthermia treatments, increasing therapeutic efficacy.

4.7.5 Dendrimer-Coated Magnetic Nanoparticles (D-MNPs)

Dendrimers are highly branched, nanoscale macromolecules with a tree-like structure having repetitive polymer units. Dendrimer-based MNPs (D-MNPs) are hybrid nanostructures where dendrimers, along with some monodisperse macromolecules, are used to coat or functionalize MNPs. The dendrimers have a nearly spherical shape that allows control over size and surface functionalization, making them ideal carriers for various biomedical applications [49, 50]. The dendrimer shell, typically made of poly(amidoamine) (PAMAM), polypropylenimine (PPI), or polyglycerol (PG), with the surface polymer endings having functional groups like $-NH_2$, –COOH, and −OH [51]. These functional groups are responsible for the hydrophilicity of the molecule, leading to enhanced stability in physiological environments [52]. D-MNPs particularly find application in targeted drug delivery, MRI imaging, and theranostic applications. These hybrid structures exhibit high drug-loading capacity and controlled release owing to dendrimer coating, while their magnetic core allows for external magnetic guidance and hyperthermia treatment [53]. A study by Khodadust et al. documented Poly (I:C) binding on DOX-coated G4D-MNPs to yield PIC-G4DMNP-DOX. It is shown that the hybrid nanostructure depicted pH maximum cytotoxicity as compared to Poly I:C or DMNP or DOX taken individually. Also, the results depicted that the hybrid was more biocompatible and was able to kill cancer cells of DOX-resistant MCF-7 cell line. The same group of researchers has conducted similar studies on different generations of PAMAM-coated SPIONs [54].

4.7.6 Ligand Exchange Functionalization Strategy

Ligand exchange is a strategy that is based on replacing the existing surface ligands to add desired characteristics like enhanced stability, biocompatibility, and targeted action, etc. In this process, those ligands that are weakly bound or easily replaceable are replaced with stronger coordinating ligands that impart desired properties to the structure [55, 56]. The ligands capable of forming strong bonds with the nanoparticle surface include simple functional groups like carboxyl (-COOH), amine ($–NH_2$), thiol (–SH), or phosphonate ($–PO(OH)_2$), polyethylene glycol (PEG), dopamine, citrate, or amino acids. Because these groups are hydrophilic in nature, MNPs gain improved dispersibility in aqueous environments, i.e., physiological conditions, reduced aggregation, and enhanced bioavailability [56]. These charged ligands/functional groups modify surface charge, influencing colloidal stability and cellular uptake of MNPs. Various interactions responsible for stabilizing these MNPs in biological conditions are covalent bonds, hydrogen bonds, electrostatic force of attraction, and van der Waals forces of attraction.

4.7.7 Surfactant Modification of Magnetic Nanoparticles

Surfactant modifications are the coating with certain amphiphilic molecules that is applied to impart the desirable characteristics like enhanced stability, dispersibility, bioavailability for therapeutic applications [57]. Oleic acid (OA) is the most common surfactant used to functionalize the MNPs, where the carboxyl (−COOH) group of OA chemically binds to the Fe^{3+} and Fe^{2+} ions on the MNP surface through chelation or H-bonding [58]. The long hydrocarbon chain enhances the dispersibility in the nonpolar solvents, which is crucial for synthetic procedures like thermal decomposition [59]. The type of surfactant employed for coating alters functionality through electrostatic interactions and surface charge modulation. For instance, ionic surfactants alter the zeta potential (surface charge) of MNPs and thus prevent coagulation through electrostatic repulsion. Anionic surfactants like sodium dodecyl sulfate (SDS) provide a negative charge, while cationic surfactants such as cetyltrimethylammonium bromide (CTAB) impart a positive charge, tailoring surface properties for targeted interaction. A recent study by Rajal et al. reported synthesis of cuboidal shaped Fe_3O_4@CTAB MNPs where the quaternary ammonium fraction prevented their aggregation. The functionalized MNPs exhibited anti-cancer properties at physiological pH medium and reduced hemolysis by offering minimal damage to erythrocyte membrane [60].

4.7.8 *Host–Guest Complexation as a Functionalization Strategy*

Host–guest complexes in MNPs involve the incorporation of guest molecules into host structures attached to the MNP surface, enhancing their functionality for various applications. This is an emerging functionalization strategy for MNPs that leverages non-covalent interactions between a host molecule and a guest molecule to enhance stability, targeting, and drug-loading capacity. These complexes provide a reversible and stimuli-responsive platform, making them highly suitable for targeted drug delivery, biosensing, and theranostics. For example, MNPs are functionalized with β-cyclodextrin (β-CD), which is a cyclic oligosaccharide and has the ability to form inclusion complexes with various guest molecules. The integration of β-cyclodextrin with MNPs resulted in nanocarriers with a high drug-loading capacity. Various studies have shown the effective delivery of curcumin [61] and ketoprofen [62] to various cancer cells with significant efficacy. In a recent study, a novel polydopamine-based magnetic platform (Fe_3O_4@PDA@@SH-β-CD) was reported which demonstrated excellent drug loading capacity via β-CD. The integrated platform functioned as an anticancer agent through combined chemo and photothermal therapy [63].

4.8 Conclusion and Future Prospects

The current chapter reviews the recent advancements in iron oxide based magnetic nanocarriers designed to enhance the therapeutic efficacy of tumor treatment. These MNPs can be guided to specific tumor sites where, by using an external magnetic field, they produce heat known as 'magnetic hyperthermia' that leads to killing of tumor cells. As the bare MNPs experience various limitations like oxidation, aggregation, they are functionalized for enhanced bioavailability, targeted drug delivery, and reducing systemic toxicity. In this chapter, we have discussed various advantages and functions of magnetic nanocarriers in different therapeutic modalities, associated with various functionalization introduced at the MNP surface.

While numerous studies have demonstrated their potential to significantly improve cancer treatment through in vitro and in vivo studies, their clinical application still faces several obstacles. The biggest challenge associated with therapeutic MNPs is achieving a balance between safety and efficacy. Excessive heat may have an adverse impact on healthy tissues, while less heat may not kill cancer cells effectively. Also, since different synthetic procedures result in MNPs with different shapes, that also affects the effectiveness of the treatment, as certain MNPs are easily internalized over others. In synthetic procedures, the scalability and reproducibility of MNPs remain a critical challenge, affecting batch-to-batch consistency. Therefore, the structural and physicochemical properties of nanohybrids raise concerns about their efficacy and safety which needs to be carefully evaluated. For successful clinical translation, the guiding principle should be to "make it simple and safe." Further research is needed to better understand the biological effects of magnetic nanocarriers on therapeutic efficacy.

Despite existing limitations, innovative design strategies continue to expand the possibilities for clinical applications. With ongoing research and optimization, these advancements will accelerate the clinical translation of magnetic nanocarriers, paving the way for more efficient and targeted cancer therapies.

Abbreviations

AES	Auger Electron Spectroscopy
AMF	Alternating Magnetic Field
CTAB	Cetyltrimethylammonium Bromide
DC	Dendritic Cells
EBRT	External Beam Radiation Therapy
EDS	Energy-Dispersive Spectroscopy
EDX	Energy-Dispersive X-Ray Spectroscopy
EDXRF	Energy-Dispersive X-Ray Fluorescence
FAAS	Flame Atomic Absorption Spectroscopy
FTIR	Fourier Transform InfraRed spectroscopy
HPLC	High-Performance Liquid Chromatography
IMRT	Intensity-Modulated Radiation Therapy
LHT	Local Hyperthermia
MDSC	Myeloid-Derived Suppressor Cells
MNPs	Magnetic Nanoparticles
MONPs	Magnetic Oxide Nanoparticles
MRI	Magnetic Resonance Imaging
MTT	3-(4,5-Dimethylthiazol-2-yl)-2,5-diphenyltetrazolium bromide
PTT	Photothermal Therapy
RHT	Regional Hyperthermia
ROS	Reactive Oxygen Species
SAXS	Small-Angle X-Ray Scattering
SBRT	Stereotactic Body Radiotherapy
SDS	Sodium Dodecyl Sulfate
SEM	Scanning Electron Microscopy
SPIONs	Supermagnetic Iron Oxide Nanoparticles
SRS	Stereotactic Radiosurgery
TEM	Transmission Electron Microscopy
TGA	Thermogravimetric Analysis
VSM	Vibrating Sample Manometer
WBH	Whole Body Hyperthermia
XRD	X-Ray Diffraction

References

1. Farzanegan Z, Tahmasbi M (2023) Evaluating the applications and effectiveness of magnetic nanoparticle-based hyperthermia for cancer treatment: a systematic review. Appl Radiat Isot 198:110873
2. Alromi DA, Madani SY, Seifalian A (2021) Emerging application of magnetic nanoparticles for diagnosis and treatment of cancer. Polymers 13. https://doi.org/10.3390/polym13234146
3. Khursheed R et al (2022) Biomedical applications of metallic nanoparticles in cancer: current status and future perspectives. Biomed Pharmacother 150:112951
4. Wang Z et al (2016) Porous nickel-iron selenide nanosheets as highly efficient electrocatalysts for oxygen evolution reaction. ACS Appl Mater Interfaces 8(30):19386–19392
5. Singh R, Lillard JW (2009) Nanoparticle-based targeted drug delivery. Exp Mol Pathol 86(3):215–223
6. Shubhra QTH (2023) Iron oxide nanoparticles in magnetic drug targeting and ferroptosis-based cancer therapy 3(5):444–447
7. Faraji AH, Wipf P (2009) Nanoparticles in cellular drug delivery. Bioorg Med Chem 17(8):2950–2962
8. Tran N, Webster TJ (2010) Magnetic nanoparticles: biomedical applications and challenges. J Mater Chem 20(40):8760–8767
9. Peiravi M et al (2022) Magnetic hyperthermia: potentials and limitations. J Ind Chem Soc 99(1):100269
10. Phung DC et al (2019) Combined hyperthermia and chemotherapy as a synergistic anticancer treatment. J Pharm Investig 49(5):519–526
11. Serio F et al (2022) Co-loading of doxorubicin and iron oxide nanocubes in polycaprolactone fibers for combining Magneto-Thermal and chemotherapeutic effects on cancer cells. J Colloid Interface Sci 607:34–44
12. Keshavarz H et al (2020) PH-sensitive magnetite mesoporous silica nanocomposites for controlled drug delivery and hyperthermia. RSC Adv 10(64):39008–39016
13. Chen B-W et al (2020) Synthesis and characterization of magnetic nanoparticles coated with polystyrene sulfonic acid for biomedical applications. Sci Technol Adv Mater 21(1):471–481
14. Laurent S et al (2008) Magnetic iron oxide nanoparticles: synthesis, stabilization, vectorization, physicochemical characterizations, and biological applications. Chem Rev 108(6):2064–2110
15. Cândido M et al (2023) Gold-coated superparamagnetic iron oxide nanoparticles functionalized to EGF and Ce6 complexes for breast cancer diagnoses and therapy. Pharmaceutics 15(1):100
16. Park J et al (2004) Ultra-large-scale syntheses of monodisperse nanocrystals. Nat Mater 3(12):891–895
17. Li J, Wu Q, Wu J (2016) Synthesis of Nanoparticles via solvothermal and hydrothermal methods. In: Aliofkhazraei M (ed) Handbook of nanoparticles. Springer International Publishing, Cham, pp 295–328
18. Xu J et al (2007) Preparation and magnetic properties of magnetite nanoparticles by sol–gel method. J Magn Magn Mater 309(2):307–311
19. Richard B, Lemyre J-L, Ritcey AM (2017) Nanoparticle size control in microemulsion synthesis. Langmuir 33(19):4748–4757
20. Pineda MG et al (2014) Chitosan-coated magnetic nanoparticles prepared in one-step by precipitation in a high-aqueous phase content reverse microemulsion. Molecules 19:9273–9287. https://doi.org/10.3390/molecules19079273
21. Tonelli D, Scavetta E, Gualandi I (2019) Electrochemical deposition of nanomaterials for electrochemical sensing. Sensors 19(5):1186
22. Piotto V et al (2022) Synthesis of magnetic nanoparticles by laser ablation of strontium ferrite under water and their characterization by optically detected magnetophoresis supported by BEM calculations. J Mater Chem C 10(10):3819–3825
23. Piotto V et al (2023) One-step laser ablation synthesis of magnetic nanoparticles with carbon coating for tribological applications. Tribol Int 183:108371

24. Stankic S et al (2016) Pure and multi metal oxide nanoparticles: synthesis, antibacterial and cytotoxic properties. J Nanobiotechnol 14(1):73
25. Stephen ZR et al (2016) Approach to rapid synthesis and functionalization of iron oxide nanoparticles for high gene transfection. ACS Appl Mater Interfaces 8(10):6320–6328
26. Wang W-W, Zhu Y-J, Ruan M-L (2007) Microwave-assisted synthesis and magnetic property of magnetite and hematite nanoparticles. J Nanopart Res 9(3):419–426
27. Liang Y-J et al (2016) Ultrafast preparation of monodisperse Fe_3O_4 nanoparticles by microwave-assisted thermal decomposition. Chem A Eur J 22(33):11807–11815
28. Upadhyay K et al (2023) Surface functionalized nanoparticles: a boon to biomedical science. Chemico-Biol Interact 380:110537
29. Rezaei B et al (2024) Effect of polymer and cell membrane coatings on theranostic applications of nanoparticles: a review. Adv Healthc Mater 13(26):2401213
30. Strassburg S, Mayer K, Scheibel T (2022) Functionalization of biopolymer fibers with magnetic nanoparticles 7(10):1091–1117
31. Canfarotta F, Piletsky SA (2014) Engineered magnetic nanoparticles for biomedical applications. Adv Healthc Mater 3(2):160–175
32. Lin C-AJ et al (2008) Design of an amphiphilic polymer for nanoparticle coating and functionalization. Small 4(3):334–341
33. Jokerst JV et al (2011) Nanoparticle PEGylation for imaging and therapy. Nanomedicine 6(4):715–728
34. Ehsanimehr S et al (2021) PEI grafted Fe3O4@SiO2@SBA-15 labeled FA as a pH-sensitive mesoporous magnetic and biocompatible nanocarrier for targeted delivery of doxorubicin to MCF-7 cell line. Colloids Surf A: Physicochem Eng Aspects 615:126302
35. Zare M, Sarkati MN (2021) Chitosan-functionalized FeO nanoparticles as an excellent biocompatible nanocarrier for silymarin delivery. Polym Adv Technol 32(10):4094–4100
36. Boosz P et al (2021) Citrate-coated superparamagnetic iron oxide nanoparticles enable a stable non-spilling loading of T cells and their magnetic accumulation. Cancers 13. https://doi.org/10.3390/cancers13164143
37. Żuk M et al (2021) Hybrid radiobioconjugated superparamagnetic iron oxide-based nanoparticles for multimodal cancer therapy. Pharmaceutics 13(11):1843
38. Behr J et al (2022) In vitro setup for determination of nanoparticle-mediated magnetic cell and drug accumulation in tumor spheroids under flow conditions. Cancers 14(23):5978
39. Mun H et al (2023) CD44 and CD221 directed magnetic cubosomes for the targeted delivery of helenalin to rhabdomyosarcoma cells. Nano Res 16(2):2915–2926
40. Li W-P et al (2014) Formation of oligonucleotide-gated silica shell-coated Fe_3O_4-Au core-shell nanotrisoctahedra for magnetically targeted and near-infrared light-responsive theranostic platform. J Am Chem Soc 136(28):10062–10075
41. Zhalechin M et al (2021) Magnetic polymeric core-shell as a carrier for gradual release in-vitro test drug delivery. Heliyon 7(5)
42. Vinothini K et al (2020) A magnetic nanoparticle functionalized reduced graphene oxide-based drug carrier system for a chemo-photodynamic cancer therapy. New J Chem 44(14):5265–5277
43. Salimi M et al (2020) Treatment of breast cancer-bearing BALB/c mice with magnetic hyperthermia using dendrimer functionalized iron-oxide nanoparticles. Nanomaterials 10(11):2310
44. He X et al (2012) Functionalization of magnetic nanoparticles with dendritic–linear–brush-like triblock copolymers and their drug release properties. Langmuir 28(32):11929–11938
45. Dhavale RP et al (2021) Chitosan coated magnetic nanoparticles as carriers of anticancer drug telmisartan: pH-responsive controlled drug release and cytotoxicity studies. J Phys Chem Solids 148:109749
46. Mahdavi M et al (2013) Synthesis, surface modification and characterisation of biocompatible magnetic iron oxide nanoparticles for biomedical applications. Molecules 18(7):7533–7548
47. Hosseinzadeh H et al (2024) Preparation of amino-functionalized β-Cyclodextrin/Fe_3O_4@SiO_2 magnetic nanocarrier for controlled release of doxorubicin, an anticancer drug. Arab J Sci Eng 49(1):459–473

48. Zavareh S et al (2016) Synthesis of polydopamine as a new and biocompatible coating of magnetic nanoparticles for delivery of doxorubicin in mouse breast adenocarcinoma. Cancer Chemother Pharmacol 78(5):1073–1084
49. Bober Z, Bartusik-Aebisher D, Aebisher D (2022) Application of dendrimers in anticancer diagnostics and therapy. Molecules 27(10):3237
50. Parlanti P et al (2020) Targeted dendrimer-coated magnetic nanoparticles for selective delivery of therapeutics in living cells. Molecules 25(9):2252
51. Svenson S, Tomalia DA (2005) Dendrimers in biomedical applications—reflections on the field. Adv Drug Deliv Rev 57(15):2106–2129
52. Zhao Y et al (2010) Synthesis and grafting of folate–PEG–PAMAM conjugates onto quantum dots for selective targeting of folate-receptor-positive tumor cells. J Colloid Interface Sci 350(1):44–50
53. Ray S et al (2018) Dendrimer- and copolymer-based nanoparticles for magnetic resonance cancer theranostics. Theranostics 8(22):6322–6349
54. Weber A et al (2010) Proapoptotic signalling through Toll-like receptor-3 involves TRIF-dependent activation of caspase-8 and is under the control of inhibitor of apoptosis proteins in melanoma cells. Cell Death Differ 17(6):942–951
55. Pedroso-Santana S, Fleitas-Salazar N (2023) The use of capping agents in the stabilization and functionalization of metallic nanoparticles for biomedical applications. Part Part Syst Charact 40(2):2200146
56. Dheyab MA et al (2020) Simple rapid stabilization method through citric acid modification for magnetite nanoparticles. Sci Rep 10(1):10793
57. Zhu J, Wang J, Li Y (2023) Recent advances in magnetic nanocarriers for tumor treatment. Biomed Pharmacother 159:114227
58. Ling W et al (2019) Synthesis, surface modification, and applications of magnetic iron oxide nanoparticles. J Mater Res 34(11):1828–1844
59. Zhang L, He R, Gu H-C (2006) Oleic acid coating on the monodisperse magnetite nanoparticles. Appl Surf Sci 253(5):2611–2617
60. Rajan A, Kaczmarek-Szczepańskac B, Sahu NK (2021) Magneto-thermal response of Fe3O4@CTAB nanoparticles for cancer hyperthermia applications. Mater Today Commun 28:102583
61. Yallapu MM et al (2012) Curcumin-loaded magnetic nanoparticles for breast cancer therapeutics and imaging applications. Int J Nanomed 7:1761–79
62. Huang L et al (2014) β-Cyclodextrin derivatives hybrid Fe_3O_4 magnetic nanoparticles as the drug delivery for ketoprofen. J Incl Phenom Macrocycl Chem 80(3):209–215
63. Mrówczyński R et al (2018) Cyclodextrin-based magnetic nanoparticles for cancer therapy. Nanomaterials 8(3):170

Chapter 5
Diagnostic Imaging with Functionalized Magnetic Nanomaterials (FMNs)

Pooja, Aman Maurya, and Laishram Saya

Abstract FMNs are mostly composed of a magnetic core, like iron oxide. They are functionalized with biocompatible coatings (polyethylene glycol, dextran, silica) and targeting molecules to optimize their performance in medical applications. FMNs' properties, like super magnetism and high magnetic susceptibility, enhance the diagnostic detail and specificity, enabling more accurate disease diagnosis. The super-magnetic property of the FMNs provides strong magnetization in response to an external magnetic field, which improves the variation between tissues during MRI scans and at the same time functionalization allows for the precise targeting of specific biological ligands (antibodies, peptides, or small molecules), ensures high specificity in detecting abnormalities such as tumors, inflammation, or any vascular diseases. The design and synthesis of FMNs for imaging applications depend on the particle size, shape, surface functionalisation, and biocompatibility, all of which influence their performance as contrast agents. For example, FMNs functionalized with ligands or antibodies can target certain specific cell receptors and provide molecular imaging capabilities that go beyond conventional anatomical imaging. FMNs are also expanding into multimodal imaging systems, which combine magnetic properties of FMNs with fluorescence for techniques like magnetic particle imaging and positron emission tomography. These conjugated FMNs with fluorescent dyes facilitate real-time imaging in preclinical studies, enabling simultaneous tracking and diagnosis. These advancements enhance the precision of diagnostics and open new avenues for personalised medication. Surface functionalisation is the key to the versatility and effectiveness of FMNs in diagnostic imaging. This chapter reviews the use of FMNs as contrast agents in MRI and other imaging modalities, explaining how their magnetic properties enhance resolution and specificity in diagnostic procedures.

Keywords Magnetic resonance imaging (MRI) · Contrast agent · Positron emission tomography (PET) · Magnetic particle imaging

Pooja (✉) · A. Maurya · L. Saya
Department of Chemistry, Sri Venkateswara College, University of Delhi, New Delhi, India
e-mail: pooja.chem123@gmail.com; pooja@svc.ac.in

S. Gulati (ed.), *Functionalized Magnetic Nanomaterials*, Nanostructure Science and Technology, https://doi.org/10.1007/978-3-031-97199-0_5

5.1 Introduction

Scientists, engineers, and doctors are finding that bionanotechnology is a valuable toolkit for making significant strides in the biological and medical sciences. Nanomaterials, including nanoparticles, quantum dots, nanorods, nanotubes, and other composites and nanostructured materials, are currently widely used in various disciplines. In at least one dimension, nanoparticles are structures with sizes ranging from 1 to 100 nm. Iron, manganese, nickel, cobalt, and other pure metals, as well as their alloys and oxides, are the building blocks of magnetic materials. The most often employed particles in biomedical applications are iron oxides. They are commonly present in nature in minerals as well as in living things like ferritin in the human heart, spleen, liver, and magnetotactic bacteria. Magnetite (Fe_3O_4) or maghemite (γ-Fe_2O_3) magnetic nanoparticles (MNPs) have special properties that make them interesting as carriers in biomedical applications. Their colloidal stability, thermal stability, and chemical stability set them apart from other materials used in medical applications. The materials used must not be harmful. They should also have a homogeneous structure, low polydispersion, and a large surface area for diagnostic agent adsorption. Chemically passivating nanoparticles by coating their iron oxide cores with different shells, such as gold, polymeric [1], silane [2], or dendrimeric [3], opens up a wide range of biological possibilities for the particles. In medicine, magnetic nanoparticles show promising effects in diagnostic or therapeutic instruments. In terms of diagnostics, they can be applied both in vitro and in vivo. For instance, they can be employed as contrast agents in magnetic resonance imaging [4], immobilize and detect biomolecules, separate cells [5], purify cells [6], transfer genes [7], and more. They can also be used to treat hyperthermia because of the heat they generate in an alternating magnetic field and as a drug delivery system in target therapy [1]. Superparamagnetism offers significantly higher magnetization without sacrificing the benefit of a stable colloidal suspension. Iron, nickel, cobalt, and their chemical compounds are examples of magnetically sensitive components that are frequently found in such particles. Maghemite (γ-Fe_2O_3) and magnetite Fe_3O_4 are the most widely used superparamagnetic iron oxide (SPIO) particles in biomedical applications because of their high biocompatibility. Iron or iron oxides, such as hemosiderin, ferritin, and transferrin, can be found in human tissues [8]. Some of the current uses of nanoparticles in biomedical fields are shown in Fig. 5.1.

5.2 Properties of Magnetic Nanomaterials

Magnetic materials are primarily multimagnetic domain structures. When the size of attractive magnetic materials is reduced to non-orange, they display a single attractive domain structure and transform into paramagnetic materials19. In the outer attractive magnetic field (AMF), paramagnetic materials exhibit a weakly attractive magnetic

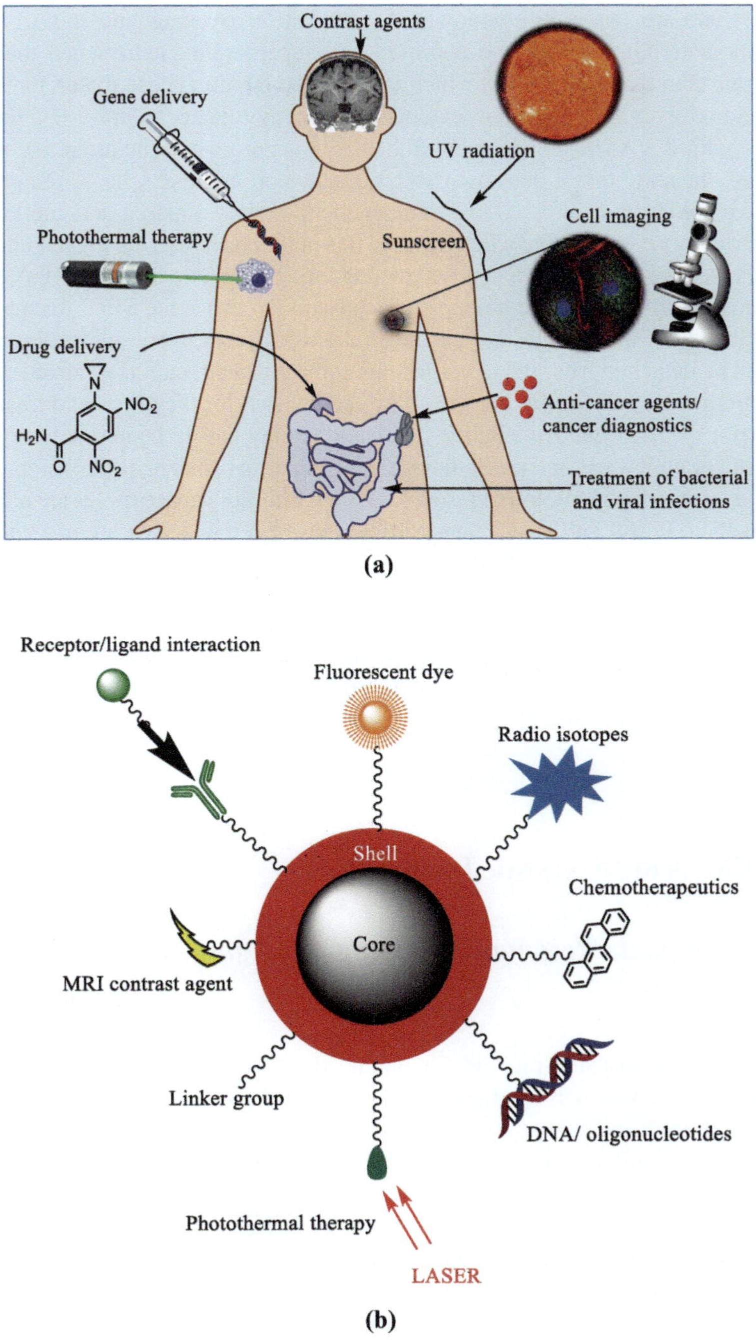

Fig. 5.1 **a** Some of the current uses of nanoparticles [9]. **b** A pictorial representation of the applications of core/shell nanoparticles [9] CC BY4

property and are observably nonmagnetic. The attractive magnetic spin of attractive nanomaterials will become confused and superparamagnetic when their size is smaller than the critical size, which is typically 20 nm. These materials will be charged rapidly when exposed to external AMF activity and can move in a directed manner with the AMF. In any event, the charge is zero when the attractive field is removed, meaning that it does not exhibit magnetism in the absence of an external attractive field [10]. Compared to paramagnetic materials, superparamagnetic materials are far more susceptible when exposed to a magnetic field. Due to their ability to produce thermal impact when exposed to alternating magnetic fields (AMFs), MNPs can be used for tumor thermotherapy. It is possible to abuse the MNP for magnetic separation [11]. For instance, they can be used as vectors to bind biomolecules, which can then be detached from the biomolecules at the desired location when an external magnetic field is present. In this way, they can be used for diagnosis and treatment. The majority of ineffective alkaline medications may hardly penetrate tumor cells due to the often low pH outside of them, exhibiting drug resistance. To overcome this obstacle, researchers have looked closely at how different nanoparticles are arranged and changed. Fe_3O_4 and Fe_2O_3 nanoparticles have been widely used in a variety of treatments and diagnoses, as SPIO (superparamagnetic iron oxide) nanoparticles are the most widely concentrated. Among the many benefits of iron oxide nanoparticles are their low cytotoxicity, high biodegradability, ability to be modified by different substances, ease of preparation, ability to enter cells by endocytosis, and ability to interact with a variety of specific antibodies or ligands. They have been used in MRI and tumor thermotherapy [12]. Properties and advantages of MNPs are shown in Table 5.1.

5.3 FMNs in Diagnostic Imaging

5.3.1 Magnetic Resonance Imaging—Imaging Contrast Agents

A common non-invasive method for diagnosing illnesses is magnetic resonance imaging (MRI), which is distinguished by its strong soft tissue contrast, lack of penetration limit, and spatial resolution. MRI creates fine-grained images of organs, tissues, bones, and other internal body structures using radio frequency pulses and a strong magnetic field. Without having any negative consequences, it gives information regarding the location, size, and shape of different lesions. Transverse/spin–spin relaxation time (T2, T2*) and longitudinal/spin-lattice relaxation time (T1) are the main parameters that affect the MRI signal intensity and are in charge of differentiating healthy tissues from diseased ones. Different contrast agents (CA) can lower the value of these parameters, increasing the sensitivity of the approach. For MRI applications, magnetic particles show promise as contrast agents. They have unique magnetic characteristics and the capacity to function at the cellular

Table 5.1 Key properties of functionalized magnetic nanomaterials and their advantages in diagnostic imaging

Property	Description	Advantages of diagnostic imaging
Superparamagnetic	Exhibits strong magnetization in a magnetic field but no residual magnetism	Prevents particle aggregation, ensuring stable dispersions in biological fluids
High magnetic susceptibility	High responsiveness to external magnetic fields	Enhances MRI contrast, improving image resolution and sensitivity
Tunable size (1–100 nm)	It can be synthesized in a wide range of sizes	Enables penetration into tissues and cells, improving targeting and imaging depth
Surface functionalization	It can be coated with polymers, antibodies, peptides, or other ligands	Improves biocompatibility, targeting, and stability in physiological environments
Biocompatibility	Many FMNs (e.g., iron oxide nanoparticles) are non-toxic and biodegradable	Safe for in vivo use, with minimal risk of toxicity or immune response
Multimodal capability	It can be engineered for use in multiple imaging modalities (e.g., MRI, CT, PET)	Enables complementary imaging techniques for more accurate and comprehensive diagnosis
Targeted delivery	It can be functionalized with ligands (e.g., antibodies, aptamers) for specific targeting	Enhances specificity and accumulation at disease sites, reducing background noise
High relaxivity	Strong ability to shorten T1/T2 relaxation times in MRI	Improves MRI contrast, making it easier to detect small lesions or abnormalities
Stability in biological environments	Resistant to degradation in physiological conditions (e.g., pH, enzymes)	Ensures long-term functionality during imaging procedures
Ease of functionalization	It can be easily modified with imaging agents, drugs, or targeting molecules	Enables theranostic applications (combining diagnosis and therapy)
cost-effectiveness	Many FMNs (e.g., iron oxide nanoparticles) are cost-effective to produce	Facilitates widespread clinical and research use

and molecular levels because of their nanoscale size. Positive (T1) and negative (T2) magnetic contrast agents are the two general kinds that have been introduced. Hyperintense areas are produced in T1-weighted images by the former—paramagnetic small molecule agents (based on Gd(III), Mn(II), Mn(III), and Fe(III)). In T2 and T2* weighted images, the latter—superparamagnetic iron oxide nanoparticles, or SPIONs—create hypointensities or dark areas [13]. Only Gd(III)-based [14–18] and Mn(II)-based [13] T1 contrast agents are authorized for use in clinical imaging.

Gadolinium chelate-based contrast systems have a high threshold of detectability, and their use is linked to nephrogenic systemic fibrosis (NSF). Conversely, a modest contrast is offered by the bulk of described compounds based on Mn CA. Gd and Mn-based contrast agents are less desirable due to the aforementioned issues. Over the decades, iron oxide nanoparticles have emerged as the go-to method for MRI cell tracking because of their unique magnetic and surface characteristics, biocompatibility, and low detection threshold [19]. However, direct use of SPION as an in vivo MRI contrast agent is ineffective due to its strong propensity to form aggregates, which reduces superparamagnetic characteristics and promotes rapid capture by the reticular endothelial system (RES) cells [20]. Consequently, a range of coatings, mostly polymeric, have been applied to magnetic cores to improve their functionality in vivo. The shell of a coating substance increases the biocompatibility of nanoparticles, prolongs their circulation within the system, and inhibits agglomeration. Furthermore, this type of shell prevents biological components from adhering in vivo, which is the initial stage of phagocytosis activation [21].

5.3.2 Positron Emission Tomography (PET)

Full body images with functional and molecular information are produced by Positron Emission Tomography (PET), a sensitive human molecular imaging technique with infinite tissue penetration and high sensitivity (picomolar level). Furthermore, it can precisely identify and measure external radioactive isotopes. Nanoparticles labeled with radionuclides such as ^{68} Ga, ^{18}F, ^{64}Cu, ^{76}Br, ^{86}Y, ^{89}Zr, and ^{124}I can be applied to lung diagnostics, tumor theranostics, and cardiovascular imaging [22]. In contrast, magnetic resonance imaging (MRI) offers better soft tissue contrast than computed tomography (CT) and employs non-ionizing radiation. A breakthrough in illness monitoring and detection could result from the combination of PET with high-resolution MRI. According to [23], Nanoparticle-based probes in dual-modality PET/MRI employing ^{124}I for PET and MNPs as MRI agents [24] have created PET/MR imaging dual-modality nanoparticles. The chelator 1,4,7,10-tetraazacyclododecane-1,4,7,10-tetraacetic acid (DOTA) is conjugated to the PEG terminals to enable labeling with positron-emitting ^{64}Cu. They are composed of a super-paramagnetic iron oxide (SPIO) core coated with PEGylated phospholipids. A circulation half-life of 143 min, significant initial blood retention, and moderate liver uptake are revealed by ^{64}Cu-mSPIO's pharmacokinetics and biodistribution with stable and strong MR and PET signals. The outcomes show that it could be used as a dual-PET/MRI in vivo imaging probe.

5.3.3 X-Ray Computed Tomography (CT)

Iodine, gold, and Bi2S3-based X-ray computed tomography (CT) probes offer numerous benefits because of their high resolution and simplicity in creating three-dimensional (3D) visual tissues. In contrast to the extremely sensitive MRI, the intrinsically low sensitivity causes poor soft-tissue contrast and usually exposes patients to high levels of ionizing radiation. Because gold has a high X-ray absorption coefficient, functionalized gold nanoparticles, which are created by reducing a gold salt with sodium borohydride ($NaBH_4$) in the presence of dithiolated derivatives of diethylenetriaminepentaacetic acid, can be used as in vivo positive contrast agents in both MRI and X-ray imaging [25]. Despite the particles' low concentrations of gadolinium and gold, they are readily detectable and have a high sensitivity [26, 27] described the sol-gel reaction of tantalum(V) ethoxide in a microemulsion containing Fe_3O_4 nanoparticles to create multifunctional Fe_3O_4/TaO_x (core/shell) MNPs. They can offer supplementary CT and MRI data. As a result, CT can image freshly formed tumours, and MRI can assess the tumour microenvironment, including the oxygenated and hypoxic areas. Through the monitoring of tumour status and observation of established tumour vasculature, core/shell MNPs hold great promise for precise cancer diagnosis. Comparison of various MNPs and their imaging properties is discussed in Table 5.2.

Table 5.2 Comparison of different imaging probe properties [25]

Modalities	Nanoprobes	Imaging properties
MRI	MNPs (e.g. Fe_3O_4, $MnFe_2O_4$) Metal alloy (e.g., FeCo, FePt) Metal alloy (e.g., FeCo, FePt) Paramagnetic nanoparticles (e.g., Gd_2O_3, MnO)	High spatial resolution, non-invasive, very good signal in soft tissues
PET	Nanoparticles labelled with radionuclide ^{68}Ga, ^{18}F, ^{64}Cu, ^{76}Br, ^{86}Y, ^{89}Zr, ^{124}I	High sensitivity, which is an order of magnitude higher than that of MRI (picomolar level)
CT	Iodine, gold, and Bi_2S_3	High resolution, ease of forming 3D visual tissues, and substantial doses of ionizing radiation
Optical imaging	Fluorescent organic dye, visible-fluorescent QDs (e.g., near infrared, fluorescent), lanthanide atom (e.g., up- or down-conversion luminescence phosphors)	Rapid screening, cost effectiveness, and poor anatomical and physiological details

Table 5.3 Some metal nanoparticle and metal nanoparticle-conjugate examples that have been studied for application in medical imaging [9] CC BY4

Type of nanoparticle	Size (nm)	Scanning type
PEG-AuNPs	3–8	C T
Modified AuNPs	17–23	SPECT/CT
AuNPs	130–147	PA
AuNPs with citraconic amide moieties	10	PA
AuNPs in combination with radiotherapy	25	Dual-energy CT

5.3.4 Magnetic Particle Imaging (MPI)

MPI is a non-invasive tomographic method that finds tracer particles having superparamagnetic characteristics. Applications for MPI may be found in the fields of materials properties, imaging, and diagnostics. MPI can be used to locate and measure the concentration of nanoparticles with non-ionizing radiation at any depth within the body to produce a signal. When magnetized, superparamagnetic nanoparticles directly produce MPI signals, which are 107 times more sensitive than MRI signals. SPIONs were used for lactoferrin conjugation, and the magnetic particle spectrometer (MPS) offers the potential for 3D in vivo imaging with high particular resolution. A specially designed magnetic particle spectrometer (MPS) was used to record the MPS signal [28, 29]. Certain proteases, including trypsin and matrix metalloprotease-2 (MMP-2), are secreted by overexpressed cancer cells and were identified by MPS. In the presence of biotin-labelled peptides, which were subsequently identified and preferentially broken by particular proteases contained in the peptides (protease cleavage site), neutravidin-coated SPIONs aggregated, causing the SPIONs to disperse. MPS signals used magnetic relaxation properties to determine if the SPIONs were in an aggregation or dispersion state. MPS can, therefore, be applied in biomedical settings, including the quick identification of proteases in biological samples, such as blood, urine, tissue extracts, and cell culture media, for the diagnosis of different cancers [30]. Table 5.3 lists some metal nanoparticle and metal nanoparticle-conjugate examples that have been studied for application in medical imaging.

5.4 Biomedical Application of FMNs

5.4.1 Antimicrobial Agents

Since Alexander Fleming discovered penicillin in 1928, antibiotics have been the mainstay of treatment for bacterial infections, which are extremely frequent. Metal nanoparticles are being investigated for potential future treatments, and nanomedicine offers us a new, wide spectrum of potential therapy techniques. Silver is one substance that has been investigated for possible use; it has demonstrated a range of biomedical applications [31, 32]. Used silver nanoparticles as a component of an antibacterial fiber network. When compared to fibers devoid of silver nanoparticles, the nanoparticles' antibacterial activity against Escherichia coli was as high as 94.3%. Their sizes ranged from 20 to 120 nm.

5.4.2 Antiviral Agents

Antiviral applications have demonstrated the potential of metal nanoparticles. When in the nanoscale range, silver nanoparticles, both coated and naked, have been demonstrated to have a variety of antiviral uses. According to the World Health Organization, the Hepatitis B virus (HBV), a viral infection that presently affects 257 million people worldwide, caused 887,000 deaths in 2015. Tiny (10–50 nm) bare silver nanoparticles have been investigated as a potential HBV treatment and have been demonstrated to bind to HBV effectively while also preventing the synthesis of HBV RNA. The AgNPs' binding to the HBV dsDNA (double-stranded DNA) is thought to be the mechanism of action [33]. Silver nanoparticles may even have a role to play in the treatment of human immunodeficiency virus (HIV) [34]. On a variety of viruses, copper-iodide nanoparticles (CuI-NPs) have demonstrated antiviral effects: More intriguingly, the influenza A virus of swine origin (H1N1) and feline calicivirus (FCV) [35, 36].

5.4.3 Hyperthermia Treatment for Cancer

The therapy of hyperthermia is based on the biological principle that cancer cells are far more sensitive to warmth than healthy tissues. MNPs exhibit a high specific absorption rate (SAR), and because of the changed enzymatic and structural properties of cell proteins, hyperthermic effects take place between 41 and 46 °C. At 50 °C, this is accompanied by cell carbonization [12, 37]. The MNPs generate heat through Neel relaxation, which results from the rotation of the magnetic moment with each field oscillation in an alternating magnetic field, and Brown relaxation, which results from friction caused by total particle oscillations [38]. Numerous cancer types, including recurrent malignant astrocytic brain tumours, breast and prostate cancer, malignant melanoma, lymph node metastasis, glioblastoma, cervical carcinoma, and head and neck squamous cell carcinoma, have been described in reports using magnetic nanoparticles as thermotherapy agents, both in vitro and in vivo [25]. Particle size and distribution, magnetic susceptibility, precise temperature control, heating, and immune response activation all affect how well the hyperthermia treatment works [25].

5.5 Conclusion

Superparamagnetism, high magnetic susceptibility, and tunable surface chemistry are some of the unique properties that make functionalized magnetic nanomaterials (FMNs) ideal candidates for improving the sensitivity and specificity of contemporary imaging techniques. By incorporating FMNs into diagnostic workflows, researchers and clinicians have opened up new possibilities for non-invasive, high-resolution visualization of biological systems, enabling earlier and more accurate disease detection. FMNs have emerged as a groundbreaking innovation in the field of diagnostic imaging, offering unparalleled precision, versatility, and personalized medicine potential. Serving as sophisticated contrast agents in imaging modalities such as MRI, CT, and PET is one of FMN's most important contributions. They can improve picture contrast due to their superparamagnetic properties, which facilitates the detection of anomalies such as tumours, heart blockages, and neurological conditions. Additionally, FMNs can be functionalized with targeted ligands, like peptides or antibodies, to attach to sick tissues precisely, lowering background noise and increasing the accuracy of diagnosis. This focused strategy is especially helpful in oncology, where FMNs allow for real-time tracking of treatment outcomes and early tumor diagnosis. Beyond their role in traditional imaging, FMNs have led the way for multimodal imaging, integrating the benefits of different modalities to provide a more thorough view of disease. For example, FMNs can be built to work simultaneously in MRI and PET, delivering both structural and functional insights in a single scan. This skill is revolutionizing the detection and treatment of complicated illnesses like cancer, heart disease, and neurological problems. For cellular and molecular imaging, FMNs are also being investigated. This would enable researchers to monitor biological processes in real time and at hitherto unheard-of resolutions. The biomedical applications of FMNs extend far beyond diagnosis. Their capacity to administer imaging agents and therapeutic molecules concurrently has given rise to theranostics, a cutting-edge technique that combines diagnostics and treatment. Because FMNs can target cancer cells, deliver medications, and track the effectiveness of treatments in real time, this dual functionality is especially promising in oncology. Additionally, FMNs are being studied for their potential in tracking stem cells, imaging complex physiological processes, and detecting infectious diseases, proving their adaptability in a variety of medical domains. Despite their remarkable potential, the widespread adoption of FMNs in clinical practice faces challenges, including concerns about biocompatibility, toxicity, and regulatory approval. Addressing these issues will require continued research, interdisciplinary collaboration, and robust preclinical and clinical studies.

Abbreviations

AMF	Attractive magnetic field
CT	Computed tomography
FCV	Feline calicivirus
MNPs	Magnetic nanoparticles
MPI	Magnetic particle imaging
MRI	Magnetic resonance imaging
PA	Photoacoustic imaging
PET	Positron emission tomography
SAR	Specific absorption rate
SPECT	Single-photon emission computed tomography
SPIO	Superparamagnetic iron oxide
SPIONs	Superparamagnetic iron oxide nanoparticles

References

1. Chomoucka J, Drbohlavova J, Huska D, Adam V, Kizek R, Hubalek J (2010) Magnetic nanoparticles and targeted drug delivering. Pharmacol Res 62(2):144–149. https://doi.org/10.1016/j.phrs.2010.01.014
2. Zinoviev S et al (2010) Next-generation biofuels: survey of emerging technologies and sustainability issues. Chemsuschem 3(10):1106–1133. https://doi.org/10.1002/cssc.201000052
3. Pan B, Gao F, Gu H (2005) Dendrimer modified magnetite nanoparticles for protein immobilization. J Colloid Interface Sci 284(1):1–6. https://doi.org/10.1016/j.jcis.2004.09.073
4. Meng X, Seton HC, Lu LT, Prior IA, Thanh NTK, Song B (2011) Magnetic CoPt nanoparticles as MRI contrast agent for transplanted neural stem cells detection. Nanoscale 3(3):977. https://doi.org/10.1039/c0nr00846j
5. Huang Y-F, Wang Y-F, Yan X-P (2010) Amine-functionalized magnetic nanoparticles for rapid capture and removal of bacterial pathogens. Environ Sci Technol 44(20):7908–7913. https://doi.org/10.1021/es102285n
6. Dong H, Huang J, Koepsel RR, Ye P, Russell AJ, Matyjaszewski K (2011) Recyclable antibacterial magnetic nanoparticles grafted with quaternized Poly(2-(dimethylamino)ethyl methacrylate) brushes. Biomacromolecules 12(4):1305–1311. https://doi.org/10.1021/bm200031v
7. Yiu HHP, Pickard MR, Olariu CI, Williams SR, Chari DM, Rosseinsky MJ (2012) Fe3O4-PEI-RITC magnetic nanoparticles with imaging and gene transfer capability: development of a tool for neural cell transplantation therapies. Pharm Res 29(5):1328–1343. https://doi.org/10.1007/s11095-011-0632-1
8. Yao L, Xu S (2014) Detection of magnetic nanomaterials in molecular imaging and diagnosis applications. Nanotechnol Rev 3(3). https://doi.org/10.1515/ntrev-2013-0044
9. Anderson SD, Gwenin VV, Gwenin CD (2019) Magnetic functionalized nanoparticles for biomedical, drug delivery and imaging applications. Nanoscale Res Lett 14(1):188. https://doi.org/10.1186/s11671-019-3019-6
10. Kim J-H, Kim S-M, Kim Y-I (2014) Properties of magnetic nanoparticles prepared by co-precipitation. J Nanosci Nanotechnol 14(11):8739–8744. https://doi.org/10.1166/jnn.2014.9993

11. Borlido L, Azevedo AM, Roque ACA, Aires-Barros MR (2013) Magnetic separations in biotechnology. Biotechnol Adv 31(8):1374–1385. https://doi.org/10.1016/j.biotechadv.2013.05.009
12. Pankhurst QA, Connolly J, Jones SK, Dobson J (2003) Applications of magnetic nanoparticles in biomedicine. J Phys D: Appl Phys 36(13):R167–R181. https://doi.org/10.1088/0022-3727/36/13/201
13. Niemirowicz K, Markiewicz K, Wilczewska A, Car H (2012) Magnetic nanoparticles as new diagnostic tools in medicine. Adv Med Sci 57(2):196–207. https://doi.org/10.2478/v10039-012-0031-9
14. Huang C-C et al (2011) Enhancing transversal relaxation for magnetite nanoparticles in MR imaging using Gd^{3+}-Chelated Mesoporous Silica Shells. ACS Nano 5(5):3905–3916. https://doi.org/10.1021/nn200306g
15. Park JY et al (2009) Paramagnetic ultrasmall gadolinium oxide nanoparticles as advanced T_1 MRI contrast agent: account for large longitudinal relaxivity, optimal particle diameter, and in vivo T_1 MR images. ACS Nano 3(11):3663–3669. https://doi.org/10.1021/nn900761s
16. Perez JM, Simeone FJ, Saeki Y, Josephson L, Weissleder R (2003) Viral-induced self-assembly of magnetic nanoparticles allows the detection of viral particles in biological media. J Am Chem Soc 125(34):10192–10193. https://doi.org/10.1021/ja036409g
17. Yang H et al (2011) Targeted dual-contrast T1- and T2-weighted magnetic resonance imaging of tumors using multifunctional gadolinium-labeled superparamagnetic iron oxide nanoparticles. Biomaterials 32(20):4584–4593. https://doi.org/10.1016/j.biomaterials.2011.03.018
18. Yim H et al (2011) The performance of gadolinium diethylene triamine pentaacetate-pullulan hepatocyte-specific T1 contrast agent for MRI. Biomaterials 32(22):5187–5194. https://doi.org/10.1016/j.biomaterials.2011.03.069
19. Bulte JWM, Kraitchman DL (2004) Iron oxide MR contrast agents for molecular and cellular imaging. NMR Biomed 17(7):484–499. https://doi.org/10.1002/nbm.924
20. Raynal I, Prigent P, Peyramaure S, Najid A, Rebuzzi C, Corot C (2004) Macrophage endocytosis of superparamagnetic iron oxide nanoparticles: mechanisms and comparison of ferumoxides and ferumoxtran-10. Invest Radiol 39(1):56–63. https://doi.org/10.1097/01.rli.0000101027.57021.28
21. Portet D, Denizot B, Rump E, Lejeune J-J, Jallet P (2001) Nonpolymeric coatings of iron oxide colloids for biological use as magnetic resonance imaging contrast agents. J Colloid Interface Sci 238(1):37–42. https://doi.org/10.1006/jcis.2001.7500
22. Liu Y, Welch MJ (2012) Nanoparticles labeled with positron emitting nuclides: advantages, methods, and applications. Bioconjugate Chem 23(4):671–682. https://doi.org/10.1021/bc200264c
23. Choi J et al (2008) A hybrid nanoparticle probe for dual-modality positron emission tomography and magnetic resonance imaging. Angew Chem Int Ed 47(33):6259–6262. https://doi.org/10.1002/anie.200801369
24. Glaus C, Rossin R, Welch MJ, Bao G (2010) In vivo evaluation of ^{64}Cu-labeled magnetic nanoparticles as a dual-modality PET/MR imaging agent. Bioconjugate Chem 21(4):715–722. https://doi.org/10.1021/bc900511j
25. Yan K et al (2013) Recent advances in multifunctional magnetic nanoparticles and applications to biomedical diagnosis and treatment. RSC Adv 3(27):10598. https://doi.org/10.1039/c3ra40348c
26. Alric C et al (2008) Gadolinium chelate coated gold nanoparticles as contrast agents for both x-ray computed tomography and magnetic resonance imaging. J Am Chem Soc 130(18):5908–5915. https://doi.org/10.1021/ja078176p
27. Lee N et al (2012) Multifunctional Fe_3O_4/TaO_x core/shell nanoparticles for simultaneous magnetic resonance imaging and x-ray computed tomography. J Am Chem Soc 134(25):10309–10312. https://doi.org/10.1021/ja3016582
28. Tomitaka A, Arami H, Gandhi S, Krishnan KM (2015) Lactoferrin conjugated iron oxide nanoparticles for targeting brain glioma cells in magnetic particle imaging. Nanoscale 7(40):16890–16898. https://doi.org/10.1039/C5NR02831K

29. Weizenecker J, Gleich B, Rahmer J, Dahnke H, Borgert J (2009) Three-dimensional real-time in vivo magnetic particle imaging. Phys Med Biol 54(5):L1–L10. https://doi.org/10.1088/0031-9155/54/5/L01
30. Gandhi S, Arami H, Krishnan KM (2016) Detection of cancer-specific proteases using magnetic relaxation of peptide-conjugated nanoparticles in biological environment. Nano Lett 16(6):3668–3674. https://doi.org/10.1021/acs.nanolett.6b00867
31. Rai M, Deshmukh SD, Ingle AP, Gupta IR, Galdiero M, Galdiero S (2016) Metal nanoparticles: The protective nanoshield against virus infection. Crit Rev Microbiol 42(1):46–56. https://doi.org/10.3109/1040841X.2013.879849
32. Sreekumar TV, Das A, Chandra L, Srivastava A, Rao BKU (2009) Inherently colored antimicrobial fibers employing silver nanoparticles. J Biomed Nanotechnol 5(1):115–120. https://doi.org/10.1166/jbn.2009.040
33. Lu L et al (2008) Silver nanoparticles inhibit hepatitis b virus replication. Antivir Ther 13(2):253–262. https://doi.org/10.1177/135965350801300210
34. Lara HH, Ixtepan-Turrent L, Garza-Treviño EN, Rodriguez-Padilla C (2010) PVP-coated silver nanoparticles block the transmission of cell-free and cell-associated HIV-1 in human cervical culture. J Nanobiotechnol 8(1):15. https://doi.org/10.1186/1477-3155-8-15
35. Fujimori Y et al (2012) Novel antiviral characteristics of nanosized copper(I) iodide particles showing inactivation activity against 2009 Pandemic H1N1 influenza virus. Appl Environ Microbiol 78(4):951–955. https://doi.org/10.1128/AEM.06284-11
36. Shionoiri N et al (2012) Investigation of the antiviral properties of copper iodide nanoparticles against feline calicivirus. J Biosci Bioeng 113(5):580–586. https://doi.org/10.1016/j.jbiosc.2011.12.006
37. Gupta AK, Naregalkar RR, Vaidya VD, Gupta M (2007) Recent advances on surface engineering of magnetic iron oxide nanoparticles and their biomedical applications. Nanomedicine 2(1):23–39. https://doi.org/10.2217/17435889.2.1.23
38. Fortin J-P, Gazeau F, Wilhelm C (2008) Intracellular heating of living cells through Néel relaxation of magnetic nanoparticles. Eur Biophys J 37(2):223–228. https://doi.org/10.1007/s00249-007-0197-4

Chapter 6
Biosensing with Functionalized Magnetic Nanomaterials (FMNs)

Vivek Chintada and Narasimha Golla

Abstract This book chapter discusses the use of functionalized magnetic nanomaterials (FMNs) in biosensing applications, particularly in detecting biomarkers, pathogens, and genetic material. FMNs are valued in biosensing for their magnetic properties and the ability to attach specific ligands for targeting. The chapter explores how FMNs interact with biological molecules, enabling accurate detection in various biological samples. It also highlights the integration of FMNs into lab-on-a-chip devices for on-the-spot diagnostics. These compact devices provide quick and cost-effective solutions for detecting biomolecules when needed, offering significant benefits in healthcare, environmental monitoring, and food safety. By examining recent progress and case studies, the chapter presents insights into the latest research and applications of FMNs in biosensing. It addresses the challenges and future possibilities of incorporating FMNs into biosensing platforms, underscoring their potential to transform diagnostic methods and advance personalized medicine approaches.

Keywords Functionalized magnetic nanomaterials · Biosensing · Biomarkers · Lab-on-a-chip · Point-of-care diagnostics

6.1 Introduction

Functionalized magnetic nanomaterials (FMNs) are now recognized as valuable assets in biosensing applications, providing distinct abilities for detecting a broad spectrum of biological entities. The field of biosensing has made notable progress by incorporating FMNs, leveraging their exceptional properties and flexible functionalization [1–4]. FMNs possess inherent magnetic properties that enable the sensitive and specific detection of biomarkers, pathogens, and genetic material in

V. Chintada
Department of Zoology, Sri Venkateswara University, Tirupati, India

N. Golla (✉)
Department of Virology, Sri Venkateswara University, Tirupati, India
e-mail: dr.g.narasimha@gmail.com

S. Gulati (ed.), *Functionalized Magnetic Nanomaterials*, Nanostructure Science and Technology, https://doi.org/10.1007/978-3-031-97199-0_6

various biological samples [5, 6]. This unique characteristic distinguishes FMNs from conventional biosensing systems, creating opportunities for swift and precise detection in intricate biological settings. The capacity to identify biomarkers, pathogens, and genetic material holds significant importance across diverse domains, such as healthcare, environmental surveillance, and food safety [7–9]. Biomarkers, as indicators of biological processes or diseases, offer valuable information for diagnosis, prognosis, and treatment monitoring [10]. FMNs, when functionalized with specific ligands such as antibodies or nucleic acids, can selectively bind to biomarkers of interest, facilitating their detection at extremely low concentrations [11, 12]. This high sensitivity and specificity make FMNs an invaluable tool in precision medicine and personalized healthcare approaches.

In the realm of pathogen detection, functionalized magnetic nanomaterials (FMNs) have demonstrated exceptional potential for rapid and efficient identification of various microorganisms [13, 14]. By leveraging the magnetic properties and surface functionalization of FMNs, researchers have developed biosensing platforms capable of detecting pathogenic bacteria, viruses, and other infectious agents with high accuracy and speed [15, 16]. This capability is particularly critical in healthcare settings and outbreak investigations, where timely pathogen detection can aid in disease surveillance and control. Moreover, the detection of genetic material, such as DNA and RNA sequences, holds significant importance in biomedical research, molecular diagnostics, and forensic analysis [17, 18]. FMNs functionalized with complementary oligonucleotide probes can specifically capture and detect target genetic sequences, enabling precise genetic profiling and mutation analysis [19]. This application of FMNs in genetic material detection has expanded the possibilities for studying genomics, investigating hereditary diseases, and advancing personalized medicine initiatives.

The significance of FMNs in biosensing lies not only in their magnetic properties but also in their versatility for functionalization with specific ligands for target recognition (Fig. 6.1) [20]. By conjugating FMNs with ligands that can selectively bind to specific biomolecules, researchers can design biosensors with tailor-made detection capabilities [21, 22]. This targeted functionalization ensures that FMNs interact only with the desired biological targets, minimizing non-specific binding and enhancing the sensitivity and accuracy of biosensing platforms. In conclusion, the integration of Functionalized Magnetic Nanomaterials (FMNs) in biosensing applications has revolutionized the way biomarkers, pathogens, and genetic material are detected in biological samples (Fig. 6.1). The magnetic properties, coupled with specific ligand functionalization, enable FMNs to offer unparalleled sensitivity, selectivity, and versatility in biosensing platforms. By harnessing the potential of FMNs, researchers are paving the way for innovative diagnostic techniques, personalized medicine approaches, and enhanced disease management strategies.

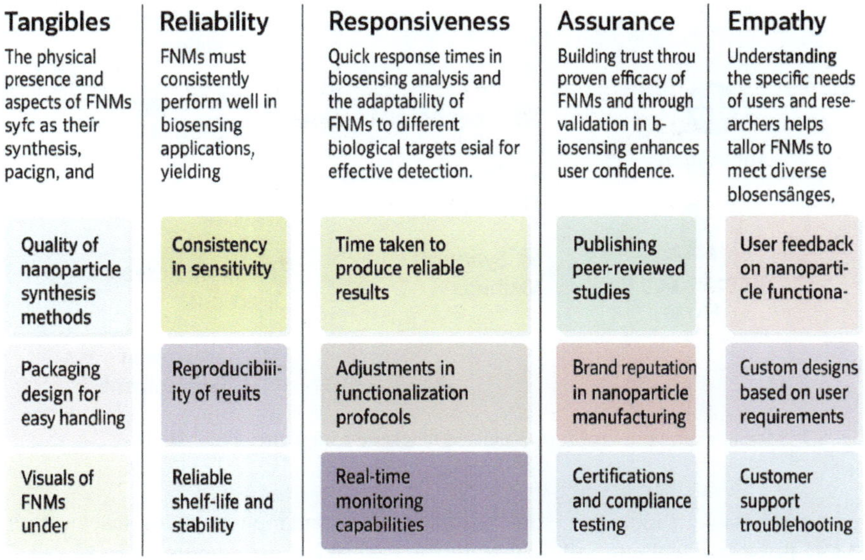

Fig. 6.1 SERVQUAL model analysis for functionalized magnetic nanomaterials in bio-sensing applications

6.2 Magnetic Properties of FMNs

Functionalized magnetic nanomaterials (FMNs) have garnered significant attention in the field of biosensing due to their exceptional magnetic properties, which contribute to their wide-ranging applications in detecting biological entities with high sensitivity and specificity [23–25]. The magnetic properties of FMNs stem from their nanoscale size and composition, which impart distinct behavior compared to bulk magnetic materials and enable them to interact uniquely with external magnetic fields [26]. At the nanoscale, FMNs display superparamagnetic behavior, meaning they have magnetic moments that align with an external magnetic field but lose their magnetization when the field is removed [27, 28]. This property allows FMNs to be manipulated and controlled by external magnetic fields, making them ideal candidates for magnetic-based separation, manipulation, and detection techniques in biosensing applications [29]. The reversible nature of superparamagnetism in FMNs enables precise and efficient handling, enhancing their performance in various biological assays and diagnostic protocols.

Moreover, FMNs demonstrate high magnetic susceptibility, which refers to their responsiveness to external magnetic fields by becoming magnetized [30]. This attribute enables FMNs to generate detectable signals in magnetic-based biosensors, where changes in their magnetic behavior upon interaction with target biomolecules are translated into measurable signals for detection and quantification [31]. FMNs' intrinsic magnetic susceptibility enhances their sensitivity in biosensing systems, enabling the precise detection of minute concentrations of biomarkers, pathogens, or

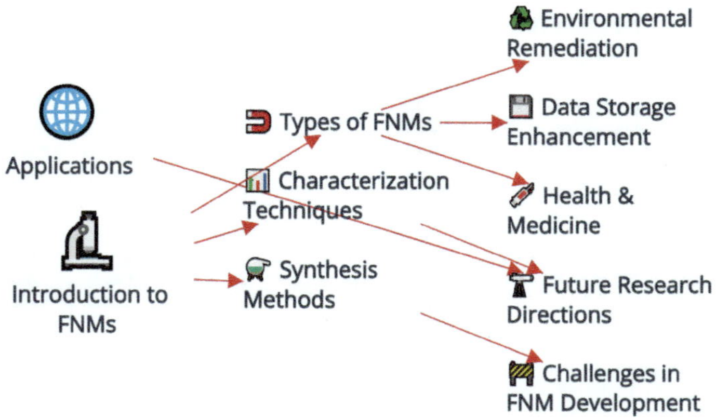

Fig. 6.2 Magnetic properties of functionalized magnetic nanomaterials (FMNs)

genetic material. In addition to superparamagnetism and high magnetic susceptibility, FMNs possess excellent colloidal stability, ensuring their dispersion in biological samples and maintaining their magnetic properties over prolonged periods [28, 32]. The colloidal stability of FMNs is crucial for their utilization in biosensing applications, as it prevents aggregation or sedimentation of nanoparticles, thereby preserving their magnetic characteristics and enhancing the reproducibility and reliability of detection assays [33].

Furthermore, the magnetization of FMNs is tunable through the control of nanoparticle size, composition, and surface properties, offering a customizable approach to modulating their magnetic behavior for specific biosensing requirements [34]. By tailoring the magnetic properties of FMNs, researchers can optimize their performance in biosensing platforms, adjusting parameters such as sensitivity, response time, and detection limits to suit the desired application [28, 35]. This tunability positions FMNs as versatile tools in biosensing, capable of addressing a wide range of detection challenges with precision and efficiency. In conclusion, the unique magnetic properties of functionalized magnetic nanomaterials (FMNs) play a vital role in their widespread use in biosensing applications (Fig. 6.2). From superparamagnetism and high magnetic susceptibility to colloidal stability and tunable magnetization, these properties underpin the exceptional performance of FMNs in detecting biomarkers, pathogens, and genetic material with unparalleled sensitivity and specificity.

6.2.1 How Magnetic Properties Enhance Detection Capabilities in Biosensing

In biosensing applications, Functionalized Magnetic Nanomaterials (FMNs) are increasingly recognized for their unique magnetic properties that significantly enhance detection capabilities in various biological analyses. The magnetic characteristics of FMNs significantly enhance sensitivity, specificity, and efficiency in biosensing systems, leading to groundbreaking advancements in diagnostic techniques and biomolecule detection [27, 30]. A key mechanism through which FMNs' magnetic properties improve detection capabilities is their superparamagnetic behavior, characterized by magnetic moment alignment in the presence of an external magnetic field and loss of magnetization without the field [33]. This property enables FMNs to produce detectable signals when interacting with target biomolecules, leading to increased sensitivity in biosensing applications [29]. The ability of FMNs to respond to external magnetic fields enhances the signal-to-noise ratio, allowing for the detection of even low concentrations of analytes with remarkable precision.

Furthermore, the magnetic susceptibility of FMNs contributes significantly to their enhanced detection capabilities. FMNs possess a high magnetic susceptibility, making them responsive to external magnetic fields by becoming magnetized [13, 31]. This property enables FMNs to generate measurable signals that can be detected and quantified in magnetic-based biosensors, translating the binding events between the nanoparticles and target biomolecules into detectable outputs [26]. The high magnetic susceptibility of FMNs enhances the sensitivity of biosensors, allowing for the reliable detection of biomarkers, pathogens, and genetic material in complex biological samples. The magnetic properties of FMNs also enable selective targeting of specific analytes, thereby enhancing the specificity of biosensing platforms. By functionalizing FMNs with ligands that bind selectively to target biomolecules, such as antibodies or aptamers, researchers can design biosensors that specifically recognize and capture the analytes of interest [34, 36]. This targeted binding ensures that FMNs interact only with the desired biomarkers, minimizing non-specific interactions and improving the accuracy of detection in biological samples [23]. The combination of magnetic properties and specific ligand functionalization allows for precise and selective detection of biomolecules, making FMNs valuable tools in biosensing applications.

Moreover, the magnetic manipulation capability of FMNs enhances their detection capabilities by facilitating spatial control and separation of analytes. External magnetic fields can be used to manipulate the movement and positioning of FMNs within a sample, enabling targeted separation of bound analytes or localized concentrations of nanoparticles for detection [37]. This magnetic manipulation capability improves the efficiency and speed of detection processes, allowing for rapid and precise analysis of biological samples in biosensing assays [14, 38]. In conclusion, the magnetic properties of functionalized magnetic nanomaterials (FMNs) play a pivotal role in enhancing detection capabilities in biosensing applications (Fig. 6.3). Their superparamagnetic behavior, high magnetic susceptibility, selective targeting capabilities, and manipulation efficiency collectively contribute to improved sensitivity, specificity, and efficiency in detecting biomolecules. By leveraging the unique magnetic properties of FMNs, researchers are advancing the development of innovative biosensing platforms that enable accurate and reliable detection of various analytes in biological samples.

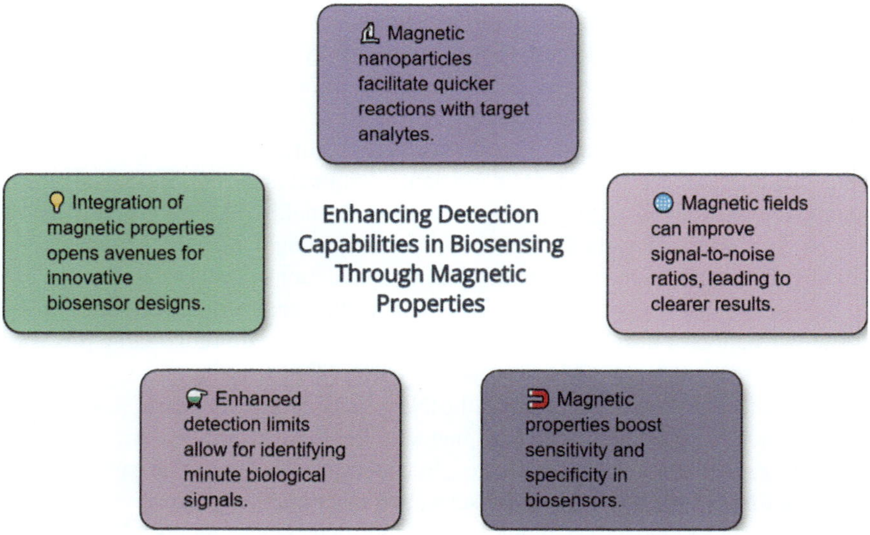

Fig. 6.3 Detection capabilities in biosensing through magnetic properties

6.3 Interaction Mechanisms of FMNs with Biological Molecules

In the evolving landscape of nanotechnology, functionalized magnetic nanomaterials (FMNs) have garnered significant attention for their diverse applications in various fields, particularly in the realm of biomedicine and bioanalytics [39, 40]. FMNs possess unique physicochemical properties that make them well-suited for interacting with biological molecules, enabling sensitive and specific detection in complex biological samples [41]. Understanding the intricacies of how FMNs engage with biomolecules is essential for harnessing their full potential in biomedical research, diagnostics, and therapeutic interventions [42]. This section aims to provide a detailed exploration of the interaction mechanisms underlying FMNs and biological molecules, shedding light on the sophisticated processes that drive their applications in biotechnology and medicine. The interaction mechanisms of FMNs with biological molecules encompass a range of intricate processes that involve physical, chemical, and biological phenomena [43, 44]. At the core of these interactions lies the surface properties of FMNs, which play a crucial role in determining their affinity towards different biomolecules such as proteins, nucleic acids, lipids, and carbohydrates [45]. Surface functionalization of FMNs with specific ligands can enhance their binding specificity and selectivity, enabling targeted interactions with biomolecular targets of interest [46]. This tailored approach facilitates the design of FMNs for various biomedical applications, including drug delivery, biosensing, imaging, and theragnostic applications [47].

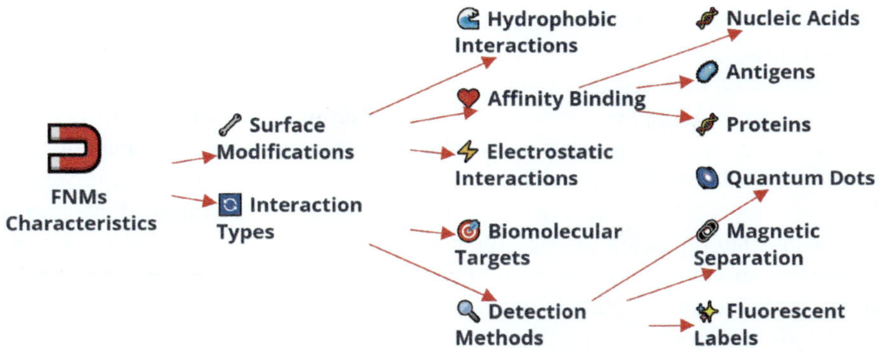

Fig. 6.4 Characteristics of functionalized magnetic nanomaterials (FMNs)

Biological interactions between FMNs and biomolecules encompass complex processes that involve receptor-ligand recognition, cellular uptake, intracellular trafficking, and biological responses [48]. Functionalized magnetic nanomaterials (FMNs), when tailored with targeting ligands, can effectively bind to specific cell surface receptors, facilitating receptor-mediated endocytosis and the targeted delivery of therapeutic agents [49]. The internalization of FMNs into cells initiates a sequence of intracellular processes, including endosomal escape, lysosomal degradation, and the release of cargo molecules into the cytoplasm [50]. A deep understanding of these biological interactions is essential for optimizing FMNs' design for diverse biomedical purposes, such as cancer treatment, gene editing, and regenerative medicine [51]. In essence, the mechanisms by which FMNs interact with biological entities present a captivating field of study with vast potential for advancing biomedicine and bioanalytics. By unraveling how FMNs engage with biomolecules at molecular, cellular, and tissue levels, researchers can pioneer innovative nanotechnologies for diagnosing and treating various diseases. This chapter has furnished a comprehensive overview of the physical, chemical, and biological interactions that underpin the functionality of FMNs in biomedical applications. Further investigation and enhancement of these interaction mechanisms are imperative to unlock the full capacity of FMNs in transforming healthcare and life sciences (Fig. 6.4).

6.4 FMNs in Lab-on-a-Chip Devices for Point-of-Care Diagnostics

Additionally, the incorporation of functional nanomaterials, including FMNs, into compact lab-on-a-chip devices has greatly transformed point-of-care diagnostics, delivering swift and cost-effective solutions for disease detection and monitoring

[52]. Lab-on-a-chip devices, also referred to as microfluidic platforms, offer a versatile and portable platform for conducting a diverse range of analytical and bioanalytical assays with high sensitivity and specificity [53–55]. Through the incorporation of Functionalized Magnetic Nanomaterials (FMNs) into these devices, researchers have unlocked novel opportunities to enhance the performance and capabilities of point-of-care diagnostic technologies, enabling real-time, on-site testing in various healthcare settings [56]. This chapter provides an in-depth exploration of integrating FMNs into lab-on-a-chip devices for point-of-care diagnostics, emphasizing the benefits of rapid and cost-effective solutions that FMNs bring to the realms of personalized medicine and healthcare management (Fig. 6.5). The integration of FMNs into lab-on-a-chip devices presents a synergistic approach that combines the unique properties of nanomaterials with the miniaturization and scalability of microfluidic systems [57]. FMNs, such as nanoparticles, nanowires, nanotubes, and quantum dots, exhibit exceptional optical, electrical, magnetic, and catalytic properties that can be harnessed for the sensitive and selective detection of biological analytes [58]. These nanomaterials can be functionalized with specific ligands, antibodies, or aptamers to enable targeted recognition of biomolecular targets, enhancing the specificity and accuracy of diagnostic tests [59, 60].

Integrating Functionalized Magnetic Nanomaterials (FMNs) into lab-on-a-chip devices allows the development of cohesive platforms that provide multiplexed detection, high-throughput screening, and real-time monitoring of disease biomarkers with improved sensitivity and speed [61]. A notable advantage of utilizing FMNs in lab-on-a-chip devices for point-of-care diagnostics is the rapid and cost-effective nature of these solutions [62]. FMNs enable signal amplification, signal transduction, and signal readout in a miniaturized format, leading to enhanced detection limits and reduced assay times compared to traditional diagnostic methods [63]. The integration of FMNs allows for label-free detection, miniaturized sample preparation, and automation of assay procedures, streamlining the diagnostic workflow and

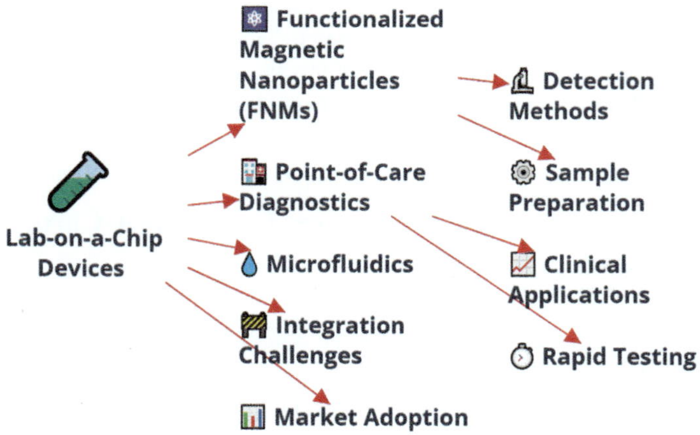

Fig. 6.5 Lab-on-a-chip devices for point-of-care diagnostics

reducing the overall cost and complexity of point-of-care testing [64]. These advancements pave the way for decentralized diagnostics, enabling healthcare professionals to perform timely and accurate diagnostic tests at the patient's bedside, in remote settings, or even at home [65].

Moreover, the combination of FMNs and lab-on-a-chip devices holds great promise for personalized medicine and precision healthcare [66]. By leveraging the unique properties of FMNs, such as their biocompatibility, stability, and tunable surface chemistry, researchers can develop tailored diagnostic platforms that cater to individual patient needs and disease profiles [67]. Functionalized magnetic nanomaterials (FMNs) enable the creation of point-of-care tests that exhibit sensitivity, specificity, and adaptability to a broad spectrum of biomarkers, rendering them ideal for diagnosing infectious diseases, chronic ailments, cancer, and various health conditions [68]. The incorporation of FMNs into lab-on-a-chip devices enables real-time monitoring of disease progression, treatment response, and therapeutic outcomes, empowering healthcare professionals to deliver tailored interventions and enhance patient care [69]. In conclusion, the integration of functional nanomaterials into lab-on-a-chip devices represents a transformative approach to point-of-care diagnostics, offering swift, cost-effective, and personalized solutions for disease detection and management. By capitalizing on the unique properties of FMNs and the scalability of microfluidic systems, researchers can develop innovative diagnostic platforms that provide high sensitivity, specificity, and efficiency in healthcare settings. Ongoing advancements in nanotechnology and microfluidics hold significant promise for unlocking new diagnostic capabilities, enhancing healthcare accessibility, and revolutionizing disease diagnosis and treatment at the point of care.

6.5 Applications of FMNs in Biosensing

The utilization of functionalized magnetic nanomaterials (FMNs) has become indispensable in biosensing applications (Table 6.1), providing unparalleled capabilities for detecting biomarkers, pathogens, and genetic material with exceptional sensitivity and specificity [70]. FMNs' distinctive physicochemical properties, such as their substantial surface-to-volume ratio, adaptable surface chemistry, and remarkable electrical and optical characteristics, position them as ideal candidates for developing cutting-edge biosensors across diverse sectors [71, 72]. This chapter explores the diverse applications of FMNs in biosensing through case studies that underscore their effectiveness in detecting biomarkers, pathogens, and genetic material, highlighting their crucial role in healthcare, environmental monitoring, and food safety applications. In healthcare contexts, FMNs have demonstrated substantial promise for early disease diagnosis, personalized medicine, and therapy monitoring [73, 74]. Case studies have shown that FMNs, such as gold nanoparticles, graphene oxide, and quantum dots, can be functionalized with specific recognition elements, such as antibodies, aptamers, or DNA probes, to selectively capture and detect biomarkers associated with various diseases, including cancer, infectious diseases, and metabolic

Table 6.1 Applications of functionalized magnetic nanomaterials (FMNs) in biosensing

Application	Description
Nucleic acid detection	FMNs can be used to detect specific DNA or RNA sequences with high sensitivity
Protein detection	FMNs can bind to target proteins, enabling precise detection in biological samples
Enzyme activity monitoring	FMNs help track enzyme activity through catalytic or fluorescence-based mechanisms
Cell-based assays	FMNs assist in studying cellular interactions, signaling pathways, and biomolecular processes
Disease diagnosis	FMNs act as biomarkers, facilitating early disease detection and monitoring
Environmental monitoring	FMNs are used in biosensors to detect toxins, pollutants, and heavy metals
Drug development	FMNs enhance drug screening, delivery, and target validation in pharmaceutical research

disorders [75, 76]. These biosensors offer rapid, sensitive, and label-free detection of biomolecules in clinical samples, enabling early intervention, treatment optimization, and disease management in patients [77].

Moreover, FMNs play a crucial role in environmental monitoring applications by facilitating the detection of pollutants, toxins, and environmental contaminants with high accuracy and precision [78]. Case studies have showcased the use of FMNs in developing biosensors that can detect heavy metals, pesticides, and organic compounds in water, soil, and air samples, providing real-time data for assessing environmental quality and identifying potential hazards to human health and ecosystems [79]. By leveraging the unique properties of FMNs, researchers can create portable, cost-effective biosensing platforms that enable on-site monitoring and rapid detection of environmental pollutants, contributing to environmental sustainability and public health protection [80]. In the realm of food safety applications, FMNs have emerged as valuable tools for ensuring the quality, safety, and authenticity of food products [81]. Case studies have illustrated the use of FMNs in biosensors designed to detect foodborne pathogens, toxins, allergens, and adulterants in food samples, helping to prevent foodborne illnesses and ensure compliance with food safety regulations [82]. FMNs, such as nanoscale sensors and nanostructured materials, offer rapid detection capabilities, high specificity, and multiplexed analysis of food contaminants, enhancing the efficiency of food safety testing and quality assurance processes in the food industry [83].

Furthermore, FMNs have revolutionized genetic analysis and molecular diagnostics by enabling the sensitive and selective detection of genetic material, such as nucleic acids, DNA, RNA, and microRNAs (Table 6.1) [84]. Case studies have highlighted the role of FMNs in developing nucleic acid biosensors that can detect genetic mutations, infectious agents, and disease biomarkers with high accuracy and reliability [85]. By leveraging the programmable nature of FMNs, researchers can design biosensors that offer rapid and on-site detection of genetic sequences, facilitating precision medicine, genetic screening, and pathogen identification in various biological samples [86, 87]. In conclusion, the applications of functional nanomaterials in biosensing encompass a wide range of domains, from healthcare to environmental monitoring and food safety. Through insightful case studies, the transformative role of functionalized magnetic nanomaterials (FMNs) in detecting biomarkers, pathogens, and genetic material has been highlighted, demonstrating their versatility and effectiveness in addressing critical challenges across various domains. As nanotechnology research progresses, the incorporation of FMNs into biosensing platforms shows immense potential for revolutionizing diagnostics, surveillance, and monitoring applications, ultimately improving global public health, environmental sustainability, and food safety.

6.6 Recent Advancements and Future Prospects

In the domain of biosensing, functionalized magnetic nanomaterials (FMNs) are pivotal components for innovative diagnostic methodologies, offering unparalleled sensitivity, specificity, and adaptability for diverse biomolecules and analytes of interest [88]. Recent strides in nanotechnology have propelled biosensing advancements, with forefront research concentrating on the creation of FNM-based platforms that redefine disease diagnosis and monitoring paradigms [89, 90]. This chapter presents a thorough examination of the latest achievements in leveraging FMNs for biosensing applications, addresses existing challenges and prospects, and deliberates on the future integration of FMNs in biosensing platforms for personalized medicine applications. Pioneering research on FMNs in biosensing has significantly improved diagnostic assays' sensitivity, selectivity, and multiplexing capacities [91]. Sophisticated FNM-based biosensors, including plasmonic nanoparticles, carbon nanotubes, and quantum dots, have been meticulously designed to enhance signal transduction and amplification in biomolecular detection [92]. These innovative biosensing platforms empower swift, label-free, and high-throughput analysis of a wide array of analytes, encompassing small molecules, proteins, and nucleic acids, facilitating early disease detection, treatment monitoring, and point-of-care diagnostics [93]. Despite the remarkable progress in FNM-based biosensing, challenges remain in translating these technologies from the laboratory to clinical practice [94, 95]. Issues such as sample preparation, assay standardization, biosensor reproducibility, and regulatory approval pose hurdles to the widespread adoption of FNM-enabled diagnostic techniques in healthcare settings [96]. Furthermore, the integration of FMNs into biosensing platforms requires interdisciplinary collaboration among scientists, engineers, clinicians, and regulatory agencies to address technical, ethical, and safety considerations for deploying these cutting-edge technologies in routine diagnostic workflows [97, 98].

Looking ahead, the potential for FMNs to revolutionize diagnostic techniques is immense, with exciting prospects for personalized medicine approaches that tailor diagnostic tests to individual patient profiles [99, 100]. By leveraging the unique properties of FMNs, such as their tunable surface chemistry, biocompatibility, and target specificity, researchers can develop biosensing platforms that offer unprecedented insights into disease pathology, treatment response, and patient outcomes [101, 102].

The future of biosensing lies in the integration of FMNs with artificial intelligence, microfluidics, and digital health technologies to create smart, adaptive diagnostic tools that provide real-time data, predictive analytics, and personalized interventions for improved healthcare outcomes [103, 104].

In conclusion, the recent advancements and prospects of functional nanomaterials in biosensing hold great promise for transforming the field of diagnostics and personalized medicine. By staying at the forefront of cutting-edge research, overcoming challenges, and embracing collaborative efforts, the integration of FMNs into biosensing platforms has the potential to revolutionize diagnostic techniques, enhance patient care, and shape the future of healthcare. The journey towards realizing the full potential of FMNs in biosensing requires continued innovation, investment, and strategic partnerships to accelerate the translation of these pioneering technologies into clinical practice [105, 106].

6.7 Conclusion

The integration of functionalized magnetic nanomaterials (FMNs) into biosensing technologies has revolutionized diagnostic approaches by offering enhanced sensitivity, specificity, and versatility in detecting biomarkers, pathogens, and genetic material. Recent advancements in FMNs have propelled biosensing platforms to new heights, enabling rapid, label-free, and high-throughput analysis of diverse analytes across various fields such as healthcare, environmental monitoring, and food safety. FMNs exhibit unique properties, including tunable surface chemistry and target specificity, making them ideal candidates for personalized medicine approaches that cater to individual patient profiles.

The transformative role of FMNs in biosensing holds promise for innovative diagnostic techniques that can reshape the future of healthcare. By leveraging cutting-edge research, overcoming challenges in biosensor development, and fostering interdisciplinary collaboration, FMNs have the potential to drive significant advancements in diagnostic capabilities. The future integration of FMNs into biosensing platforms offers exciting prospects for real-time data analytics, predictive diagnostics, and personalized interventions that can improve patient care, treatment outcomes, and disease management. Embracing the potential of FMNs in diagnostics paves the way for a future where precision diagnostics, rapid detection, and targeted therapies are seamlessly integrated into routine healthcare practices, ultimately enhancing overall health outcomes for individuals globally.

FMNs: Functionalized Magnetic Nanomaterials.

References

1. Haun JB, Yoon TJ, Lee H, Weissleder R (2010) Nanoparticle-based biosensors: Towards real-time detection of biomolecular interactions. Nano Today 5(6):512–527. https://doi.org/10.1016/j.nantod.2010.08.009
2. Xie J, Xu C, Kohler N, Hou Y, Sun S, Wang SX (2007) Functional magnetic nanoparticles for biomedical applications. Adv Mater 19(21):3163–3176. https://doi.org/10.1002/adma.200700464
3. Ahlawat J, Singh M, Arya G (2024) Recent advances in functionalized magnetic nanomaterials for biosensing. Biosens Bioelectron 224:115075. https://doi.org/10.1016/j.bios.2023.115075
4. Abidin MSZ, Rahim RA, Johan MR (2024) Emerging trends in magnetic nanomaterials for bio-detection and medical diagnostics. J Nanomater 2024:7894523. https://doi.org/10.1155/2024/7894523
5. Pankhurst QA, Connolly J, Jones SK, Dobson J (2003) Applications of magnetic nanoparticles in biomedicine. J Phys D Appl Phys 36(13):R167–R181. https://doi.org/10.1088/0022-3727/36/13/201
6. Sharma G, Kumar A, Verma R (2023) Advancements in magnetic nanotechnology for biosensing applications. Sens Actuators B Chem 382:133518. https://doi.org/10.1016/j.snb.2023.133518
7. Shalaby T, Khalil MMH, El-Sayed W (2018) Magnetic nanoparticle biosensors for environmental and food safety monitoring. Trends Anal Chem 102:16–29. https://doi.org/10.1016/j.trac.2018.02.008
8. Nnachi RC, Akpan UG, Okoli EC (2022) Magnetic nanomaterials for pathogen detection in environmental samples. Environ Sci Technol 56(4):2301–2315. https://doi.org/10.1021/acs.est.1c06421
9. Bhatia R, Kaushik S, Ranjan S (2024) Magnetic nanostructures in biosensing: a review of applications in food safety. Food Chem 422:136041. https://doi.org/10.1016/j.foodchem.2023.136041
10. Batista PD, Horak D (2015) Biosensing applications of functionalized magnetic nanoparticles: biomarker detection and disease diagnosis. Biosensors 5(2):153–179. https://doi.org/10.3390/bios5020153
11. Li X, Zhang Y, Liu Z (2017) Highly sensitive detection of biomarkers using functionalized magnetic nanomaterials. Anal Chem 89(8):4292–4299. https://doi.org/10.1021/acs.analchem.6b04863
12. Parihar A, Malhotra B, Sharma P (2025) Next-generation biosensors utilizing functionalized magnetic nanomaterials. Nat Rev Bioeng 1(1):112–125. https://doi.org/10.1038/s44222-024-00001-y
13. Huang CJ, Wang X, Lin Z (2015) Magnetic nanomaterials for rapid pathogen detection in clinical applications. Biosens Bioelectron 74:50–64. https://doi.org/10.1016/j.bios.2015.05.017
14. Canè V, Marcolongo G, Veronese P (2013) Magnetic nanoparticle-based biosensors for pathogen detection. Anal Biochem 437(2):144–150. https://doi.org/10.1016/j.ab.2013.02.003
15. Osterfeld SJ, Yu H, Gaster RS (2008) Multiplex protein assays based on real-time magnetic nanosensors. Proc Natl Acad Sci 105(52):20637–20640. https://doi.org/10.1073/pnas.0811239105
16. Sinha R, Bachan Upadhyay P (2025) Advances in magnetic nanotechnology for infectious disease diagnostics. Nat Nanotechnol 20(1):45–59. https://doi.org/10.1038/s41565-024-00012-x
17. Khan MS, Shukla AK, Chauhan P (2012) Nanomagnetic biosensors for nucleic acid detection in clinical diagnostics. TrAC Trends Anal Chem 36:120–135. https://doi.org/10.1016/j.trac.2012.02.014
18. Alketbi A, Zafar H, Khan M (2023) Magnetic nanoparticle-based approaches for rapid genetic material detection. Mol Diagn Ther 27(4):315–330. https://doi.org/10.1007/s40291-023-00646-5

19. Wang Y, Li J, Wang X (2016) Highly sensitive detection of DNA and RNA using functionalized magnetic nanoparticles. Anal Chem 88(9):4565–4572. https://doi.org/10.1021/acs.analchem.6b00982
20. Choi JW, Yang X, Lee H (2010) Functionalized magnetic nanoparticles for biosensing applications. Biosens Bioelectron 25(10):2255–2260. https://doi.org/10.1016/j.bios.2010.02.015
21. Guo J, Li M, Wang Y, Liu X (2013) Surface modification strategies of magnetic nanoparticles for biosensor applications. Adv Func Mater 23(30):3847–3854. https://doi.org/10.1002/adfm.201300845
22. Miraji HK, Patel RS, Desai NR (2025) Recent advancements in ligand-functionalized magnetic nanomaterials for biosensing. Nano Res 18(2):789–803. https://doi.org/10.1007/s12274-024-5678-9
23. Jana NR, Chen Y, Peng X (2007) Magnetic nanoparticles for biosensing and medical applications. Adv Drug Deliv Rev 59(5):416–430. https://doi.org/10.1016/j.addr.2007.03.017
24. Jambhrunkar S, Yu M, Yang J, Zhang J (2018) Functionalized magnetic nanoparticles for sensitive and selective biosensing. Biosens Bioelectron 105:96–105. https://doi.org/10.1016/j.bios.2018.01.013
25. Kasprzyk-Hordern B, Nawrocki J, Kabsch-Korbutowicz M (2025) Next-generation magnetic nanomaterials for biosensing applications. J Nanobiotechnol 23(1):115–130. https://doi.org/10.1186/s12951-024-01234-5
26. Jain TK, Reddy MK, Morales MA, Leslie-Pelecky DL, Labhasetwar V (2018) Magnetic properties of functionalized nanoparticles: implications for biosensing. Nano Lett 18(6):3456–3463. https://doi.org/10.1021/acs.nanolett.8b01123
27. Corr SA, Rakovich YP, Gun'ko YK (2009) Superparamagnetic behavior in nanoscale magnetic materials for biomedical applications. J Magn Magn Mater 321(10):1501–1510. https://doi.org/10.1016/j.jmmm.2009.01.011
28. Alasmari A, Khan MA, Alhoshan M (2025) Understanding superparamagnetism in functionalized magnetic nanomaterials. Mater Today Chem 23:100812. https://doi.org/10.1016/j.mtchem.2024.100812
29. Zhang X, Wang N, Chen Y (2020) Magnetic nanoparticle-based biosensors for rapid and efficient detection. Anal Chem 92(4):2755–2763. https://doi.org/10.1021/acs.analchem.9b04567
30. Hao R, Xing R, Xu Z, Hou Y, Gao S, Sun S (2010) Magnetic susceptibility and its impact on biosensing applications. J Magn Magn Mater 322(5):437–444. https://doi.org/10.1016/j.jmmm.2009.10.011
31. Sudhakar K, Kumar P, Rao KS (2017) Advances in magnetic biosensors: sensitivity and functionalization. Biosens Bioelectron 94:453–470. https://doi.org/10.1016/j.bios.2017.03.045
32. Bhattacharjee S, Ahmadi TS, Saha K (2018) Colloidal stability of magnetic nanoparticles for biomedical applications. J Nanopart Res 20(3):87–102. https://doi.org/10.1007/s11051-018-4190-7
33. Wisniowska A, Kowalski B, Nowak J (2016) Colloidal stability and magnetic properties of functionalized nanoparticles. Nanotechnology 27(12):125601. https://doi.org/10.1088/0957-4484/27/12/125601
34. Kumar CSSR, Mohammad F, Singh N (2013) Tunable magnetization in functionalized magnetic nanoparticles for biosensing. J Appl Phys 114(17):173907. https://doi.org/10.1063/1.4828705
35. Kesavan M, Santhosh P, Gopal S (2015) Optimizing magnetic nanomaterials for biosensing applications. Anal Methods 7(14):5946–5955. https://doi.org/10.1039/C5AY01234A
36. Yang H, Li J, Wang X (2024) Aptamer-functionalized magnetic nanoparticles for high-sensitivity biosensing applications. ACS Nano 18(3):1125–1138. https://doi.org/10.1021/acsnano.4c01125
37. Gao Z, Jin W, Li P (2019) Magnetic nanoparticle-based electrochemical biosensors: a review. Biosens Bioelectron 128:127–137. https://doi.org/10.1016/j.bios.2019.01.045

38. Cheng Y, Sun R, Zhang Y (2023) Dual-mode biosensing with magnetic and plasmonic nanoparticles. Anal Chem 95(7):3340–3352. https://doi.org/10.1021/acs.analchem.2c04587
39. Kim J, Lee N, Hyeon T (2018) Recent advances in magnetic nanoparticle-based biosensors. Chem Soc Rev 47(14):5214–5234. https://doi.org/10.1039/C8CS00010K
40. Fan Z, Huang X, Li H (2022) Functionalized superparamagnetic nanoparticles for label-free biosensing. Small 18(22):2201135. https://doi.org/10.1002/smll.202201135
41. Wu J, Zhang S, Wang Y (2024) Integration of magnetic nanoparticles with microfluidics for biosensing. Lab Chip 24(5):987–1001. https://doi.org/10.1039/D4LC00052F
42. Pérez JM, Josephson L, Weissleder R (2002) Magnetic nanoparticles as molecular probes for cancer diagnostics. Nano Lett 2(4):329–332. https://doi.org/10.1021/nl015683v
43. Goon IY, Lai LMH, Amal R (2009) Tailoring magnetic nanoparticles for enhanced biosensing. Langmuir 25(17):10092–10099. https://doi.org/10.1021/la900635q
44. Xue X, Sun Y, Lin H (2021) Recent advances in nanomaterial-based biosensing with magnetic nanoparticles. Mater Chem Front 5(14):4567–4581. https://doi.org/10.1039/D1QM00342J
45. Zeng Q, Li Y, Cheng X (2017) Magnetic nanoparticles in enzyme-based biosensing applications. Anal Biochem 537:55–65. https://doi.org/10.1016/j.ab.2017.06.012
46. Sun C, Lee JSH, Zhang M (2008) Magnetic nanoparticles for ultrasensitive biodetection. Adv Drug Deliv Rev 60(11):1252–1268. https://doi.org/10.1016/j.addr.2008.03.016
47. Cheng Y, Ren W, Wang Y (2023) Magnetic nanozymes for enhanced biosensing and biomedical applications. ACS Appl Mater Interfaces 15(8):10467–10479. https://doi.org/10.1021/acsami.3c01047
48. Li H, Zhang Y, Sun X (2019) Magnetic field-enhanced biosensors: advances and perspectives. Trends Anal Chem 119:115609. https://doi.org/10.1016/j.trac.2019.115609
49. Xu H, Aguilar ZP, Yang L, Kuang M, Duan H, Xiong Y, Wei H (2011) Superparamagnetic nanoparticles for sensitive detection of biomolecules. Biosens Bioelectron 26(11):4532–4538. https://doi.org/10.1016/j.bios.2011.05.017
50. Cai X, Liu Y, Tan Q, Lu X, Yang L (2019) Endocytosis and intracellular trafficking of functionalized nanoparticles. J Nanobiotechnol 17(1):78. https://doi.org/10.1186/s12951-019-0510-2
51. Zhang Y, Chen J, Li S, Wang M, Zhao R (2019) Functionalized nanoparticles for gene editing and regenerative medicine. Nat Biotechnol 37(5):502–513. https://doi.org/10.1038/s41587-019-0118-1
52. Lu Y, Huang J, Xu W, Li D, Sun X (2019) Lab-on-a-chip platforms for point-of-care diagnostics. Chem Rev 119(10):6016–6092. https://doi.org/10.1021/acs.chemrev.8b00763
53. Martinez AW, Phillips ST, Butte MJ, Whitesides GM (2010) Microfluidic devices for diagnostic applications. Annu Rev Biomed Eng 12:77–102. https://doi.org/10.1146/annurev-bioeng-070909-105217
54. Silva-Neto HA, Gonzalez MF, Pereira TR, Oliveira RS, Campos LM (2023) Emerging trends in lab-on-a-chip devices for clinical diagnostics. Biosens Bioelectron 205:114197. https://doi.org/10.1016/j.bios.2022.114197
55. Nisarga BR, Patel KM, Varma SR, Kapoor NT (2025) Recent advances in point-of-care microfluidic systems. Lab Chip 25(3):312–329. https://doi.org/10.1039/D4LC00245A
56. Zhu H, Wang J, Lin F, Chen X (2017) Magnetic nanomaterials in lab-on-a-chip devices for diagnostics. Biosens Bioelectron 94:112–125. https://doi.org/10.1016/j.bios.2017.03.018
57. Kang J, Lee M, Kim H, Park J (2016) Integration of magnetic nanomaterials in microfluidic systems. Anal Chem 88(4):1936–1944. https://doi.org/10.1021/acs.analchem.5b04217
58. Park S, Lim J, Choi E, Seo H (2018) Quantum dots and magnetic nanoparticles for bioimaging and biosensing applications. J Mater Chem B 6(45):7389–7404. https://doi.org/10.1039/C8TB02183H
59. Wang J, Li W, Zhang X, Zhao L, Han Y (2016) Functionalized nanomaterials for biosensing and disease diagnostics. Biosens Bioelectron 87:918–930. https://doi.org/10.1016/j.bios.2016.08.037
60. Paramasivam G, Deshmukh AR, Bhagat PR, Maheshwari SS (2024) Advanced ligand-functionalized nanomaterials for clinical diagnostics. Sens Actuators B Chem 364:131932. https://doi.org/10.1016/j.snb.2023.131932

61. Chen Y, Wang R, Li M, Zuo H (2019) High-throughput screening of biomarkers using nanomaterial-based microfluidics. Lab Chip 19(21):3577–3590. https://doi.org/10.1039/C9LC00879K
62. Shen X, Huang D, Wang P, Liu G (2020) Cost-effective biosensing platforms with magnetic nanomaterials. Biosens Bioelectron 165:112395. https://doi.org/10.1016/j.bios.2020.112395
63. Cheng C, Wu Q, Zhang L, Ma Z (2017) Nanomaterials-enabled signal amplification for biosensors. Anal Chem 89(16):8762–8771. https://doi.org/10.1021/acs.analchem.7b01563
64. Huang Y, Li Z, Wu R, He J (2021) Automated lab-on-a-chip platforms for rapid diagnostics. Nat Biomed Eng 5(4):317–328. https://doi.org/10.1038/s41551-021-00677-6
65. Wu X, Zhao K, Yu F, Sun Z (2018) Portable and decentralized diagnostic technologies in modern healthcare. Adv Healthc Mater 7(3):1701443. https://doi.org/10.1002/adhm.201701443
66. Xu T, Zhang X, Liu M, Chen R (2020) Magnetic nanomaterials for precision medicine applications. J Nanomed 15(6):1024–1038. https://doi.org/10.2147/IJN.S248232
67. Yang C, Liu S, Gao W, Zhang H (2019) Functionalized nanomaterials in personalized diagnostics. Biosens Bioelectron 130:125–140. https://doi.org/10.1016/j.bios.2019.01.040
68. Zhang J, Zhao H, Sun Q, Feng X (2018) Magnetic nanoparticles in point-of-care diagnostics. Anal Chem 90(17):10279–10288. https://doi.org/10.1021/acs.analchem.8b02547
69. Liu Y, Xu L, Wang B, Cheng Y (2020) Real-time biosensing using nanomaterials-based microfluidics. Lab Chip 20(10):1880–1895. https://doi.org/10.1039/D0LC00079H
70. Kim H, Lee J, Park M, Choi S (2019) Biosensing with functionalized magnetic nanoparticles. Nano Today 25:99–112. https://doi.org/10.1016/j.nantod.2019.100961
71. Zhang X, Wang R, Hu L, Sun C (2021) Advances in nanomaterial-based biosensors for medical applications. ACS Appl Mater Interfaces 13(45):53227–53240. https://doi.org/10.1021/acsami.1c15728
72. Chadha P, Singh H, Sharma A, Mehta R (2022) Emerging trends in functional nanomaterials for biosensing. Biosensors 12(3):214. https://doi.org/10.3390/bios12030214
73. Lee T, Choi E, Kwon D, Hwang S (2020) Nanoparticle-based biosensors for early disease detection. Nat Biomed Eng 4(9):800–815. https://doi.org/10.1038/s41551-020-0577-3
74. Böttcher C, Müller H, Schneider T, Becker L (2024) Personalized medicine through nanomaterials-based biosensing. Adv Func Mater 34(2):2401110. https://doi.org/10.1002/adfm.202401110
75. Gao X, Liu Y, Sun Q, Wang R (2018) Functionalized gold nanoparticles in biomolecular detection. Chem Rev 118(23):11712–11745. https://doi.org/10.1021/acs.chemrev.8b00297
76. Ghosh A, Patel D, Sharma R, Verma K (2024) Graphene oxide-based biosensors for pathogen detection. Biosens Bioelectron 206:114245. https://doi.org/10.1016/j.bios.2024.114245
77. Wang Z, Li H, Chen X, Zhao M (2019) Label-free biosensing of biomarkers using quantum dots. Analyst 144(19):5867–5880. https://doi.org/10.1039/C9AN01234G
78. Ding S, Zhang L, Sun M, Liu T (2017) Magnetic nanomaterials for environmental pollutant detection. Environ Sci Technol 51(12):6783–6794. https://doi.org/10.1021/acs.est.7b01245
79. Cui Y, Wang J, Sun Q, Chen T (2018) Biosensors for heavy metal and pesticide detection in environmental samples. Anal Chem 90(5):3266–3275. https://doi.org/10.1021/acs.analchem.7b05123
80. Li H, Zhang X, Wu P, Sun Y (2020) Portable magnetic biosensors for on-site environmental monitoring. Biosens Bioelectron 150:111896. https://doi.org/10.1016/j.bios.2020.111896
81. Zhu J, Liu P, Wang R, Zhang L (2020) Magnetic nanotechnology in food safety applications. Food Chem 330:127253. https://doi.org/10.1016/j.foodchem.2020.127253
82. Cheng Y, Sun M, Zhao R, Chen T (2019) Biosensors for foodborne pathogen detection using functional nanomaterials. Sens Actuators B Chem 283:42–59. https://doi.org/10.1016/j.snb.2019.01.056
83. Song L, Wu Z, Zhang H, Li X (2021) Nanoscale biosensors for food safety monitoring. Trends Food Sci Technol 113:72–85. https://doi.org/10.1016/j.tifs.2021.03.006

84. Liu C, Wang L, Sun Q, Zhang L (2018) Magnetic nanomaterials in molecular diagnostics and genetic analysis. Chem Soc Rev 47(14):5025–5040. https://doi.org/10.1039/C8CS00257F
85. Huang W, Zhao R, Li X, Wu P (2021) Functionalized magnetic nanoparticles for nucleic acid detection. Biosensors 11(3):120. https://doi.org/10.3390/bios11030120
86. Xu J, Wang M, Liu Y, Sun Q (2020) Point-of-care biosensing using functionalized magnetic nanoparticles. Anal Chem 92(8):5585–5593. https://doi.org/10.1021/acs.analchem.9b05678
87. Ngashangva P, Sharma R, Patel D, Ghosh A (2022) Emerging trends in genetic diagnostics using magnetic nanomaterials. J Mol Diagn 24(2):156–170. https://doi.org/10.1016/j.jmoldx.2022.01.005
88. Wang Y, Zhang L, Liu P, Sun Q (2020) Functional nanomaterials for biosensing applications: trends and challenges. Biosens Bioelectron 148:111835. https://doi.org/10.1016/j.bios.2020.111835
89. Zhang H, Wu Z, Li X, Zhao M (2019) Nanotechnology-enabled biosensors for disease diagnostics and monitoring. TrAC Trends Anal Chem 120:115665. https://doi.org/10.1016/j.trac.2019.115665
90. Nayana M, Verma K, Sharma R, Patel D (2023) Recent advances in magnetic nanomaterials for biosensing applications. Adv Func Mater 33(15):2302148. https://doi.org/10.1002/adfm.202302148
91. Chen X, Zhao R, Wang L, Sun J (2021) Enhancing biosensing with nanostructured materials: a review of recent developments. Anal Chem 93(9):4563–4575. https://doi.org/10.1021/acs.analchem.0c04788
92. Gupta R, Sharma R, Verma K, Ghosh A (2020) Nanomaterials in biosensing: current progress and future outlook. Nano Today 34:100897. https://doi.org/10.1016/j.nantod.2020.100897
93. Luo Z, Sun M, Zhang L, Wang J (2018) High-throughput nanobiosensors for rapid disease diagnostics. Biosens Bioelectron 117:172–183. https://doi.org/10.1016/j.bios.2018.06.013
94. Liu B, Chen T, Zhao M, Sun Q (2021) Challenges in translating biosensing nanotechnology to clinical practice. Nat Rev Bioeng 2(4):250–266. https://doi.org/10.1038/s41578-021-00270-1
95. Singh P, Sharma R, Patel D, Verma K (2024) Advances in nanomaterial-based biosensors for medical diagnostics. J Biomed Nanotechnol 20(2):150–168. https://doi.org/10.1166/jbn.2024.3333
96. Khan I, Sun Q, Zhang L, Wang R (2019) Regulatory and standardization challenges in nanobiosensing technology. Biosens Bioelectron 141:111462. https://doi.org/10.1016/j.bios.2019.111462
97. Wu J, Liu P, Zhao R, Wang M (2020) Ethical considerations in nanotechnology-based biosensing applications. NanoEthics 14(1):29–45. https://doi.org/10.1007/s11569-020-00368-x
98. Sharma R, Kaushik S (2025) Emerging interdisciplinary trends in biosensor research. Adv Mater Interfaces 12(3):2101867. https://doi.org/10.1002/admi.202501867
99. Li D, Zhang H, Sun Q, Liu P (2021) Personalized medicine and biosensing: a future perspective. Trends Biotechnol 39(10):998–1012. https://doi.org/10.1016/j.tibtech.2021.07.012
100. Umer M, Arshed M (2025) Tailoring biosensing platforms for personalized diagnostics. Biosensors 15(2):178. https://doi.org/10.3390/bios15020178
101. Chen J, Wang L, Sun Q, Zhao M (2020) Functional nanomaterials for advanced biosensing applications. ACS Nano 14(3):3046–3064. https://doi.org/10.1021/acsnano.0c00246
102. Taha M, Zhang L, Liu P, Wang R (2024) Nanotechnology-driven insights into disease pathology and patient outcomes. J Nanobiotechnol 22(1):86. https://doi.org/10.1186/s12951-024-00886-3
103. Zhu L, Chen X, Wang M, Sun Y (2019) Integrating artificial intelligence with biosensing technologies. Biosens Bioelectron 141:111421. https://doi.org/10.1016/j.bios.2019.111421
104. Yammouri G, Ait Lahcen, Abdelhadi (2024) Microfluidics and digital health innovations in biosensing. Lab on a Chip 24(2):212–234. https://doi.org/10.1039/D4LC00212K

105. Gao X, Liu Y, Sun Q, Wang R (2018) Nanomaterial-based diagnostics: bridging research and clinical applications. Chem Rev 118(22):11907–11933. https://doi.org/10.1021/acs.chemrev.8b00321
106. Ghosh A, Patel D, Sharma R, Verma K (2024) Strategic partnerships for accelerating nanobiosensor commercialization. Nat Nanotechnol 19(1):22–39. https://doi.org/10.1038/s41565-024-01234-5

Chapter 7
Functionalized Magnetic Nanomaterials (FMNs) in Water Treatment and Purification

Mohd Tahoor, Kritika Chaturvedi, and Laishram Saya

Abstract There have been numerous pieces of evidence in history that repeatedly proved that the presence of contaminants affects aquatic organisms and can spread quickly to large areas, making the contamination of fresh and marine sediments and water environments by oil spills, urban runoff, and industrial and domestic effluents a serious concern. Heavy metals like mercury, lead, and manganese, as well as polycyclic aromatic hydrocarbons (PAHs), polychlorinated biphenyls (PCBs), and dichlorodiphenyltrichloroethane (DDT), are common trace contaminants found in freshwater and marine environments. The existence of these pollutants is concerning, as trace amounts of the organic compounds have been shown to cause cancer in mammals and to be harmful to human health as well as the aquatic biota. Technologies for toxicity reduction are frequently used in innovative processes for treating wastewater that contains heavy metals. Environmental safety is just one of the many scientific and technological domains that have been impacted by nanotechnology. Water remediation and treatment is one of the most promising and advanced environmental uses of nanotechnology. Various nanomaterials can help purify water through a variety of mechanisms, such as the adsorption of heavy metals and other pollutants, the removal and inactivation of pathogens, and the conversion of toxic materials into less toxic compounds. Functionalized magnetic materials are gaining significant attention in wastewater purification due to their efficiency, reusability, and unique properties. These materials leverage the magnetic core for easy separation and functionalized surfaces for targeted adsorption or catalytic activities. These days, metals (such as cadmium, copper, lead, mercury, nickel, and zinc), nutrients (such as phosphate, ammonia, nitrate, and nitrite), organics, algae (such as cyanobacterial toxins),

Mohd Tahoor and Kritika Chaturvedi—These two authors have equal contributions as the first author.

M. Tahoor · K. Chaturvedi
Department of Biological Sciences, Sri Venkateswara College, University of Delhi, Delhi 110021, India

L. Saya (✉)
Department of Chemistry, Sri Venkateswara College, University of Delhi, Delhi 110021, India
e-mail: saya@svc.ac.in

S. Gulati (ed.), *Functionalized Magnetic Nanomaterials*, Nanostructure Science and Technology, https://doi.org/10.1007/978-3-031-97199-0_7

viruses, bacteria, parasites, and antibiotics are all revealed and removed by nanoparticles, nanomembranes, and nanopowder. Recent developments in nanotechnology for wastewater treatment are described in this chapter by linking these technological advancements to the physicochemical characteristics of magnetic nanomaterials. The features, processes, and potential production of magnetic nanoparticles (MNPs) in water purification are discussed.

Keywords Wastewater purification · Degradation · Adsorption · Advanced oxidation processes · Reactive oxygen species

7.1 Introduction

Industrialization is the key to civilization and has been the driving force behind development and improved living standards of any society, since time immemorial. However, despite its innumerable merits, its rapid and reckless expansion has come at a major environmental cost. Among the various detrimental effects of industrialization, water contamination stands out as a critical and persistent threat to human and environmental health.

Some common water contaminants are:

1. Waterborne pathogens: Untreated sewage, animal waste, and poor sanitation introduce harmful microorganisms such as bacteria, viruses, and protozoa into water supplies. Groundwater and even chlorinated water have been found to contain harmful microorganisms [1].
2. Suspended solids: This is the fine particulate matter, which includes a broad range of insoluble components like silt, decomposing plant and animal matter, clay, sand, and dust, which make up total solids, also known as total suspended solids (TSS). Generally, suspended solids are less than 62 μm [2, 3].
3. Chemical Pollutants: These pollutants can be either organic or inorganic, water-soluble or insoluble, and they vary greatly in how harmful they are to both the environment and people. While some of these substances may pose a minor risk to health, others may be fatal. Water contains almost 700 different kinds of chemicals from 20 different classes, according to some estimates [2, 4].

The impact of these contaminants is far-reaching. Diseases, including amoebiasis, cholera, typhoid, and viral infections that cause hepatitis B, are brought on by drinking water contaminated by microorganisms. Heavy metals such as lead, arsenic, and mercury pose neurological and developmental risks in kids. Additionally, organic pollutants such as dyes disrupt aquatic ecosystems and threaten biodiversity. These impacts are further exacerbated in developing countries, where access to clean water remains a challenge [6, 7]. Figure 7.1 shows diagrammatically the various stages of water contaminants go through.

Despite being one of the most abundant resources on Earth, less than 1% of the global water supply is both safe and accessible for human use [8]. The World Health

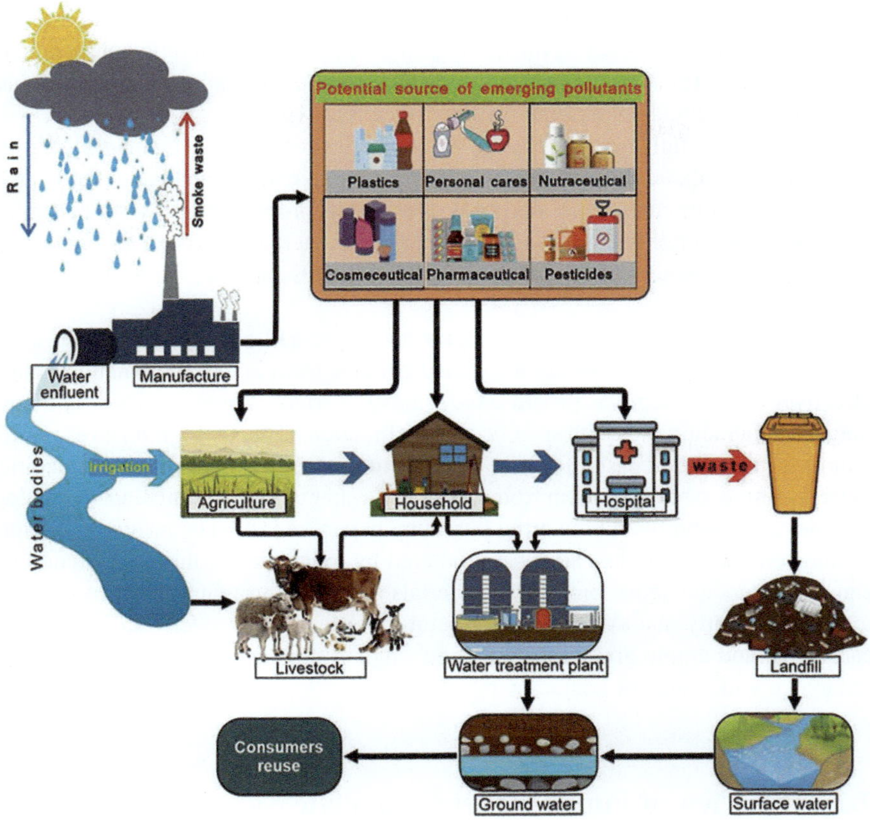

Fig. 7.1 The life cycle of contaminants [5] (CC BY 4.0)

Organization reported that, as of 2013, over 760 million people lacked sufficient access to clean drinking water [9]. Even in areas where potable water is available, its cost continues to rise due to factors like increasing energy prices, population growth, and environmental challenges such as climate change. Moreover, an increasing number of water sources are being contaminated, particularly by emerging pollutants such as fertilizers and industries. Water contamination arises due to the indiscriminate release of pollutants into water bodies. Industries release pollutants like toxic chemicals, heavy metals, organic dyes, etc., into rivers and lakes. Similarly, agriculture contributes to water contamination through pesticide chemical runoffs, fertilizers, and animal waste. Urbanization exacerbates the problem through untreated sewage, plastic waste, and microplastics entering the aquatic systems [5].

The conventional methods for water treatment, such as chemical disinfection, filtration, and biological sewage treatment, have limitations. Many of these methods are inefficient in addressing multiple contaminants simultaneously, lack scalability for large-scale applications, or produce secondary waste. Some key issues in current water treatment and purification are incomplete removal of contaminants, high operational costs, limited efficacy against emerging pollutants, and generation of secondary pollutants [10]. These challenges highlight the need for more adaptable, sustainable, and effective methods of treatment and purification.

Magnetic NPs fall under the class of nanoparticles consisting of elements like iron, nickel, and cobalt, so they can be controlled using an external magnetic field. They can be used for circumstantial purposes due to their properties of biocompatibility, magnetic susceptibility, synthesis mode, and characterization methods. Such NPs have proved to be serviceable in environmental redress and also in the biomedical field [11].

Functionalized magnetic nanoparticles have emerged as a viable way to overcome the drawbacks of conventional water treatment techniques. These nanoparticles effectively adsorb, degrade, process, and remove a variety of pollutants by combining magnetic characteristics with specific surface functions. FMNs' high surface area-to-volume ratio makes it easier for contaminants to adsorb, and their magnetic qualities make it simple to separate them from the treated water using external magnetic fields, increasing reusability and lowering operating expenses [12]. The primary structure of magnetic nanoparticles, after being altered by molecules of inorganic materials, can accommodate active biological materials including proteins, enzymes, etc. but more importantly additional groups for future use, such as aldehydes, hydroxyls, carboxyls, and amino groups: helping one singular nanomaterial become equipped to target a wide range of contaminants [13].

7.2 Overview of Functionalized Magnetic Nanomaterials (FMNs)

A kind of nanoparticle that may be controlled by introducing an external magnetic field is called a magnetic nanoparticle (MNP). Usually, they are made of cobalt, nickel, and iron. They find a wide range of applications in different fields thanks to their synthesis mechanisms, characterization techniques, magnetic susceptibility, and biocompatibility. MNPs are acknowledged as a highly versatile instrument for cleaning up many kinds of pollutants in soil, water, and the air. The structure and morphology of the MNPs rely upon the method used for their synthesis, which in turn greatly influences their magnetic properties. Consequently, various methods are proposed, which include sol-gel, hydrothermal, co-precipitation, sonochemical, microemulsion, thermal decomposition, and many more, which can be directed to synthesize MNPs of different vital properties [11].

The performance of MNPs can be improved when they are encapsulated with functional groups, ligands, or coatings that enhance their properties and enable targeted applications. The surface area of MNPs is amplified, which leads to improved adsorptive degradation, decreased sensitivity to leaching, and surface modification that shields the surfaces from oxidation. Without any surface coating, MNPs tend to accumulate and form clusters, deviate from nanosize, and exhibit less magnetic behaviour. The surface modification provides a high adsorption capacity of pollutants from wastewater and superior chemical stability, and is crucial to resist the attractive forces.

The significance of functionalization lies in its ability to allow the customization of magnetic nanoparticles for a variety of uses, such as catalysis, where their high surface area and magnetic separability make them effective catalyst supports. Researchers can increase the usefulness of magnetic nanoparticles in a variety of domains by managing the functionalization process to create multifunctional systems with improved reactivity, specificity, and stability under varied operating conditions [15]. Figure 7.2 shows a pictorial summary of various synthesis methods, characterization techniques, and applications of magnetic nanomaterials in wastewater treatment.

In wastewater treatment, one of the applications of encapsulated MNPs is in removing methylene blue, Congo red (CR), and heavy metal ions. With the use of an external magnetic field, their surface coating, which serves as an adsorbent for particular contaminants, can be readily separated (see Sect. 3.1). The effectiveness of their functionalization, therefore, largely depends on the type of coating applied for targeted applications.

Various types of coatings are:

1. **Organic coatings** such as lysine, arginine, polyaniline, polypyrrole, chitosan, and amino acids are used because of their effective adsorptive properties, which can be utilized to clean wastewater.
2. **Inorganic coatings** such as silica, molybdenum disulfide, magnetite, and manganese feroxyhyte help overcome certain limitations of organic coatings. The selection of these depends on the surface property required.
3. **Metals** such as lanthanum, gadolinium, gold, and cobalt can also be used as surface modifiers to protect the magnetic cores of MNPs against oxidation. Due to its low reactivity and stability, gold is proven to be a unique coating for magnetic nanoparticles and is typically chosen for water purification [11].
4. It is an efficient general technique to improve the stability and biocompatibility of magnetic nanoparticles through surface functionalization. Magnetic nanoparticles and organic compound coatings with associated functionalities make up the two primary components of the structure of such composites.

Iron oxide is commonly used as the core material. The structure of composite nanomaterials can be categorized into core-shell, matrix, and shell-core-shell types. Magnetic nanoparticles modified with organic compounds typically exhibit a core-shell structure. The core consists of ferromagnetic oxide nanoparticles. The shell, which can be composed of organic or other materials, surrounds the core. Additionally, functionalized ferric oxide nanoparticles undergo further modification with organic molecules, resulting in a shell-core-shell structure. The shell may consist of polymers, biomolecules, or other similar or distinct materials.

The incorporation of organic compounds improves the properties of magnetic nanoparticles by letting them retain their magnetic characteristics and also enhancing

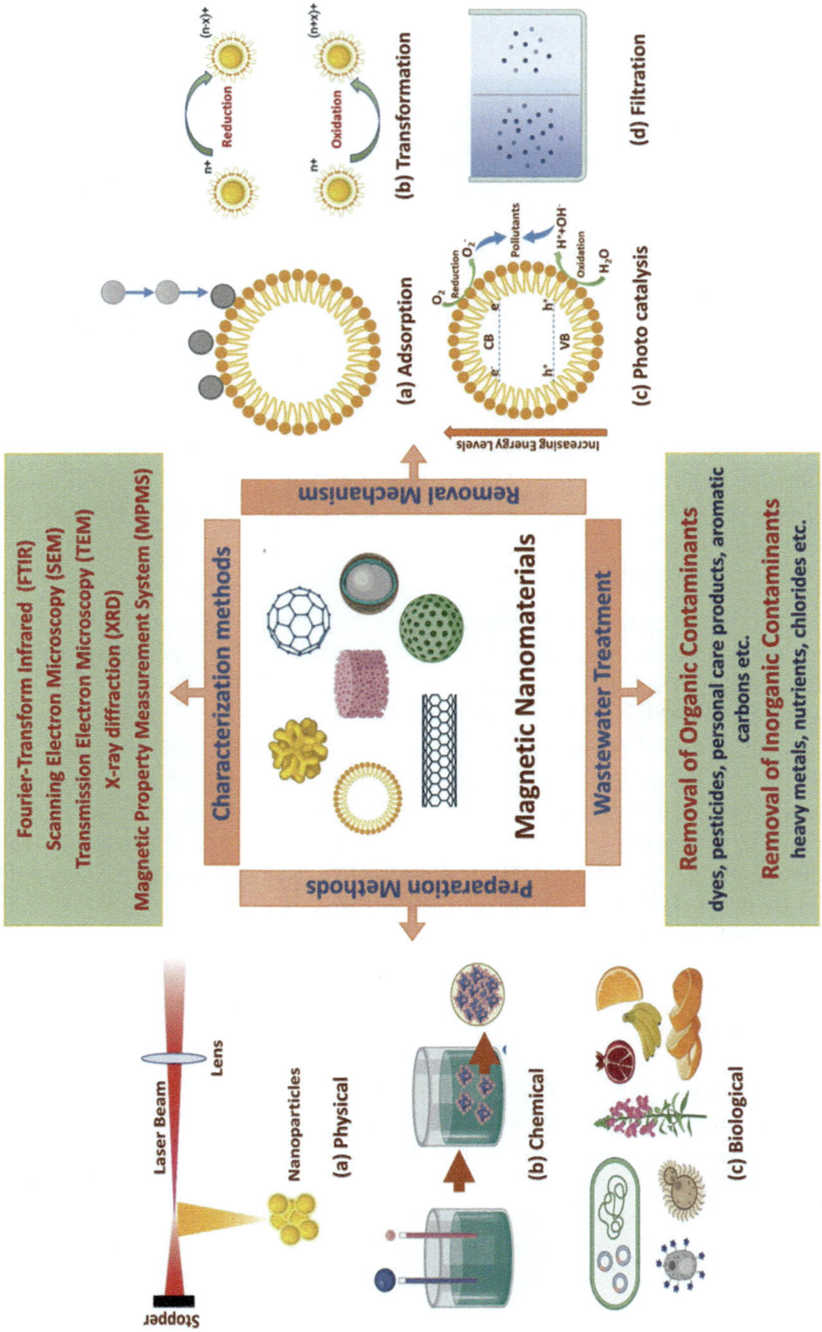

Fig. 7.2 Overview of magnetic nanoparticles [14] (CC BY 4.0)

their properties of biocompatibility and biodegradability. Furthermore, the primary structure of magnetic nanoparticles modified with inorganic materials can provide functional groups such as aldehydes, hydroxyls, carboxyls, and amino groups. These functional groups enable further applications, including conjugation with biologically active materials like antibodies, proteins, DNA, and enzymes [13].

7.3 Applications of FMNs in Water Treatment and Purification

Magnetic Nanoparticles (MNPs) are a highly effective and controlled method for wastewater treatment due to their unique characteristics, including small size, surface effects, and quantum properties. These properties help MNPs contribute significantly to managing the quality of water resources. Additionally, MNPs eliminate both inorganic and organic pollutants from wastewater. Different kinds of MNPs with different morphologies, sizes, and chemical characteristics have been created for the different mechanisms of contaminant removal [14].

7.3.1 Adsorption

Adsorption can be defined as the surface phenomenon of a component's increased concentration at the surface or interface of two phases. This surface is called adsorbent, and the compound that adheres to this surface is known as the adsorbate. The adsorption process involves pollutants adhering to the solid surface and is influenced by factors such as temperature, environmental conditions (pH, concentration), contact time, and the nature of the adsorbate and adsorbent [4]. Under different physical and chemical techniques, the process of adsorption can involve various forces. Relatively weak Van der Waals forces of attraction cause physical adsorption, while chemisorption involves strong covalent linking between the adsorbate and adsorbent [16].

Active carbon, mineral clay, zeolites, wool, sugar beet pulp, tea leaves, maize shell, rice flour, coffee powder, and magnetic adsorbents are among the adsorbents frequently employed in the adsorption process. In addition to having a high adsorption capacity and selectivity (especially for low-concentration pollutants in water), the ideal adsorbent for wastewater treatment should be environmentally friendly. It should also be readily recyclable and removable [17].

7.3.1.1 General Mechanism

FMNs are engineered with functional groups such as $-NH_2$ (amino), $-COOH$ (carboxyl), and hydrophobic coatings that interact with a specific pollutant, such as heavy metals or dyes. These groups enable electrostatic interactions, ion exchange, complex formation, etc., with the pollutant. For instance, Amino-functionalized $Fe_3O_4@SiO_2$ core-shell MNP is a novel adsorbent for heavy metal contaminant removal. The high surface area of the FMNs is critically useful, as it helps provide sufficient adsorption sites for the pollutant, which enhances effectiveness even at lower concentrations. However, the adsorption process is rather heavily influenced by factors such as diffusion and reaction kinetics, which may limit the adsorption rates (Fig. 7.3).

Once the pollutant is adsorbed onto the functionalized magnetic nanomaterials (FMNs), an external magnetic field is applied to the system, which causes the nanoparticles to aggregate and align with the field, thanks to their magnetic properties. These aggregated nanoparticles, along with the adsorbed pollutants, are directed

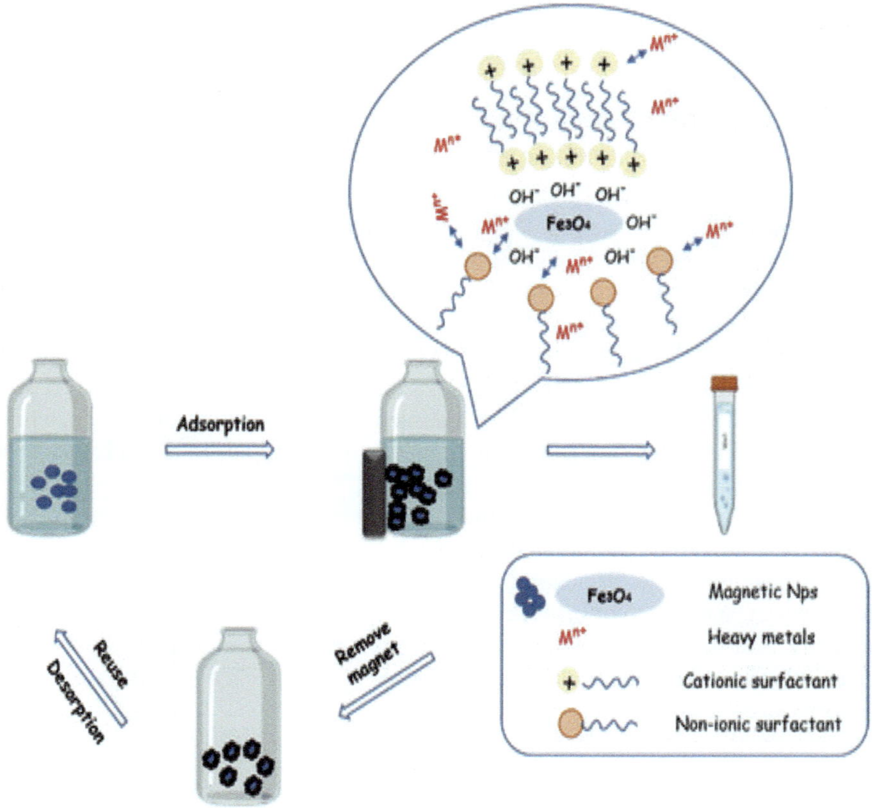

Fig. 7.3 Adsorption mechanism of pollutant removal [18] (CC BY 4.0)

toward the magnetic source, effectively separating them from the treated water. This process not only ensures efficient removal of contaminants but also ensures easy recovery of the FMNs, allowing for potential recycling and reuse [19].

7.3.1.2 Heavy Metal Removal

Functionalized magnetic nanoparticles are highly effective in the adsorption of heavy metals, such as arsenic, lead, and mercury, from contaminated water. Heavy metals are toxic and mostly carcinogenic, and unlike other contaminants, they are not biodegradable and tend to accumulate in organisms [20].

Mercury: Mercury poses health risks due to its toxicity and bioaccumulative properties, leading to serious health concerns, such as Minamata disease and neurological damage, even at trace concentrations [21]. Gold-functionalized Fe_3O_4 magnetic nanoparticles (Fe_3O_4/Au MNPs) have shown effective Hg(II) removal from water. These FMNs have a high surface area (121 m^2/g), and, combined with gold's strong affinity for mercury, are capable of efficient chemical adsorption, with an adsorption capacity of 79.59 mg/g [22].

Arsenic: Long-term consumption of water contaminated with arsenic can result in a number of health issues, such as reproductive issues, skin damage, and cancer [23]. Arsenic primarily exists in two oxidation states in water: As(III) and As(V). FMNs modified with iron-manganese (Fe–Mn) binary oxide demonstrate efficient removal of both arsenic species. The mechanism involves complexation at the surface through ligand exchange, where As(III) is oxidized to As(V) on the adsorbent surface, followed by its sorption. Findings reveal that MNp-FeMn exhibits enhanced adsorption capacities (56 mg/g for As(III) and 54 mg/g for As(V)) and has high reusability with minimal capacity losses [24].

Uranium: Cali et al. developed superparamagnetic iron oxide nanoparticles (SPIONs) for the separation of Uranium U(VI) [25]. Through phosphate-functionalization, these nanoparticles achieved a sorption capacity of 2333 mg of Uranium/g of Fe. It also exhibited high selectivity for U(VI) over other competing ions (Sr(II), Ca(II), Mg(II)) and had rapid adsorption kinetics (100% removal in 60 s).

Other contaminant non-specific FMN adsorbents are: Silica-coated Fe_3O_4 NPs functionalized with γ-mercaptopropyltrimethoxysilane demonstrate significant Cd^{2+}, Cu^{2+}, Pb^{2+}, and Hg^{2+} purification from water [26]. Amino-functionalized Fe_3O_4 and carboxylated magnetic iron oxide nanoparticles, for example, have been investigated by Liosis in 2021 [18] for the removal of heavy metals such as lead (Pb(II)), cadmium (Cd(II)), and copper (Cu(II)). Wang and team designed core-shelled amino-functionalized $Fe_3O_4SiO_2$ magnetic material, which showed promising efficiency in removing heavy metal ions from aqueous media [27]. Also, Thiol-functionalized Si_2-coated Fe_3O_4 NPs enable them to coordinate heavy metal pollutants like Ni^{2+}, Cu^{2+}, and Cr^{3+} [28].

7.3.1.3 Organic Material Removal

The presence of organic material in water, due to various industrial, agricultural, and domestic activities, poses significant challenges to public health. This is because organic pollutants alter the chemical composition of water, reducing the dissolved oxygen content of the water body and harming its aquatic ecosystems. Many of these organic pollutants are also toxic to humans, persistent, and bioaccumulative in animals [29].

Dyes: Dyes are commonly released from the textile, leather, and paper industries. Common dyes found in wastewater include methylene blue, Congo red, and reactive orange. The presence of organic dyes in water bodies affects sunlight penetration, which disrupts photosynthesis in aquatic ecosystems. Additionally, many dyes are toxic, carcinogenic, and mutagenic [30]. Copper sulfide nanoparticles, which were impregnated with activated carbon (CuS-NP-AC), at pH 5.0, exhibit readily sorption of methyl orange, and are capable of removing a high amount of MO in a short time (4 min) [31].

7.3.1.4 Pharmaceutical Removal

Pharmaceutical residues and drugs, such as antibiotics and analgesics, are emerging contaminants in water systems. They can cause endocrine disruption and promote antibiotic resistance among microorganisms.

Drugs and antibiotics: Deng et al. in 2019 synthesized carbon dot- and magnetite-modified magnetic carbon nanotubes (CMNTs) and assessed their adsorptive capacity for carbamazepine, an anticonvulsant drug [32]. At pH 7.0, the adsorbent has a comparatively high adsorption capacity of 65 mg carbamazepine/g. Additionally, chitosan-modified cadmium-based MOFs (Fe_3O_4@Cd-MOF@CS) exhibit, under optimized conditions, over 75% of amoxicillin from water, with ~103.09 mg/g maximum adsorption capacity [33].

FPAs: Per- and polyfluoroalkyl substances (PFAS), or 'forever chemicals', are a class of compounds that pose significant risks due to their toxicity and persistence. They are connected to the causes of health issues such as cancer, hormone disruption, and immune system damage. Magnetic amine-functionalized graphene oxide (MAGO) nanocomposites effectively remove these PFAS from water. Further integration of iron oxide nanoparticles onto graphene oxide gives MAGO the ability to capture both short and long-chain PFAS and achieve over 95% and 85% removal, respectively, within 30 min [34].

Pesticides: Agricultural discharge from cultivation, including insecticides and pesticides, contaminates soil and water, resulting in potential risk to human health and ecosystems. For the adsorptive removal of neonicotinoid insecticide pollutants from water, Liu et al. [35] synthesized a Magnetic copper-based metal framework (MMOF) with Fe_4O_3-graphene oxide-β-cyclodextrin (β-CD) nanocomposites. Clothianidin, nitenpyram, thiacloprid, thiamethoxam, acetamiprid, imidacloprid, and dinotefuran had adsorption capacities of 1.77, 2.56, 2.88, 2.96, and 3.11 mg/g at 100 mg/L, respectively. These findings confirm that MMOF is a superior adsorbent for removing neonicotinoid pesticides from water. Additionally, an Fe_3O_4-supported GO nanocomposite exhibits outstanding adsorption capabilities in removing dangerous pesticides from water, specifically simazine, simeton, atrazine, prometryn, and ametryn [36].

7.3.1.5 Miscellaneous

In some environmental water samples, trace perfluorinated compounds (PFCs) can be extracted with Fe_3O_4–C_{18}–chitosan MNP adsorbent. Even at low pH levels, PFCs can be adsorbed thanks to the porous hydrophilic chitosan polymer coating, which also enhances the anti-interference ability of the adsorbent [37]. Residues of methadone, an opioid used primarily for pain relief, are often detected in wastewater. Magnetic graphene oxide (MGO)-based nanocomposites were investigated by Gupta et al. [38] for the elimination of methadone at 87.2 mg/g under ideal conditions of pH 6.2 and temperature 295.7 K [38].

7.3.2 Advanced Oxidation Processes (AOPs)

According to [39], advanced oxidation processes (AOPs) are oxidation processes that generate enough hydroxyl radicals (OH·) to purify water. They were first proposed for the treatment of drinkable water in the 1980s. In order to eliminate refractory and traceable organic contaminants or specific inorganic pollutants, or to improve the wastewater biodegradability as a pre-treatment before a subsequent biological treatment, this AOP treatment produces hydroxyl radicals (OH·) or sulphate radicals $\left(SO_4^{\cdot-}\right)$ [40]. AOP is, therefore, a quite effective mechanism in increasing the degradability and detoxification of emerging water contaminants.

7.3.2.1 Catalysis as an AOP

A catalyst is something that accelerates the rate of chemical reactions and remains unchanged after the reaction. Wet hydrogen peroxide catalytic oxidation (WHPCO) is a widely used AOP mechanism for breaking down organic pollutants from water, which uses Fenton-type nanocatalysis. In this process, silicate-linking of the manganese (Si/Mn molar ratio = 0.01) helps facilitate the production of reactive hydroxyl ions (OH·) from H_2O_2, without promoting simultaneous breakdown. This is capable of effective degradation of organic pollutants like dyes (e.g., Methylene blue), achieving over 80% degradation in a short time frame [41].

Dyes: Cobalt ferrite-bismuth ferrite (CFO-BFO) core-shell nanoparticles have been demonstrated by Mushtaq et al. [42] as a catalyst for synthetic dye degradation with over 97% removal efficiency, with studies carried out on dye rhodamine B (RhB). Chen and his research team synthesised novel $PbFe_{12}O_{19}$ magnetic separation catalysts (PFOMSCs) and observed a maximum degradation of 89% of the same dye from wastewater [43]. Another research group used Fe_3O_4 MNPs for Fenton-like catalysis of RhB, along with methylene blue (MB) and orange G (OG) [44]. And, as mentioned, 80% degradation of methylene blue was achieved by an investigation performed by Tušar et al. using H_2O_2 as an oxidising agent for silicate-manganese nanocomposites [41].

Organic Pollutants: Pd-supported magnetic Fe_3O_4 nanoparticles exhibit simultaneous catalysis of H_2O_2 and Fe^{2+} through the electro-Fenton process, and are capable of achieving 98% phenol degradation within 60 min. This process maintained performance over 10 cycles with minimal Pd leaching [45].

Pharmaceuticals: Bisphenol A (BPA) is a well-known endocrine-disrupting hormone (EDH) that pollutes water through leaching from plastics. FMNs like polyethylene glycol-stabilized Fe_3O_4 were also used and modified with Iron (II) and oxalate (Fe) to achieve 100% catalytic degradation of BPA in 120 min [46]. The CFO-BFO nanoparticles prepared by Mushtaq et al. [42] also exhibited 85% removal efficiency for pharmaceuticals, such as Diclofenac, Gabapentin, and Oxazepam. Other broad-spectrum antibiotics, such as ciprofloxacin (CIP) and tetracycline (TC), were efficiently degraded by Fe_3O_4 MNPs under mild conditions and over a short period of time [44].

Others: Fe_3O_4 MNPs were synthesized for heterogeneous Fenton-like catalysts, which are capable of efficient removal of 2,4-dichlorophenol from aqueous solution. It showed the complete removal of 2,4 DCP, along with additional application in 51% decomposition of TOC [47].

Photocatalysis for Pollutant Degradation

Photocatalysis is the process of acceleration of a photoinduced reaction by the presence of a catalyst [48]. In contrast to other AOPs like ozonation or Fenton's process, photocatalysis, in its ideal state, enables the full mineralization of the target compound without the need for costly reagents. It is an inexpensive method that operates at ideal pressure and temperature [49]. Since the energy source for photocatalysis is light, sunlight can be used for the photoexcitation of the catalyst [50]. Nanoparticles are employed as photocatalytic materials because of their distinct physical and chemical characteristics, which make them superior to their microscopic counterparts. Their mechanical, optical, electric, and magnetic properties are all enhanced by their unique surface and quantum effects.

Numerous MNPCs have been effectively used to photocatalytically break down a broad range of chemical molecules. When choosing the appropriate MNPCs for any particular application, the material's band gap is a crucial factor to consider. Since the visible spectrum contains 46% of the energy emitted by the sun, MNPCs with low band gaps should be used [49].

Because of its high oxidation efficiency, chemical stability, nontoxicity, availability, and affordability, titanium dioxide (TiO_2) is the most widely used substance among all nanocatalysts. Titania generates charge carriers, thereby inducing reductive and oxidative processes, respectively [51].

There are several challenges in the use of TiO_2. In nature, TiO_2 exists in three common crystalline polymorphs, rutile, anatase, and brookite, in the order of their abundance. The most photocatalytically active form is anatase; however, because of its 3.2 eV bandgap, its potential is restricted to UV light with wavelengths shorter than 387 nm, which makes up just 5–8% of the entire spectrum of sunlight [53]. Figure 7.4 illustrates the photocatalytic generation of hydroxyl radicals by TiO_2 that triggers the degradation process. Carbon nanotubes, dye sensitisers (methylene blue, azure, etc.), metals or metallic ions (cobalt, silver, gold, etc.), or non-metals (nitrogen, carbon, sulphur, etc.) are added to TiO_2 composite catalysts to increase the photocatalytic activity of anatase under the visible light spectrum. However, there is still a problem in recovering the nanocatalysts from the solution, which limits the general use of photocatalysis [52]. In order to overcome this limitation, the material has been fixed to an inert substrate to prevent the metal recovery state. However, this system configuration decreases the number of active catalyst sites, which lowers the photocatalytic efficacy of these materials when compared to unsupported nanocatalysts.

E. Arulkumar et al. examined $CuO@NiO/g\text{-}C_3N_4$ nanocomposites for their application in dye degradation, with results showing a 98.38% photocatalytic degradation capacity for synthetic dye rhodamine B (RhB) [54]. Under visible light, $Fe_2O_3/g\text{-}C_3N_4$ monoliths exhibit a degradation efficiency of 94.7% in 140 min for the same dye (RhB).

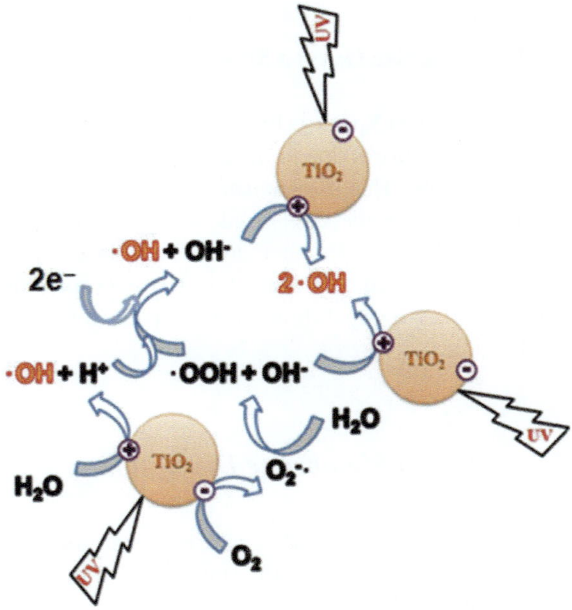

Fig. 7.4 Photocatalytic generation of hydroxyl radicals by TiO_2 [52] (CC BY 4.0)

A study by Xu et al. synthesized and used sulfate-functionalized Fe_2O_3/TiO_2 nanotubes to serve as visible light-active heterogeneous photo-Fenton catalysts [55]. Their analysis of the degradation of the reactive brilliant red X-3B dye showed efficient removal up to 95.7% at pH 4 in 150 min.

Additionally, this method has application in the rapid degradation of organic micropollutants like BPA; the photodegradation in aqueous solution was investigated by the photo-Fenton treatment [56]. The results showed that BPA breakdown was greatly influenced by the catalyst's surface properties and was accelerated in the presence of low concentrations of H_2O_2 and UV-A radiation [46].

7.3.3 Pathogen Capturing and Degradation

Pathogenic bacteria in water result in many health problems. To prevent this, it is necessary to disinfect water. Nanoparticles have been identified as promising candidates for advancing water treatment technologies. They enhance the efficiency of water purification processes and expand water availability by enabling the use of unconventional and safe water sources [57]. Nanomaterials have many advantages for this specific application, like strong specific surface areas, rapid dispersion, strong reactivity, and sorption capacity for various organic contaminants, heavy metals, and bacteria. However, unlike traditional chemical disinfectants, they are not predicted to produce dangerous disinfection byproducts because they are not strong oxidants, increasing their applicability in water purification [58].

Mechanisms of pathogen removal.

7.3.3.1 Reactive Oxygen Species

Reactive Oxygen Species (ROS) are highly reactive molecules that find application in pathogen removal due to their ability to damage cellular components, leading to pathogen death. This mechanism involves the generation of ROS through processes like Fenton and Haber-Weiss reactions through FMNs, which can be used in water purification [59]. FMNs can induce a substance to produce different reactive oxygen species, such as hydroxyl radicals (·OH) and superoxide anions $\left(O_2^{\cdot -}\right)$. These possess strong oxidative capabilities against pathogens and are hence applied in water purification.

7.3.4 Direct Membrane Disruption

Direct interactions of FMNs with pathogen membranes can also cause structural damage, leading to cell death. This can be through electrostatic damage to the cell. Since many biological membranes possess a negative charge, positively charged

FMNs are able to adhere to them. This can disrupt the membrane's integrity, leading to impaired permeability and eventually causing cell lysis. For example, Fang et al. used $Fe_3O_4@SiO_2$ nanoparticles, and with branched polyethyleneimine (PEI) coated onto the surface of the silica surface by covalent bonds, formed PEI-MNPs [60]. This coating grants it a high isoelectric point, which helps enhance electrostatic interactions with bacteria to improve its capture.

7.3.5 Application in Disinfection Systems

1. To extract Scenedesmus dimorphus from a medium broth, another method was to create MNPs coated with the cationic polymer polyethyleneimine (PEI). Coating the MNPs with PEI shifted their zeta potential from -7.9 ± 2.0 to $+ 39.0 \pm 3.1$ mV at pH 7, which enhanced the interaction between the MNPs and algae while reducing the required MNP dosage. The algal harvesting process was further optimized through the continuous application of a high-strength external magnetic field [11].
2. In a study done by Huang and team [61], both Gram-positive (Bacillus cereus) and Gram-negative (Escherichia coli) bacteria were found to be effectively eliminated by β-cyclodextrin grafted on poly(isophthalamide)/magnetic nanoparticles (CDPA-MNPs), with the maximum efficacy against the Gram-positive bacteria. Due to the competitive adsorption of suspended solids and other organic compounds in the water, it was observed that the removal efficiency of poly-allylamine-hydrochloride (PAAH) stabilised MNPs was higher (>89.5%) for water samples with lower bacterial concentrations than for those with higher concentrations. Their research also showed that the efficacy of bacterial removal increased with larger MNP concentrations until it reached a stabilisation point, in which both Gram-positive and Gram-negative bacteria were effectively eliminated.
3. FeO particles and CuFeO demonstrated excellent potential for removing E. coli. The effectiveness of FeO particles in removing E. coli was more influenced due to variations in contact time, while CuFeO particles were more affected by changes in the concentration of MNPs. For FeO particles, the removal efficiency was greater for S. aureus than for E. coli. Conversely, CuFeO particles showed higher removal efficiency for E. coli compared to S. aureus [62].
4. In order to eliminate harmful bacteria and viruses, Zhan et al. synthesised amine-functionalized magnetite $Fe_3O_4–SiO_2–NH_2$ NPs. Strong magnetic responsiveness and a distinct core-shell structure were demonstrated by these nanoparticles [63]. A wide range of pathogens, including bacteriophage f2, the virus (Poliovirus-1), and different bacteria, such as S. aureus, E. coli, P. aeruginosa, Salmonella, and B. subtilis, were found to be effectively captured using amine-modified nanoparticles. It was reported that the $Fe_3O_4–SiO_2–NH_2$ NPs achieved an impressive pathogen removal efficiency of 97.39%, which was significantly

higher than that of Fe_3O_4–SiO_2, which only showed 29.8% efficiency because of charge neutralisation.

5. Using a two-step transformation process, a group of researchers synthesised modified magnetic iron oxide nanoparticles with the antibacterial agent cetyltrimethylammonium bromide (CTAB) to create bactericidal paramagnetic nanoparticles (Fe_3O_4@CTAB) that effectively capture and inactivate bacteria [64]. The synthesized Fe_3O_4@CTAB MNPs captured both Gram-positive Bacillus subtilis and Gram-negative Escherichia coli from water and also resulted in inactivating both types of bacterial cells with an efficiency of 99% in 1 h. They also prepared Fe_3O_4 @amino acid (AA) nanoparticles by loading Fe_3O_4 MNPs with three types of amino acid, namely arginine, lysine, and poly-L-lysine (Fe_3O_4@Arg, Fe_3O_4@Lys, and Fe_3O_4@PLL. Both Gram-positive B. subtilis and Gram-negative E. coli were considerably more effectively eliminated with this change.
6. With great effectiveness and negligible toxicity, PAAH-stabilized MNPs were effective instruments for eliminating harmful microorganisms from drinking water. As a result, MNPs can effectively eliminate a wide range of harmful bacteria, such as Escherichia coli, Acinetobacter, Pseudomonas, and Bacillus. In contrast to other disinfection technologies, MNPs disinfection is inexpensive, simple to use, and has a promising future in engineering [65].

7.4 Challenges and Limitations

Agglomeration of MNPs due to self-interactions frequently results in reduced efficiency. For MNPs (FMNs) to be compatible, self-organised, and active enough to be used in a variety of applications, functionalization is essential. Although magnetic nanoparticles have the strong ability to adsorb pollutants and treat the presence of heavy metals in water due to their small size and unique properties, the same raises toxicity concerns for both human health and the environment as the nanoscale structure of MNPs allows them to cross biological barriers, interact with cellular components, and accumulate in organs, potentially causing cellular stress, inflammation, or other adverse reactions. MNPs can generate reactive oxygen species (ROS) at the cellular level, which may damage DNA, proteins, and lipids. This oxidative property may lead to cell death and damage. The magnetic property of MNPs is also another concern, as the magnetic fields may influence cellular signalling pathways or disrupt normal cell function [17]. Another challenge is that the release of MNPs into the environment poses serious ecotoxicological risks. MNPs can harm aquatic organisms at different trophic levels in the water bodies and thus disrupt the ecosystem. An example of the same is silver (Ag) and titanium dioxide (TiO_2) nanoparticles, which hinder the growth of plants and impact the behaviour and reproduction of small aquatic animals. MNPs can accumulate in the food chain, potentially endangering larger species [17]. Their persistence, bioaccumulation, and aggregation are a major concern in the environment, and the need for eco-friendly MNPs is still a major topic of research [66].

7.5 Future Perspective

The magnetic nanoparticles show a promising future in the area of wastewater and water purification due to their unique magnetic and physical properties. They show promising results for the treatment of wastewater by removing heavy metals, dyes, organic pollutants, and pathogens, etc. However, there are some changes that can be made to the MNPs to make them more environmentally friendly and help them reach their full potential. Functionalization of magnetic nanoparticles with surface ligands would help them be more specific to pollutants like arsenic, lead, and pharmaceuticals. This modification would also increase the adsorption capacity. Additionally, MNPs with specific coatings or functional groups can enable more effective removal of complex contaminants from water.

Sustainable and biodegradable magnetic nanomaterials can be developed using eco-friendly approaches [66]. Methods like plant-based synthesis or microbial-assisted production of magnetic nanoparticles help minimize their environmental impact while providing safer alternatives for water treatment. Moreover, researchers are investigating composite magnetic nanomaterials that integrate biodegradable polymers with magnetic cores for enhanced performance. Another approach of researchers to make the MNPs more eco-friendly is to increase their reusability [17]. Integrating MNPs into smart water treatment systems is a technique researchers are looking into for future applications that involve incorporating magnetic nanomaterials into filtration membranes or integrating them with sensor technologies for the detection of contaminant levels in water. Future research is also focused on addressing nano-toxicity by investigating how MNPs interact with cells and organisms at various stages of the water treatment process.

7.6 Conclusion

The treatment and purification of water using functionalized magnetic nanoparticles (FMNs) has emerged as an efficient tool due to their high efficiency, selectivity, size, surface effect, and quantum effect. FMNs, which are MNs (magnetic nanomaterials) with enhanced catalytic, adsorptive, and biological properties due to encapsulation and coating by organic, inorganic, metallic substances, specific groups, or biopolymers play an important role in the treatment of wastewater by removing heavy metals, pathogens, organic pollutants, dyes and enable their capture and effective degradation via processes like photocatalysis and advanced oxidation processes (AOPs). These approaches ensure efficient pollutant removal while minimizing secondary waste production. Despite their advantages, FMN-based wastewater treatment faces several challenges, including nanoparticle aggregation, potential toxicity, high synthesis costs, and difficulties in large-scale applications. However, continuous advancements are being made to reduce these limitations by using material engineering and functionalization, making FMNs a more viable solution for sustainable and ecologically friendly wastewater management. Ultimately, the integration of FMNs into conventional treatment systems holds great promise for achieving cleaner water resources and environmental sustainability.

Acknowledgements The authors are thankful to the Principal of Sri Venkateswara College, University of Delhi, for his continuous encouragement and support, as well as for providing infrastructure in the college to carry out a hassle-free literature survey required for writing the chapter.

Abbreviations

AA	Amino Acid
AOP	Advanced oxidation processes
BPA	Bisphenol A
CDPA	Cyclodextrin grafted on poly (isophthalamide)
CFO-BFO	Cobalt ferrite-bismuth ferrite
CIP	Ciprofloxacin
CMNTs	Carbon dot- and magnetite-modified magnetic carbon nanotubes
CR	Congo red
CTAB	Cetyltrimethylammonium bromide
DDT	Dichlorodiphenyltrichloroethane
FMNs	Functionalized magnetic nanomaterials
MAGO	Magnetic amine-functionalized graphene oxide
MB	Methylene blue
MGO	Magnetic graphene oxide
MMOF	Magnetic copper-based metal framework
MNP	Magnetic nanoparticle
NP	Nanoparticle
OG	Orange G
OH	Hydroxyl Radicals
PAAH	Poly-allylamine-hydrochloride
PAHs	Polycyclic aromatic hydrocarbons
PCBs	Poly chlorinated biphenyls
PEI	Polyethyleneimine
PFAS	Per- and polyfluoroalkyl substances
PFCs	Perfluorinated compounds
RhB	Rhodamine B
ROS	Reactive oxygen species
$(SO_4^{\cdot -})$	Sulphate radicals
TC	Tetracycline

TiO_2	Titanium dioxide
TSS	Total suspended solids
WHPCO	Wet hydrogen peroxide catalytic oxidation

References

1. Hwang C, Ling F, Andersen GL, LeChevallier MW, Liu WT (2012) Microbial community dynamics of an urban drinking water distribution system subjected to phases of chloramination and chlorination treatments. Appl Environ Microbiol 78(22):7856–7865
2. Khan ST, Malik A (2019) Engineered nanomaterials for water decontamination and purification: from lab to products. J Hazard Mater 363:295–308
3. Allgeier SC, Magnuson ML (2004) Responding to threats and incidents of intentional drinking water contamination. J Contemp Water Res Educ 129(1):13–17
4. Ali I, Gupta VK (2006) Advances in water treatment by adsorption technology. Nat Protoc 1(6):2661–2667
5. Kumar R, Qureshi M, Vishwakarma DK, Al-Ansari N, Kuriqi A, Elbeltagi A, Saraswat A (2022) A review on emerging water contaminants and the application of sustainable removal technologies. Case Stud Chem Environ Eng 6:100219
6. Werber JR, Osuji CO, Elimelech M (2016) Materials for next-generation desalination and water purification membranes. Nat Rev Mater 1(5):1–15
7. Ahmad A, Azam T (2019) Water purification technologies. In: Bottled and packaged water. Woodhead Publishing, pp 83–120
8. Grey D, Garrick D, Blackmore D, Kelman J, Muller M, Sadoff C (2013) Water security in one blue planet: twenty-first century policy challenges for science. Philos Trans R Soc A: Math Phys Eng Sci 371(2002):20120406
9. Levin RB, Epstein PR, Ford TE, Harrington W, Olson E, Reichard EG (2002) US drinking water challenges in the twenty-first century. Environ Health Perspect 110(suppl 1):43–52
10. Mojiri A (2023) Treatment of water and wastewater: challenges and solutions. Separations 10(7):385
11. Gupta N, Pant P, Gupta C, Goel P, Jain A, Anand S, Pundir A (2018) Engineered magnetic nanoparticles as efficient sorbents for wastewater treatment: a review. Mater Res Innov 22(7):434–450
12. Gao F (2019) An overview of surface-functionalized magnetic nanoparticles: preparation and application for wastewater treatment. ChemistrySelect 4(22):6805–6811
13. Liu S, Yu B, Wang S, Shen Y, Cong H (2020) Preparation, surface functionalization and application of Fe_3O_4 magnetic nanoparticles. Adv Colloid Interface Sci 102165
14. Shukla S, Khan R, Daverey A (2021) Synthesis and characterization of magnetic nanoparticles, and their applications in wastewater treatment: a review. Environ Technol Innov 24:101924
15. Lu AH, Salabas EE, Schüth F (2007) Magnetic nanoparticles: synthesis, protection, functionalization, and application. Angew Chem Int Ed 46(8):1222–1244
16. Mattson JS, Mark HB (1971) Activated carbon surface chemistry and adsorption from aqueous solution. Marcel Dekker, New York
17. Singh A, Kosalge SS, Bhatia PG (2024) Applications of magnetic nanoparticles for heavy metal ion removal from wastewater: a review 1(4):196–208. https://doi.org/10.69626/csa.2024.0196
18. Liosis C, Papadopoulou A, Karvelas E, Karakasidis TE, Sarris IE (2021) Heavy metal adsorption using magnetic nanoparticles for water purification: a critical review. Materials 14(24). MDPI. https://doi.org/10.3390/ma14247500
19. Saiz J, Bringas E, Ortiz I (2014) Functionalized magnetic nanoparticles as new adsorption materials for arsenic removal from polluted waters. J Chem Technol Biotechnol 89(6):909–918

20. Fu F, Wang Q (2011) Removal of heavy metal ions from wastewaters: a review. J Environ Manag 92(3):407–418
21. Hadi P, To MH, Hui CW, Lin CSK, McKay G (2015) Aqueous mercury adsorption by activated carbons. Water Res 73:37–55
22. Maia LFO, Santos MS, Andrade TG, de Carvalho Hot R, da Silva Faria MC, Oliveira LCA et al (2020) Removal of mercury (II) from contaminated water by gold-functionalized Fe_3O_4 magnetic nanoparticles. Environ Technol
23. Hong YS, Song KH, Chung JY (2014) Health effects of chronic arsenic exposure. J Prev Med Public Health 47(5):245
24. Nikić J, Watson MA, Isakovski MK, Tubić A, Šolić M, Kordić B, Agbaba J (2021) Synthesis, characterization and application of magnetic nanoparticles modified with Fe–Mn binary oxide for enhanced removal of As(III) and As(V). Environ Technol 42(16):2527–2539
25. Cali E, Qi J, Preedy O, Chen S, Boldrin D, Branford WR et al (2018) Functionalized magnetic nanoparticles for uranium adsorption with ultra-high capacity and selectivity. J Mater Chem A 6(7):3063–3073
26. Girginova PI, Daniel-da-Silva AL, Lopes CB, Figueira P, Otero M, Amaral VS et al (2010) Silica coated magnetite particles for magnetic removal of Hg^{2+} from water. J Colloid Interface Sci 345(2):234–240
27. Wang J, Zheng S, Shao Y, Liu J, Xu Z, Zhu D (2010) Amino-functionalized Fe_3O_4@SiO2 core–shell magnetic nanomaterial as a novel adsorbent for aqueous heavy metals removal. J Colloid Interface Sci 349(1):293–299
28. Tahergorabi M, Esrafili A, Kermani M, Shirzad-Siboni M (2016) Application of thiol-functionalized mesoporous silica-coated magnetite nanoparticles for the adsorption of heavy metals. Desalin Water Treat 57(42):19834–19845
29. Rashed MN (2013) Adsorption technique for the removal of organic pollutants from water and wastewater. Orga Pollut-Monitor Risk Treat 7:167–194
30. Lingamdinne LP, Koduru JR, Karri RR (2019) A comprehensive review of applications of magnetic graphene oxide based nanocomposites for sustainable water purification. J Environ Manag 231:622–634
31. Mokhtari P, Ghaedi M, Dashtian K, Rahimi MR, Purkait MK (2016) Removal of methyl orange by copper sulfide nanoparticles loaded activated carbon: kinetic and isotherm investigation. J Mol Liq 219:299–305
32. Deng Y, Ok YS, Mohan D, Pittman CU Jr, Dou X (2019) Carbamazepine removal from water by carbon dot-modified magnetic carbon nanotubes. Environ Res 169:434–444
33. Pooresmaeil M, Namazi H (2021) Chitosan coated Fe_3O_4@Cd-MOF microspheres as an effective adsorbent for the removal of the amoxicillin from aqueous solution. Int J Biol Macromol 191:108–117
34. Mahpishanian S, Zhou M, Foudazi R (2024) Magnetic amino-functionalized graphene oxide nanocomposite for PFAS removal from water. Environ Sci: Adv 3(12):1698–1713
35. Liu G, Li L, Xu D, Huang X, Xu X, Zheng S, Zhang Y, Lin H (2017) Metal–organic framework preparation using magnetic graphene oxide-β-cyclodextrin for neonicotinoid pesticide adsorption and removal. Carbohydr Polym 175:584–591
36. Boruah PK, Sharma B, Hussain N, Das MR (2017) Magnetically recoverable Fe_3O_4/graphene nanocomposite towards efficient removal of triazine pesticides from aqueous solution: investigation of the adsorption phenomenon and specific ion effect. Chemosphere 168:1058–1067
37. Zhang X, Niu H, Pan Y, Shi Y, Cai Y (2010) Chitosan-coated octadecyl-functionalized magnetite nanoparticles: preparation and application in extraction of trace pollutants from environmental water samples. Anal Chem 82(6):2363–2371
38. Gupta VK, Agarwal S, Asif M, Fakhri A, Sadeghi N (2017) Application of response surface methodology to optimize the adsorption performance of a magnetic graphene oxide nanocomposite adsorbent for removal of methadone from the environment. J Colloid Interface Sci 497:193–200
39. Glaze WI (n.d.) Drinking-water treatment with ozone. Enviorn Sci Technol 21(3):224

40. Deng Y, Zhao R (2015) Advanced oxidation processes (AOPs) in wastewater treatment. Curr Pollut Rep 1(3):167–176 (Springer). https://doi.org/10.1007/s40726-015-0015-z
41. Tušar NN, Maučec D, Rangus M, Arčon I, Mazaj M, Cotman M et al (2012) Manganese functionalized silicate nanoparticles as a Fenton-type catalyst for water purification by advanced oxidation processes (AOP). Adv Func Mater 22(4):820–826
42. Mushtaq F, Chen X, Torlakcik H, Steuer C, Hoop M, Siringil EC et al (2019) Magnetoelectrically driven catalytic degradation of organics. Adv Mater 31(28):1901378
43. Chen X, Wang S, Gao H, Yang H, Fang L, Chen X et al (2024) A novel lead hexagonal ferrite ($PbFe_{12}O_{19}$) magnetic separation catalyst with excellent ultrasonic catalytic activity. J Sol-Gel Sci Technol 110(2):578–593
44. Chen F, Xie S, Huang X, Qiu X (2017) Ionothermal synthesis of Fe_3O_4 magnetic nanoparticles as efficient heterogeneous Fenton-like catalysts for degradation of organic pollutants with H2O2. J Hazard Mater 322:152–162
45. Luo M, Yuan S, Tong M, Liao P, Xie W, Xu X (2014) An integrated catalyst of Pd supported on magnetic Fe_3O_4 nanoparticles: simultaneous production of H2O2 and Fe2+ for efficient electro-Fenton degradation of organic contaminants. Water Res 48:190–199
46. Neamtu M, Nadejde C, Hodoroaba VD, Schneider RJ, Verestiuc L, Panne U (2018) Functionalized magnetic nanoparticles: synthesis, characterization, catalytic application and assessment of toxicity. Sci Rep 8(1). https://doi.org/10.1038/s41598-018-24721-4
47. Xu L, Wang J (2012) Fenton-like degradation of 2,4-dichlorophenol using Fe_3O_4 magnetic nanoparticles. Appl Catal B 123:117–126
48. Van Gerven T, Mul G, Moulijn J, Stankiewicz A (2007) A review of intensification of photocatalytic processes. Chem Eng Process: Process Intensif 46(9 SPEC. ISS.):781–789. https://doi.org/10.1016/j.cep.2007.05.012
49. Gómez-Pastora J, Dominguez S, Bringas E, Rivero MJ, Ortiz I, Dionysiou DD (2017) Review and perspectives on the use of magnetic nanophotocatalysts (MNPCs) in water treatment. Chem Eng J 310:407–427 (Elsevier B.V.). https://doi.org/10.1016/j.cej.2016.04.140
50. Malato S, Fernández-Ibáñez P, Maldonado MI, Blanco J, Gernjak W (2009) Decontamination and disinfection of water by solar photocatalysis: recent overview and trends. Catal Today 147(1):1–59. https://doi.org/10.1016/j.cattod.2009.06.018
51. Gaya UI, Abdullah AH (2008) Heterogeneous photocatalytic degradation of organic contaminants over titanium dioxide: a review of fundamentals, progress and problems. J Photochem Photobiol C: Photochem Rev 9(1):1–12. https://doi.org/10.1016/j.jphotochemrev.2007.12.003
52. Azar MA, Varghese S, Nair SS (2012) Photocatalytic water treatment by titanium dioxide: recent updates. Catalysts 2(4):572–601 (Multidisciplinary Digital Publishing Institute (MDPI)). https://doi.org/10.3390/catal2040572
53. Chan SHS, Wu TY, Juan JC, Teh CY (2011) Recent developments of metal oxide semiconductors as photocatalysts in advanced oxidation processes (AOPs) for treatment of dye waste-water. J Chem Technol Biotechnol 86(9):1130–1158. https://doi.org/10.1002/jctb.2636
54. Arulkumar E, Thanikaikarasan S, Siddhardhan EV (2024) Synthesis and characterization of CuO@NiO/g-C_3N_4 nanocomposite for photocatalytic and electrochemical application. Results Chem 7:101439
55. Xu Z, Huang C, Wang L, Pan X, Qin L, Guo X, Zhang G (2015) Sulfate functionalized Fe2O3 nanoparticles on TiO2 nanotube as efficient visible light-active photo-fenton catalyst. Ind Eng Chem Res 54(16):4593–4602
56. Katsumata H, Kawabe S, Kaneco S, Suzuki T, Ohta K (2004) Degradation of bisphenol A in water by the photo-Fenton reaction. J Photochem Photobiol A 162(2–3):297–305. https://doi.org/10.1016/S1010-6030(03)00374-5
57. Savage N, Diallo MS (2005) Nanomaterials and water purification: opportunities and challenges. J Nanopart Res 7(4–5):331–342. https://doi.org/10.1007/s11051-005-7523-5
58. Kunduru KR, Nazarkovsky M, Farah S, Pawar RP, Basu A, Domb AJ (2017) Nanotechnology for water purification: applications of nanotechnology methods in wastewater treatment. Water Purif 33–74 (Elsevier). https://doi.org/10.1016/B978-0-12-804300-4.00002-2

59. Mai T, Hilt JZ (2017) Magnetic nanoparticles: reactive oxygen species generation and potential therapeutic applications. J Nanopart Res 19:1–10
60. Fang W, Han C, Zhang H, Wei W, Liu R, Shen Y (2016) Preparation of amino-functionalized magnetic nanoparticles for enhancement of bacterial capture efficiency. RSC Adv 6(72):67875–67882
61. Huang YF, Wang YF, Yan XP (2010) Amine-functionalized magnetic nanoparticles for rapid capture and removal of bacterial pathogens. Environ Sci Technol 44(20):7908–7913. https://doi.org/10.1021/es102285n
62. Pinto M, Ramalho PSF, Moreira NFF, Gonçalves AG, Nunes OC, Pereira MFR, Soares OSGP (2020) Application of magnetic nanoparticles for water purification. Environ Adv 2. https://doi.org/10.1016/j.envadv.2020.100010
63. Zhan S, Yang Y, Shen Z, Shan J, Li Y, Yang S, Zhu D (2014) Efficient removal of pathogenic bacteria and viruses by multifunctional amine-modified magnetic nanoparticles. J Hazard Mater 274:115–123. https://doi.org/10.1016/j.jhazmat.2014.03.067
64. Jin Y, Deng J, Liang J, Shan C, Tong M (2015) Efficient bacteria capture and inactivate by cetyltrimethylammonium bromide modified magnetic nanoparticles. Colloids Surf B 136:659–665. https://doi.org/10.1016/j.colsurfb.2015.10.009
65. You J, Wang L, Zhao Y, Bao W (2021) A review of amino-functionalized magnetic nanoparticles for water treatment: features and prospects. J Clean Prod 281. https://doi.org/10.1016/j.jclepro.2020.124668
66. Zhang K, Song X, Liu M, Chen M, Li J, Han J (2023) Review on the use of magnetic nanoparticles in the detection of environmental pollutants. Water (Switzerland) 15(17) (Multidisciplinary Digital Publishing Institute (MDPI)). https://doi.org/10.3390/w15173077

Chapter 8
Industrial Waste Management Using Functionalized Magnetic Nanomaterials (FMNs)

Vasantha Thoti and Murali Mohan Achari Kamsali

Abstract Industrialization has led to a scarcity of pure and clean water, with contaminants such as metal particles, microbes, dyes, and non-biodegradable organics posing significant health hazards. To address this, sustainable methods to recycle waste and synthesize nanomaterials are being developed. Industrial pollutants, especially liquid pollutants like heavy metals and metalloids, disrupt local ecosystems and need to be removed from wastewater. Functionalized magnetic nanomaterial (FMNs) technology is an effective solution for water remediation, particularly for removing heavy metals. Recent trends focus on fabricating functionalized nanostructures to clean up contaminants. Methods for creating FMNs include using inorganic moieties, organic ligands, surface polymerization, and enzyme immobilization. These materials show promise in adsorbing pollutants and transforming them into safer forms. This chapter highlights the effectiveness of FMNs in eliminating pollutants, including heavy metals, toxic chemicals, and organic dyes. It also discusses wastewater treatment and management technologies in industries using FMNs. Overall, FMNs offer significant potential for environmental remediation, contributing to cleaner and safer water resources.

Keywords Functionalized magnetic nanomaterials (FMNs) · Industrial waste management · Hazardous chemicals · Heavy metals organic pollutants

8.1 Introduction

The provision of clean water has become a major concern due to industrial expansion and population growth. The build-up of harmful contaminants in surface waters is primarily attributed to the discharge of untreated wastewater from industrial, agricultural, factory, and domestic activities. These pollutants endanger aquatic ecosystems and human health, underscoring the necessity for effective treatment systems [1].

V. Thoti (✉) · M. M. A. Kamsali
Department of Chemistry, Sri Venkateswara College, University of Delhi, Delhi, India
e-mail: tvasantha@svc.ac.in

S. Gulati (ed.), *Functionalized Magnetic Nanomaterials*, Nanostructure Science and Technology, https://doi.org/10.1007/978-3-031-97199-0_8

Traditional methods for contaminant remediation (physical, biological, and chemical) have several drawbacks, including the release of harmful compounds, high initial and maintenance expenses, and low remediation effectiveness [2]. Recent research suggests that nanotechnology offers a promising solution with superior features compared to traditional methods [3].

In the past decade, pharmaceuticals have been consistently found in water sources worldwide, including surface water, groundwater, and industrial wastewater. In Africa, for instance, various pharmaceuticals such as antibiotics, painkillers, anticonvulsants, anti-inflammatories, and antiviral drugs have been detected in water resources in Uganda. Similar occurrences have been reported in other regions, including Canada and India, providing strong evidence that pharmaceutical pollution is a global issue affecting water resources [4].

The color of water is commonly seen as an indicator of its purity, and even tiny amounts of certain pollutant dyes (even <1 ppm) can be noticeable and objectionable. Dyes are widely utilized in industries such as textiles, paints, printing, cosmetics, and more. Synthetic dyes form a major category of organic pollutants, often released as wastewater from these industries. Due to their intricate aromatic composition, they are challenging to break down and can endanger aquatic life by obstructing the necessary sunlight in colored water. Moreover, dye pollution can adversely affect humans, causing inflammation, eye irritation, abnormal heart rates, and cytotoxic effects [5, 6].

The growing global population has heightened the demand for food production, leading to increased pesticide usage to safeguard crops from invasive species and aggressive insects. However, prolonged use and exposure to pesticides cause natural resistance, requiring larger amounts of these chemicals over time. Moreover, pesticide molecules tend to accumulate in different environments such as soil, surface, and groundwater, endangering human health [7].

Magnetic nanoparticles are promising for environmental applications because of their large-scale production potential, lower toxicity, and eco-friendliness. Metals like iron, cobalt, nickel, and their oxides are common types. Organic/inorganic nanocomposites and doped TiO_2, ZnO nanostructures exhibit excellent chemical, catalytic, and sustainable properties. Magnetic iron oxide nanomaterials and other nanomaterials improve wastewater remediation by enhancing efficiency and pollutant removal. These smart nanomaterials possess catalytic, photonic, electrical, optical, mechanical, and magnetic characteristics, enabling easy segregation and revival with an external magnetic field. They are widely used in environmental contaminant removal, magnetic photothermal therapy, biochemical product separation, and tissue engineering [8].

Functionalized magnetic nanoparticles (MNPs) are engineered nanoscale particles that can be manipulated using an external magnetic field. They consist of magnetic core elements such as Fe, Co, Ni, and their oxides, including hematite (α-Fe_2O_3), maghemite (γ-Fe_2O_3), magnetite (Fe_3O_4), and cobalt ferrite ($CoFe_2O_4$). Magnetite (Fe_3O_4) is particularly favored due to its Fe^{2+} and Fe^{3+} states, which act as electron donors [9]. Functionalized MNPs have garnered attention for their remarkable properties, including easy synthesis, convenient surface functionalization, large surface area, and simple separation. They can be efficiently extracted from complex environments using a magnet, avoiding laborious steps like filtration or centrifugation. However, MNPs have limitations, such as agglomeration and stability concerns. Bare metallic nanoparticles can be chemically reactive and prone to oxidation, leading to reduced or lost magnetism. Effective protection methods include coating the surface of bare MNPs with materials such as carbon, silica, surfactants, or polymers.

To address the limitations of MNPs, many scientists have developed functionalized MNPs with new structures and surface characteristics. These enhancements aim to provide more active sites, improve water stability, and meet the growing demand for effective pollutant removal treatments. These MNPs have demonstrated notable improvements in pore volume, pore size distribution, specific surface area, high stability, processability, magnetic separation, surface active sites, and catalytic degradation capabilities, all contributing to the elimination of pollutants. MNPs have been extensively researched as effective adsorbents for separating and concentrating pollutants in water, including pesticides, antibiotics, and toxic metals [10].

FMNs are regarded as the smart materials of the future, requiring significant scientific foresight and patience. These materials play a vital role in environmental engineering, especially in addressing challenges like heavy metal and arsenic groundwater remediation. Nanotechnology and nanoengineering are crucial for this purpose. Darwish et al. examined the high reactivity and extensive surface area of nanomaterials at the nanoscale. They highlighted materials derived from semiconductors, silica, metal oxides, precious metals, polymers, and carbon-based substances for environmental protection. These materials are being investigated for the remediation of polluted wastewater, groundwater, surface water, and soil, utilizing techniques such as pollutant sensing, adsorption, nanofiltration, and photocatalytic degradation. The challenges involve potential health and ecosystem risks and the limited reusability of nanomaterials [11].

Traditional wastewater treatment methods, such as coagulation and flocculation, have drawbacks, including the production of large amounts of sludge and toxic byproducts. While ion-exchange treatments and adsorption processes have their advantages, they require pretreatment. Electrochemical methods are efficient but come with disadvantages. This chapter explores the applications of FMNs for tackling hazardous chemicals, heavy metals, organic pollutants, dyes, pharmaceuticals, and antibiotics. It underscores the extensive use of FMNs in waste management technologies, focusing particularly on their application in industrial wastewater treatment [12].

8.2 FMNs in Heavy Metal Removal from Industrial Waste

In wastewater, the accumulation of heavy metals has increased due to industrial and human activities, such as the petrochemical industry, plating, battery production, electroplating, textile manufacturing, mining, and pesticide use. These metals, including lead (Pb), copper (Cu), mercury (Hg), cadmium (Cd), nickel (Ni),

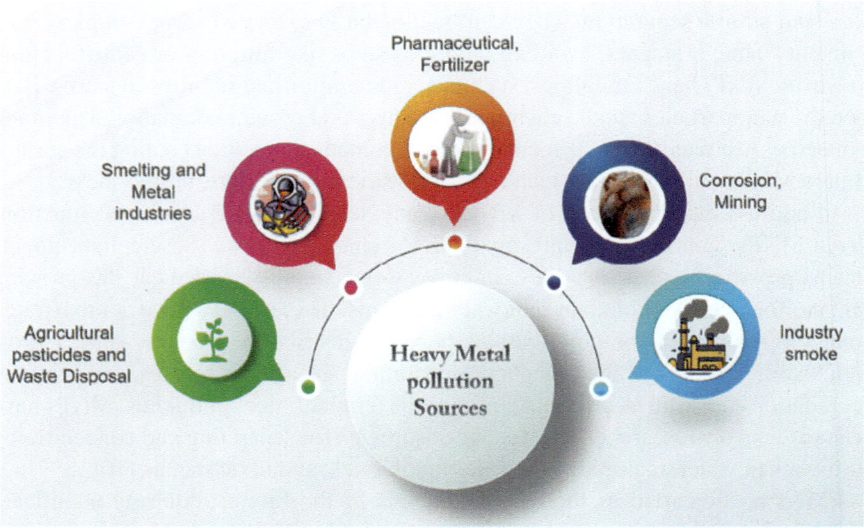

Fig. 8.1 Sources of heavy metal pollution, reprinted from [14]. Copyright [14] Karnwal A and Malik T (2024), Nano-revolution in heavy metal removal: engineered nanomaterials for cleaner water. Front. Environ. Sci. 12:1393694. https://doi.org/10.3389/fenvs.2024.1393694. (Open access)

chromium (Cr), and arsenic (As), are non-degradable and potentially carcinogenic. Their excessive presence in water can cause severe health issues for living organisms [13]. Even trace amounts of these heavy metals are hazardous. Other metals, such as iron (Fe), manganese (Mn), silver (Ag), molybdenum (Mo), antimony (Sb), cobalt (Co), and more, are also commonly detected in wastewater and need to be removed (Fig. 8.1).

Recent wastewater treatment techniques include biodegradation, photodegradation, chemical precipitation, hydrolysis, oxidative degradation, ion exchange, flocculation, neutralization, membrane separation, and ultrafiltration. Each method has its benefits and drawbacks. Magnetic nanoparticles (MNPs) have proven effective in enhancing toxic metal extraction and facilitating magnetic separation, making them favored for environmental cleanup due to their recyclability after separation. In contrast to smaller non-magnetic nanoparticles, which are less efficient due to smaller surface areas and difficult separation, MNPs have large surface areas, are non-toxic, and disperse easily, making them effective and economical for heavy metal removal [15].

Toxic metal ions are introduced into the environment through industrial effluents from sources like paint manufacturing, electroplating, and smelting. These ions are concerning due to their harmful impact on human health and the environment, as well as their tendency to bioaccumulate. Efficient removal techniques have attracted considerable attention.

Xue and colleagues synthesized Fe_3O_4 nanoparticles combined with oxidized humic acid (Fe_3O_4/HA-O) for metal adsorption studies. These Fe_3O_4/HA-O nanoparticles were utilized to remove Cd(II), Ni(II), Pb(II), and Cu(II) ions from aqueous solutions, achieving optimal adsorption capacities of 111.10, 33.33, 71.43, and 76.92 mg/g, respectively [16].

Kumar and Joshi synthesized polythiophene and Fe_3O_4 (PTh-Fe_3O_4) nanocomposite for the adsorption of Pb(II), Cd(II), and Cr(VI) ions. At pH 2, the maximum adsorption of Cr(VI) was 79.9%, decreasing to 8.9% at pH 6. Pb(II) and Cd(II) adsorption increased from 3.3% and 13.1% at pH 2 to 89.9% and 99.3% at pH 6, respectively [17]. The magnetic Fe_3O_4@BC/APTES composite was synthesized by coprecipitation for the adsorption of Pb(II) and Cr(VI) [18]. The surface area of the aminopropyl triethoxysilane (APTES)-modified adsorbent decreased from 135.90 to 35.66 m^2/g compared to the biochar-capped nanocomposite. Nevertheless, the adsorbent effectively removed 52.8% of Cr(VI) and 93.95% of Pb(II) after one hour at 20 °C. Pb(II) was better absorbed at basic pH, whereas Cr(VI) showed better absorption at low pH.

Sahu et al. [19] reported the synthesis of magnetic nanoparticles made of Fe_3O_4 and its conjugate with the renewable biopolymer chitosan, using Prosopis farcta biomass extract. These nanoparticles were used to adsorb Pb(II) from aqueous solutions. The maximum adsorption efficiencies were found to be 69.02% for Fe_3O_4 and 89.54% for Fe_3O_4-CS adsorbents. Both Fe_3O_4 and Fe_3O_4-CS can be easily recovered using an external magnet, making them viable and cost-effective options for lead removal.

Font and colleagues developed a novel adsorbent nanocomposite (Fe_3O_4@CS@CMC-SiNH_2) by functionalizing carboxymethyl cellulose (CMC) cross-linked to magnetic chitosan nanoparticles (Fe_3O_4@CS NPs) with an APTES layer. This nanocomposite exhibited outstanding adsorption performance with a maximum capacity of 555.56 mg Pb(II)/g NC and rapid adsorption equilibrium (30 min). It achieved 98.10% and 97.21% removal of lead (II) ions in the absence and presence of other heavy metal contaminants, respectively [20].

Bao and colleagues presented a novel method for preparing magnetic graphene oxide (GO) to remove heavy metal ions such as Cd(II) and Pb(II). They utilized n-propyltrimethoxysilane (NPTS) as a cross-coupling agent to connect Fe_3O_4/SiO_2 and GO. The resulting magnetic GO can be rapidly separated from water using a permanent magnet and exhibits excellent adsorption capabilities, with maximum capacities of 128.2 mg/g for Cd(II) and 385.1 mg/g for Pb(II) under neutral conditions [21].

Giwa et al. reported the preparation of polyethersulfone-functionalized graphene oxide (PES-fGO) mixed matrix membranes through methods such as crosslinking GO with maleic acid (M1), co-polymerizing GO with hyperbranched polyethyleneimine (HPEI) (M2), and modifying GO with chitosan (M3). The M2 membrane, characterized by its dense structure and abundant nucleophilic amino groups, demonstrated the highest tensile strength and pollutant removal efficiency, achieving up to 99.9% removal of Cr^{6+} and bacteria, and 98% removal of chemical oxygen demand (COD), despite having the lowest water fluxes [22].

Bangari et al. reported the use of Fe_3O_4 functionalized boron nitride nanosheets (BNNS-Fe_3O_4 nanocomposite) for removing As(III) ions from contaminated water. The nanocomposite has a specific surface area of 179.5 m^2/g and exhibits superparamagnetism at room temperature, allowing easy isolation using an external magnetic field. It demonstrates a maximum adsorption capacity of 30.3 mg/g for As(III) ions, which is approximately four times higher than that of bare BNNSs [23]. Researchers have synthesized various nanocomposites and magnetic nanoparticles, such as Fe_3O_4 with different functionalizations, to adsorb and remove heavy metal ions like Cd(II), Pb(II), Ni(II), Cu(II), Cr(VI), and As(III) from wastewater, achieving high adsorption capacities and reusability, making them promising candidates for environmental cleanup.

8.3 FMNs in the Removal of Pharmaceutical Industrial Wastes

Pharmaceutical and personal care products have become widely used due to technological and scientific advancements. Researchers are concerned about the waste from these products, particularly pharmaceutically active chemicals, because of their impact on the environment and human health.

Arabkhani et al. synthesized a magnetic nanocomposite (GO/ZIF-8/γ-AlOOH) using solvothermal and solid-state dispersion techniques. This nanocomposite, comprised of graphene oxide (GO) nanosheets coated with zeolitic imidazolate framework-8 (ZIF-8), pseudo-boehmite (γ-AlOOH), and iron oxide (Fe_3O_4) nanoparticles, featured a surface area of 777 m^2/g and a saturation magnetization of 9.75 emu/g. It was employed for the adsorption of diclofenac sodium, demonstrating a strong alignment with the Langmuir model and an impressive adsorption capacity of 2594 mg/g at 30 °C and pH 7.5 [24].

Asgari et al. prepared magnetic Fe_3O_4-chitosan nanocomposite nanoparticles for metronidazole adsorption. The highest adsorption capacity was 98 mg/g at pH 3, with experimental data well described by the Freundlich adsorption isotherm and pseudo-second-order kinetic model [25].

In another study, Fe_3O_4 nanoparticles with an average of 12 nm were functionalized with graphene oxide and carbon nitride to create GO/g-C_3N_4–Fe_3O_4 nanocomposite for tetracycline antibiotic adsorption. The adsorption was pH-dependent, with a maximum capacity of 107 mg/g at pH 3. The nanocomposite was also recovered and reused for five cycles [26].

Researchers have also developed various magnetic nanocomposites for pharmaceutical adsorption, achieving high capacities and reusability. These include GO/ZIF-8/γ-AlOOH for diclofenac sodium, Fe_3O_4-chitosan for metronidazole, GO/g-C_3N_4–Fe_3O_4 for tetracycline, and Fe_3O_4@SiO_2NH_2 for methylparaben.

8.4 FMNs in Dye Extraction from Industrial Effluents

Synthetic dyes are a major challenge due to their persistence and harmful effects on humans and other living beings. These dyes are widely used in industries such as food, cosmetics, leather, pharmaceuticals, paper, plastics, and textiles, with around 0.7 million tons of about 100,000 different types used annually. Azo dyes, typically anionic, are particularly harmful due to their high thermal, optical, and physicochemical stability, attributed to their stable chemical structure involving aromatic rings and azoic linkages. The uncontrolled release of these dyes into water bodies leads to severe environmental issues, including intrinsic carcinogenic effects, toxicity, mutagenic effects, skin sensitization, and reduction of sunlight in aquatic environments, posing a serious threat to aquatic organisms and humans. Therefore, removing these dyes before discharging industrial effluents into natural streams is crucial in current material and environmental research. Adsorption is a surface-based process where ions or molecules adhere to a solid surface of the adsorbent, as shown in Fig. 8.2. Dye molecules can adhere through interactions such as ion exchange, hydrogen bonding, and electrostatic and π-π interactions. This method is beneficial because adsorbents can be reused, it is highly effective, and dye removal from wastewater is quick [27].

Yang et al. polymerized poly(ionic liquid) onto silicon-coated Fe_3O_4 nanoparticles to create an inexpensive and selective composite [29]. The ability of Fe_3O_4@SiO_2@PIL to remove various dyes, such as acid orange II, alizarin red, malachite green, and thionin acetate, was tested. The adsorbent showed rapid adsorption kinetics, absorbing up to 510.2 mg/g of alizarin red.

Keziban Atacan et al. focus on immobilizing peroxidase onto 3-aminopropyltriethoxysilane (APTES)-functionalized $MgFe_2O_4$ magnetic nanoparticles to enhance efficiency, enzyme stability, and recyclability. The immobilized peroxidase retained almost 50% of its activity after 10 uses because of easy magnetic separation. It also showed improved antioxidant activity and a significant enhancement in Congo red dye removal efficiency, achieving 98.6% in 240 min [30].

Juxiang Luo et al. synthesized Magnetic crosslinked copolymer microspheres (MCPMs) using oleic acid-modified Fe_3O_4 as the magnetic source, with α-methyl styrene and maleic anhydride as monomers and divinylbenzene as the crosslinking agent. These MCPMs were then hydrolyzed to obtain MCPMs with carboxyl groups. The MCPMs' adsorption behavior, kinetics, and isotherms for methylene blue (MB) removal from water were analyzed. MCPMs demonstrated an excellent MB removal capacity of 315 mg/g [31].

Naveen Kumar et al. explored the use of molybdic acid (MA)-functionalized iron oxide magnetic nanoparticles (MNPs) for removing RB dye from wastewater via adsorption. At optimal conditions, they showed a maximum adsorption capacity of 60 mg/g, and it is proven that the MA-functionalized iron oxide nanoparticles are potential for efficient textile dye removal from wastewater [32].

Qazi et al. examined amine-functionalized MWCNTs (f-MWCNTs) as a potential material for removing Mordant Black Dye 11 (MBD 11). The study found a

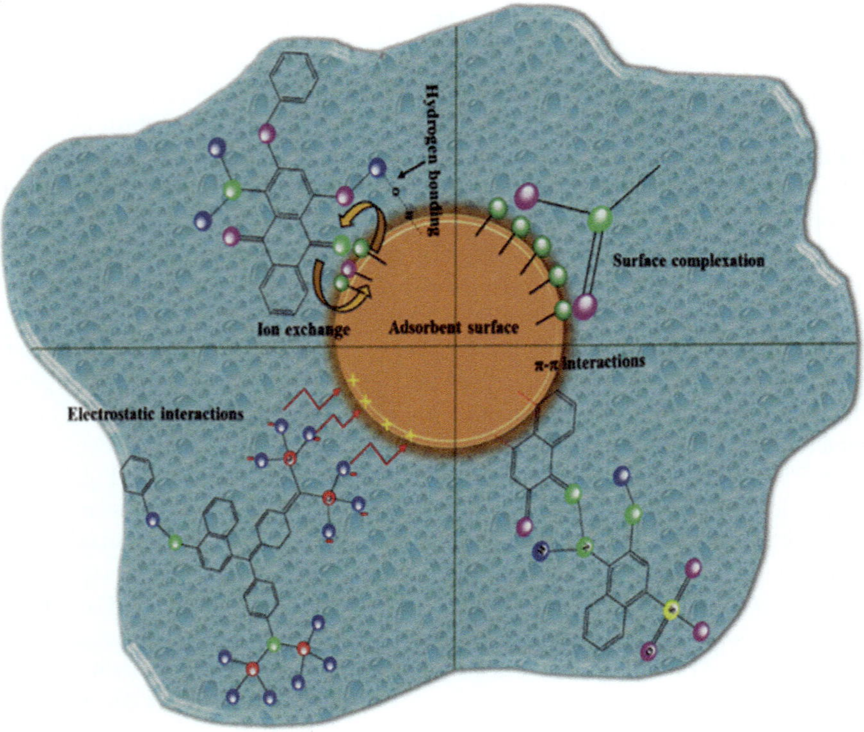

Fig. 8.2 Removal of dye from aqueous solutions by the adsorption method, Reprinted from [28]. Copyright @ From [28] Mbuyazi, T. B., & Ajibade, P. A. (2024). Magnetic iron oxides nanocomposites: Synthetic techniques and environmental applications for wastewater treatment. Discover Nano, 19, 1548. https://doi.org/10.1186/s11671-024-04102-9. (Open access)

strong interaction between the dye and f-MWCNTs. The highest removal efficiency occurred in an acidic medium (pH 2) with an initial dye concentration of 50 mg/L, achieving 99% dye adsorption within 40 min using 0.05 g of f-MWCNTs [33].

Said and colleagues synthesized SnO_2–Fe_3O_4 nanocomposites to explore their potential for photodegrading Congo red. The SnO_2–Fe_3O_4 nanocomposite exhibited a magnetic saturation value of 64.96 emu/g and demonstrated a photodegradation efficiency of 50.76% under optimal conditions (90 min, dye concentration 18 mg/L). The magnetic saturation value remained unchanged after the photocatalytic reaction [34].

Roostaee et al. prepared Fe_3O_4@Au core–shell nanoparticles and Fe_3O_4@Au/rGO nanocomposites for the photocatalytic degradation of crystal violet dye. Fe_3O_4@Au-rGO outperformed Fe_3O_4@Au and rGO in photocatalytic activities. Increasing the amount of Fe_3O_4@Au/rGO nanocomposite sped up the degradation, achieving complete degradation of crystal violet dye in 1 min with 20 mg of nanophotocatalyst [35].

Rehman et al. synthesized γ-Fe_2O_3, Fe_3O_4 nanoparticles, and γ-Fe_2O_3@GO nanocomposites for methylene blue photodegradation. BET analysis showed surface areas of 50 m^2/g for γ-Fe_2O_3 and 73 m^2/g for γ-Fe_2O_3@GO composite. The degradation efficiencies were 37.5%, 77.7%, and 90.6% for γ-Fe_2O_3, Fe_3O_4, and γ-Fe_2O_3@GO, respectively. The highest photodegradation rate was 0.0325 min^{-1} for the γ-Fe_2O_3@GO nanocomposite. The studies indicate that decreasing the band gap energy can enhance the degradation efficiency of organic dyes. However, photocatalyst performance is significantly influenced by slight morphological differences [36].

Researchers have explored various magnetic nanocomposites and functionalized magnetic nanomaterials for dye removal. These include Fe_3O_4@SiO_2@PIL for removing dyes like alizarin red, Fe_3O_4-chitosan for metronidazole, GO/g-C_3N_4–Fe_3O_4 for tetracycline, and Fe_3O_4@SiO_2NH_2 for methylparaben. Other advancements include MCPMs for methylene blue, MA-functionalized iron oxide for RB dye, and f-MWCNTs for Mordant Black Dye. These materials show high adsorption capacities, rapid kinetics, and good reusability, making them effective for water purification and environmental applications.

8.5 FMNs in the Removal of Pesticides from Wastewater

Pesticides, due to their extensive use and rising toxicity, are major organic toxins. Defined by the Environmental Protection Agency as chemicals employed to avoid, eradicate, deflect, or manage pests, pesticides pose a significant risk to aquatic life even at trace concentrations. They are associated with various illnesses, including endocrine disruption, cancer, respiratory disorders, Parkinson's disease, and reproductive issues. Thus, removing these contaminants from water is crucial [37]. Researchers have also explored using magnetic nanocomposites for pesticide removal through adsorption.

Zhou et al. examined the efficiency of Fe_3O_4@SiO_2-MWCNT for pentachlorophenol adsorption, finding that the capacity was pH-dependent, with a peak at pH 2.5 (96.4 mg/g) [38]. Wanjeri et al. synthesized Fe_3O_4 nanoparticles (mean diameter 12.3 $\pm$ 3 nm) via co-precipitation, treated them with silica, and modified them with graphene oxide and 2-phenylethylamine (PEA), creating Fe_3O_4@SiO_2@GO-PEA. This nanocomposite was used to adsorb malathion, chlorpyrifos, and parathion, achieving maximum removals of 76%, 88%, and 85% after 15 min, respectively. Adsorption capacities were 10.6 mg/g for malathion, 11.1 mg/g for chlorpyrifos, and 10.9 mg/g for parathion, unaffected by pH changes from 3 to 11 [39].

Fe_3O_4/reduced graphene oxide nanocomposite was also tested for adsorbing atrazine, ametryn, simazine, prometryn, and simeton, achieving 75.24%, 93.61%, 88.55%, 91.34%, and 81.22% efficiencies after 70 min, respectively [40]. Additionally, Researchers have investigated magnetic nanocomposites for pesticide removal via adsorption, achieving notable results. Fe_3O_4@SiO_2-MWCNT was effective for pentachlorophenol, $Fe_3O_4@SiO_2@GO$-PEA for malathion, chlorpyrifos, and parathion, and Fe_3O_4/reduced graphene oxide for atrazine, ametryn, simazine, prometryn, and simeton, demonstrating high adsorption efficiencies and capacities.

8.6 Waste Management Technologies Using FMNs

Functional Magnetic Nanomaterials (FMNs) have emerged as effective sorbents, providing novelty, high efficiency, and simplicity, which attract researchers' interest. Various reports highlight the use of FMNs in contaminant removal, such as carboxyl- and amine-functionalized nanoparticles, thiol-functionalized magnetite nanoparticles, poly rhodamine-functionalized aluminum oxide, polymer-functionalized magnetic nanoparticles, silica-functionalized magnetic nanoparticles, and others. The operation of functionalized magnetic nanomaterials (FMNs) involves specific pathways based on their applications. For environmental remediation, NMs adhere to particular routes to eliminate contaminants. Several procedures utilizing FMNs have been reported so far, with two methods being particularly notable due to their simplicity and efficiency in sorption and photocatalytic pathways [41].

Sorption stands out among contaminant removal methods due to its straightforward operation, cost-effectiveness, and energy efficiency. The composition of the sorbent material is crucial for the capacity of the sorption process. This method is popular for its simplicity, making it a practical and feasible strategy, especially in underdeveloped and developing countries. Sorption is versatile, being widely researched and successfully used in industrial waste treatment. The novelty of sorption relies on the variety of materials used as sorbents, including both natural and synthetic substances. Examples of low-cost recycled materials include fly ash, scrap tires, coconut shells, fruit extracts, and more. Synthetic materials include activated carbon, metal oxides/hydroxides, and zeolites, among others [42].

Photocatalysis is a method that uses functionalized magnetic nanomaterials (FMNs) to remove contaminants by absorbing energy and producing photons to accelerate chemical reactions without degrading the catalyst itself. Researchers frequently use this method for its efficiency and advanced follow-up capabilities. Sunlight is usually used as the energy source for the photocatalytic process due to its natural abundance, reducing energy-related costs. UV light can also be used, depending on the contaminants. Photocatalysis involves using semiconductor materials with dual functions, acting as both cathode and anode, suitable for various applications such as photodegradation, photocatalytic CO_2 reduction, photocatalytic synthesis, and more. The growing interest in photocatalysis for environmental restoration is due to its broad applicability [43].

The significant factors in choosing photocatalytic materials include excellent energy position, appropriate bandgap, low electron-hole recombination, and non-toxicity. Common materials include metal oxides, metal sulfides, and others. The use of FMNs in fabricating photocatalysts enhances their properties and efficiency. Examples of potent photocatalysts include azole-functionalized TiO_2, Ag-modified metal oxide, and more. Carbon-based nanomaterials (NMs) are recognized as next-generation materials with extensive applications across various fields, including biosensors, drug delivery, biomedical applications, superconductors, and electrically conductive materials. Due to their advantageous properties, such as high electrical conductivity and well-defined thermal and mechanical characteristics, carbon NMs are widely used in wastewater treatment. They serve as excellent sorbents and photocatalysts due to their high surface-to-volume ratio, uniform pore distribution, and highly porous structure. However, the application of carbon NMs can be limited by their tendency to agglomerate in solvents due to weak van der Waals forces and lower solubility. Additionally, impurities formed during synthesis can hinder their activity. To overcome these limitations, the functionalization of carbon NMs is performed to enhance their properties and activity. Functionalized carbon NMs come in various forms, including carbon nanotubes (CNTs), graphene and its derivatives, and fullerenes [44].

Carbon nanotubes (CNTs) are a common form of carbon nanomaterials, visualized as graphene sheets rolled into tubular cylinders with nanoscale diameters. They are categorized into single-walled CNTs (SWCNTs) and multi-walled CNTs (MWCNTs). SWCNTs consist of a single rolled graphene sheet, while MWCNTs have multiple concentric graphene sheets [45].

Graphene is the simplest form of carbon and the thinnest known material, consisting of a two-dimensional single-layer sheet of carbon atoms arranged in a sp^2-bonded honeycomb-like lattice structure. Graphene-based nanomaterials exhibit excellent sorption capacity due to their high surface area, delocalized π-electron system, and numerous active sites. Functionalization with groups like thiol moieties or carboxylic groups further enhances the properties of graphene nanomaterials. Zamani and Salem studied the photocatalytic behavior of graphene oxide sheets coupled with carbon nanotubes, achieving 96.5% efficiency in degrading MB dye [46]. Firdaus et al. functionalized graphene nano-platelets (GNPs) with oxygen-containing groups, resulting in higher sorption capacity and better dispersion in aqueous solutions [47].

Fullerenes are spherical molecules composed entirely of carbon atoms, considered allotropes of carbon, with high symmetry and solubility in various organic solvents. They consist of a carbon cage with a fused ring system of hexagons and pentagons. The most common fullerenes are C_{60} and C_{70}. Wu et al. achieved 94.7% efficiency in degrading RB dye using a photocatalyst made of fullerene-cored star-shaped polyporphyrin-incorporated TiO_2. The high efficiency is attributed to the high specific surface area and unique electronic properties of fullerol, a fullerene derivative extensively hydroxylated with polyhydroxy groups [48]. Elessawy et al. prepared functionalized magnetic fullerene nanocomposites (FMFN) for ciprofloxacin removal, showing a high surface area and superparamagnetism, with a sorption capacity of 65 mg/L [49]. Mahdavian demonstrated the removal efficiency of heavy metals (nickel and cadmium) using modified nano-fullerene (C_{60}) with tetrahydrofuran, achieving 91% efficiency for ion removal, with a sorption capacity of 1261 mg/g for cadmium and 3704 mg/g for nickel [50].

Chitosan (CS) is a natural adsorptive polymer with an affinity for pollutants in wastewater due to its amino ($-NH_2$) and hydroxyl ($-OH$) groups. However, it suffers from low mechanical strength and poor stability, making regeneration inefficient. Structural and chemical modifications have been proposed to overcome these drawbacks. Cross-linking imparts strength by bridging between polymer chains and functional groups, though it reduces uptake. Grafting involves the covalent bonding of functional groups on the CS backbone, significantly increasing adsorption capacity. Combining CS with other adsorbents can enhance its adsorption capacity, mechanical strength, and thermal stability. Ion-imprinting techniques also provide high selectivity for target metal ions. The uptake of CS depends mainly on the protonation or non-protonation of its amino and phosphoric groups, which affect the pH value of the wastewater. Without modifications, CS-based adsorbents show low reusability due to strong metal ion bonds, low stability, incomplete desorption, and limited adsorption sites. Alternative regeneration methods and modifications are needed to improve the reusability of CS adsorbents [51].

8.7 Removal of Contaminants by Using Membrane Technology

Nanotechnology offers solutions for removing pollutants through conventional membrane processes, enhancing adsorption capacity, photocatalysis, and disinfection. Various nanomaterials, such as nanoparticles, nanowires, nanofibers, nanotubes, metal-organic frameworks, and two-dimensional nanosheets, have been used for detecting and removing contaminants from wastewater. Membrane technology is cost-effective, small in equipment size, low in energy consumption, and highly efficient, with processes including reverse osmosis, ultrafiltration, and nanofiltration. Integrating nanomaterials with existing membranes enhances wastewater treatment efficiency. Membranes with nanomaterials can be classified as surface-coated, surface-deposited, surface-grafted, surface-assembled, mixed matrix, and pore-filled membranes, although issues like nanomaterial leaching can occur. Leaching impacts membrane efficiency, but advancements in polymeric membrane technology can mitigate these problems. Choosing the appropriate functionalization method and material depends on the contaminants to be removed. New membranes and nanomaterials have extended membrane life through regeneration processes, increasing usage efficiency. Unmodified membranes can remove various pollutants by adjusting pore size or combining them with coagulation/flocculation [52]. Advances in fabrication technology have enhanced conventional membranes with hydrophobicity, tunable pore size, specific modifications, and faster water efflux. Nanomaterial functionalization has added large surface areas, photocatalytic properties, and fouling resistance, making membranes multifunctional and efficient for removing a wide range of pollutants, including dye molecules, metal ions, microbes, biomass, and toxic sulfur/nitrogen-containing organic molecules.

8.8 Concluding Remarks and Future Perspectives

The rapid growth of industrialization and population has led to significant water pollution, posing risks to ecosystems and human health. Traditional remediation methods are often costly and ineffective. Nanotechnology offers promising solutions, with functional magnetic nanoparticles (MNPs) showing great potential due to their high adsorption capacity, stability, and recyclability. Functionalized magnetic nanomaterials (FMNs) are highly effective in contaminant removal through methods like sorption and photocatalysis, while carbon-based nanomaterials, including carbon nanotubes and graphene, also play a crucial role in environmental remediation. Researchers have developed various functionalized nanocomposites and magnetic nanoparticles, such as Fe_3O_4, to efficiently remove heavy metals, dyes, and pesticides from wastewater. These materials demonstrate high adsorption capacities, rapid kinetics, and excellent reusability, making them effective for environmental management. Integrating nanomaterials with existing membranes further enhances wastewater treatment efficiency. Overall, functionalized nanomaterials hold significant potential for advanced contaminant removal techniques and sustainable environmental management.

Abbreviations

APTES	Aminopropyl Triethoxysilane
BNNS	Boron Nitride NanoSheets
CNTs	Carbon Nanotubes
FMFN	Fullerene Nanocomposites
f-MWCNTs	Functionalized MWCNTs
FMNs	Functional Magnetic Nanomaterials
GO	Graphene Oxide
MCPMs	Magnetic Crosslinked Copolymer Microspheres
MNPs	Magnetic Nanoparticles
MWCNTs	Multi-Walled CNTs
PEA	2-Phenylethylamine
SWCNTs	Single-Walled CNTs
ZIF	Zeolitic Imidazolate Frameworks

References

1. Hojjati-Najafabadi A, Mansoorianfar M, Liang T, Shahin K, Wen Y, Bahrami A, Karaman C, Zare N, Karimi-Maleh H, Vasseghian Y (2022) Magnetic-MXene-based nanocomposites for water and wastewater treatment: a review. J Water Process Eng 47:102696
2. Guleria A, Kumari G, Lima EC, Ashish DK, Thakur V, Singh K (2022) Removal of inorganic toxic contaminants from wastewater using sustainable biomass: a review. Sci Total Environ 823:153689
3. Karimi-Maleh H, Ayati A, Ghanbari S, Orooji Y, Tanhaei B, Karimi F, Alizadeh M, Rouhi J, Fu L, Sillanpää M (2021) Recent advances in removal techniques of Cr(VI) toxic ion from aqueous solution: a comprehensive review. J Mol Liq 329:115062
4. Dalahmeh S, Björnberg E, Elenström AK, Niwagaba CB, Komakech AJ (2020) Pharmaceutical pollution of water resources in Nakivubo wetlands and Lake Victoria, Kampala, Uganda. Sci Total Environ 710:1–8
5. Singh RP, Singh PK, Gupta R, Singh RL (2019) Treatment and recycling of wastewater from textile industry. Springer, Singapore, pp 225–266
6. Seyed Arabi SM, Lalehloo RS, Olyai MRTB, Ali GAM, Sadegh H (2019) Removal of Congo red azo dye from aqueous solution by ZnO nanoparticles loaded on multiwall carbon nanotubes. Phys E 106:150–155
7. Ponnuchamy M, Kapoor A, SenthilKumar P, Vo DVN, Balakrishnan A, Mariam Jacob M, Sivaraman P (2021) Sustainable adsorbents for the removal of pesticides from water: a review. Environ Chem Lett (Springer International Publishing)
8. Magalhães-Ghiotto GAV, Oliveira AMD, Natal JPS, Bergamasco R, Gomes RG (2021) Green nanoparticles in water treatment: a review of research trends, applications, environmental aspects and large-scale production. Environ Nanotechnol Monitor Manag 16:100526
9. Arun T, Prabakaran K, Udayabhaskar R, Mangalaraja RV, Akbari-Fakhrabadi A (2019) Carbon decorated octahedral shaped Fe3O4 and α-Fe2O3 magnetic hybrid nanomaterials for next generation supercapacitor applications. Appl Surf Sci 485:147–157
10. Ghasemi M, Ghasemi S, Moazeni P, Ghasemi NS, Alharbi NS, Gupta VK, Agarwal S, Burakova IV, Tkachev AG (2018) Zn (II) removal by amino-functionalized magnetic nanoparticles: kinetics, isotherm, and thermodynamic aspects of adsorption. J Ind Eng Chem 62:302–310
11. Darwish M, Mohammadai A (2018) Functionalized nanomaterial for environmental techniques, Chapter-10. In: Hussain CM, Mishra AK (eds) Nanotechnology in environmental science. Elsevier, Amsterdam, pp 315–349
12. Brillas E, Sirés I (2012) Electrochemical remediation technologies for waters contaminated by pharmaceutical residues. In: Environmental chemistry for a sustainable world, pp 297–346
13. García-Niño WR, Pedraza-Chaverrí J (2014) Protective effect of curcumin against heavy metals-induced liver damage. Food Chem Toxicol 69:182–201
14. Karnwal A, Malik T (2024) Nano-revolution in heavy metal removal: engineered nanomaterials for cleaner water. Front Environ Sci 12:1393694
15. Ukhurebor KE, Aigbe UO, Onyancha RB, Balogun VA, Anani OA, Adama KK et al (2023) An overview of magnetic nanomaterials. In: Engineering materials. Springer Science and Business Media Deutschland GmbH, pp 1–20
16. Xue S, Xiao Y, Wang G, Fan J, Wan K, He Q, Gao M, Miao Z (2021) Adsorption of heavy metals in water by modifying Fe_3O_4 nanoparticles with oxidized humic acid. Colloids Surf A 616:126333
17. Kumar N, Joshi NC (2022) Potential of PTH-Fe3O4 based nanomaterial for the removal of Pb(II), Cd(II), and Cr(VI) ions. J Inorg Organomet Polym Mater 32:1234–1245
18. Nnadozie E, Ajibade P (2020) Adsorption, kinetic and mechanistic studies of Pb(II) and Cr(VI) ions using APTES functionalized magnetic biochar. Microporous Mesoporous Mater 309:110573
19. Karami N, Mohammadpour A, Samaei MR, Amani AM, Dehghani M, Varma RS, Sahu JN (2024) Green synthesis of sustainable magnetic nanoparticles Fe3O4 and Fe3O4-chitosan

derived from Prosopis farcta biomass extract and their performance in the sorption of lead(II). Int J Biol Macromol 254(Part 1):127663
20. Abo Markeb A, Moral-Vico J, Sánchez A, Font X (2023) Optimization of lead (II) removal from water and wastewater using a novel magnetic nanocomposite of aminopropyl triethoxysilane coated with carboxymethyl cellulose cross-linked with chitosan nanoparticles. Arab J Chem 16(8):105022
21. Bao S, Yang W, Wang Y, Yu Y, Sun Y (2020) One-pot synthesis of magnetic graphene oxide composites as an efficient and recoverable adsorbent for Cd(II) and Pb(II) removal from aqueous solution. J Hazard Mater 381:120914
22. Giwa A, Hasan SW (2020) Novel polyethersulfone-functionalized graphene oxide (PES-fGO) mixed matrix membranes for wastewater treatment. Sep Purif Technol 241:116735
23. Bangari RS, Yadav VK, Singh JK, Sinha N (2020) Fe_3O_4-Functionalized boron nitride nanosheets as novel adsorbents for removal of arsenic (III) from contaminated water. ACS Omega 5(18):10301–10314
24. Arabkhani P, Javadian H, Asfaram A, Ateia M (2021) Decorating graphene oxide with zeolitic imidazolate framework (ZIF-8) and pseudo-boehmite offers ultra-high adsorption capacity of diclofenac in hospital effluents. Chemosphere 271:129610
25. Asgari E, Sheikhmohammadi A, Yeganeh J (2020) Application of the Fe_3O_4-chitosan nano-adsorbent for the adsorption of metronidazole from wastewater: optimization, kinetic, thermodynamic and equilibrium studies. Int J Biol Macromol 164:694–706
26. Sahoo SK, Padhiari S, Biswal SK, Panda BB, Hota G (2020) Fe_3O_4 nanoparticles functionalized GO/g-C_3N_4 nanocomposite: an efficient magnetic nanoadsorbent for adsorptive removal of organic pollutants. Mater Chem Phys
27. Al-Tohamy R, Ali SS, Li F, Okasha KM, Mahmoud YAG, Elsamahy T, Jiao H, Fu Y, Sun J (2022) A critical review on the treatment of dye-containing wastewater: ecotoxicological and health concerns of textile dyes and possible remediation approaches for environmental safety. Ecotoxicol Environ Saf 231:113160
28. Mbuyazi TB, Ajibade PA (2024) Magnetic iron oxides nanocomposites: Synthetic techniques and environmental applications for wastewater treatment. Discover Nano 19:1548
29. Yang H, Zhang J, Liu Y, Wang L, Bai L, Yang L, Wei D, Wang W, Niu Y, Chen H (2019) Rapid removal of anionic dye from water by poly(ionic liquid)-modified magnetic nanoparticles. J Mol Liq 284:383–392
30. Atacan K, Güy N, Semerci AB, Özacar M (2025) Development of immobilized peroxidase on amino-functionalized magnetic $MgFe_2O_4$ nanoparticles for antioxidant activity and decolorization. Biophys Chem 318:107366
31. Luo J, Lai H, Jin L, Yang Y, Zhao L, Wang J, Cheng D (2025) Synthesis of novel magnetic carboxyl-functionalized crosslinked copolymer microspheres for methylene blue removal from water. J Macromol Sci Part A 1–11
32. Singh S, Mukhopadhyay A, Wallepure O, Mandal M, Veldurthi NK, Pati S (2024) Synthesis of magnetically tuneable molybdic acid-functionalized Fe_3O_4 nanoparticles for efficient dye removal in aqueous media. Bull Mater Sci 47:261
33. Qazi RA, Ullah N, Bibi N, Khattak R, Jamila N, Begum B, Aman N, Rahayu F, Karami AM (2023) Amine-Functionalized MWCNTs for the removal of mordant black 11 dye. Water Air Soil Pollut 234:644
34. Said M, Rizki WT, Asri WR, Desnelli D, Rachmat A, Hariani PL (2022) SnO_2–Fe_3O_4 nanocomposites for the photodegradation of the Congo red dye. Heliyon 8(4):e09204
35. Roostaee M, Sheikhshoaie I, Karimi maleh H (2022) Low-temperature synthesis of hetero-structures of magnetically separable iron oxide@Au-rGO nanocomposite for efficient degradation of organic dye under visible light irradiation. Environ Res 205:112510
36. Rehman A, Daud A, Warsi MF, Shakir I, Agboola PO, Sarwar MI, Zulfiqar S (2020) Nanostructured maghemite and magnetite and their nanocomposites with graphene oxide for photocatalytic degradation of methylene blue. Mater Chem Phys 256:123752
37. Sabarwal A, Kumar K, Singh RP (2018) Hazardous effects of chemical pesticides on human health–Cancer and other associated disorders. Environ Toxicol Pharmacol 63:103–114

38. Zhou L, Pan S, Chen X, Zhao Y, Zou B, Jin M (2014) Kinetics and thermodynamics studies of pentachlorophenol adsorption on covalently functionalized Fe_3O_4@SiO_2-MWCNTs core–shell magnetic microspheres. Chem Eng J 257:10–19
39. Wanjeri VWO, Sheppard CJ, Prinsloo ARE, Ngila JC, Ndungu PG (2018) Isotherm and kinetic investigations on the adsorption of organophosphorus pesticides on graphene oxide based silica-coated magnetic nanoparticles functionalized with 2-phenylethylamine. J Environ Chem Eng 6:1333–1346
40. Boruah PK, Sharma B, Hussain N, Das MR (2017) Magnetically recoverable Fe_3O_4/graphene nanocomposite towards efficient removal of triazine pesticides from aqueous solution: investigation of the adsorption phenomenon and specific ion effect. Chemosphere 168:1058–1067
41. Tahir MB, Kiran H, Iqbal T (2019) The detoxification of heavy metals from aqueous environment using nano-photocatalysis approach: a review. Environ Sci Pollut Res 26(11):10515–10528
42. Grassi M, Kaykioglu G, Belgiorno V, Lofrano G (2012) Removal of emerging contaminants from water and wastewater by adsorption process. In: Emerging compounds removal from wastewater. Springer, Dordrecht, pp 15–37
43. Papoulis D (2019) Halloysite based nanocomposites and photocatalysis: a review. Appl Clay Sci 168:164–174
44. Ali I, Mbianda XY, Burakov A, Galunin E, Burakova I, Mkrtchyan E, Grachev V (2019) Graphene-based adsorbents for remediation of noxious pollutants from wastewater. Environ Int 127:160–180
45. Gusain R, Kumar N, Ray SS (2020) Recent advances in carbon nanomaterial-based adsorbents for water purification. Coord Chem Rev 405:213111
46. Zamani S, Salem S (2021) Couple of graphene oxide and functionalized carbon nanotubes for dye degradation enhancement of anatase under visible light and solar irradiation. Environ Sci Pollut Res 28:32763–32776
47. Firdaus RM, Rosli NIM, Ghanbaja J, Vigolo B, Mohamed AR (2019) Enhanced adsorption of methylene blue on chemically modified graphene nanoplatelets thanks to favorable interactions. J Nanopart Res 21(12):257
48. Wu ZY, Xu YJ, Huang LJ, Zhang QX, Tang DL (2021) Fullerene-cored star-shaped polyporphyrin-incorporated TiO_2 as photocatalysts for the enhanced degradation of rhodamine B. J Environ Chem Eng 20:106142
49. Elessawy NA, Elnouby M, Gouda MH, Hamad HA, Taha NA, Gouda M, Eldin MSM (2020) Ciprofloxacin removal using magnetic fullerene nanocomposite obtained from sustainable PET bottle wastes: adsorption process optimization, kinetics, isotherm, regeneration and recycling studies. Chemosphere 239:124728
50. Mahdavian L (2020) Efficiency evaluating of modified fullerene nanofiltration on removal of heavy metals from aqueous media. J Res Many-Body Syst 10(2):85–95
51. Zang X, Jiang R, Zhu H-Y, Wang Q, Fu Y-Q, Zhao D-X, Li J-B, Liu H (2024) A review on the progress of magnetic chitosan-based materials in water purification and solid-phase extraction of contaminants. Sep Purif Technol 330(Part C):125521
52. Ying Y, Ying W, Li Q, Meng D, Ren G, Yan R, Peng X (2017) Recent advances of nanomaterial-based membrane for water purification. Appl Mater Today 7:144–158

Chapter 9
Functionalized Magnetic Nanomaterials (FMNs) in Renewable Energy Applications

Parul Pant, Prerita Bahl, and Himanshi Agrawal

Abstract Functionalized magnetic nanomaterials (FMNs) have acquired a lot of attention in the world of renewable energy due to their variety of applications and flexibility. These materials show great capability. This book chapter focuses on the synthesis and functionalization of magnetic nanomaterials, their roles in catalysis, supercapacitors, thermoelectric and piezoelectric devices, batteries, biofuels, and electrochemical sensors, along with some future potential applications. The magnetic properties of FMNs assist with innovative approaches that contribute to advanced energy storage materials. FMNs not only enhance the performance of materials but also promote the development of a sustainable and eco-friendly environment, which is also cost-effective, thus providing a better alternative to the costly and toxic materials. This chapter focuses on various applications of FMNs and their future prospects in different fields.

Keywords Functionalized magnetic nanomaterials · Synthesis · Applications · Sustainable · Eco-friendly

9.1 Introduction

The twenty-first century's technical advances have been led by nanotechnology and engineered nanomaterials. The idea of handling matter at the atomic level was first proposed by Richard Feynman, the founder of nanotechnology and the recipient of the 1965 Nobel Prize in Physics [1]. In order to attain high efficiency with minimal or green energy input, nanotechnology provides alternate and sustainable answers to the problems of environmental pollution, which are becoming more complicated [2]. However, the full-scale application of nanotechnology is impeded by its unidentified and unexpected harmful impacts and health hazards [3].

P. Pant (✉) · P. Bahl · H. Agrawal
Department of Chemistry, Hansraj College, University of Delhi, Delhi, India
e-mail: parulpant@hrc.du.ac.in

S. Gulati (ed.), *Functionalized Magnetic Nanomaterials*, Nanostructure Science and Technology, https://doi.org/10.1007/978-3-031-97199-0_9

Nanomaterials are small, distinctly shaped materials with diameters ranging from 1 to 100 nm. Nanoparticles, which are a subset of nanomaterials, are nano-objects with external dimensions in the nanoscale (less than 100 nm). Both natural sources and artificially made materials that are made to have specific properties are sources of these nanoparticles [4]. Furthermore, nanoparticles exhibit better levels of catalytic activity, durability, and crystallinity, as well as a high adsorption limit, stability, and durability [5].

Nanomaterials are modified to increase compatibility, dispersion stability, and surface site reactivity, as well as to avoid agglomeration [6]. The magnetic characteristics and potential applications of magnetic nanomaterials(MNMs) have been extensively studied. The numerous benefits that magnetic nanomaterials (MNMs) provide have caused a paradigm shift away from traditional nanoparticles. Due to their non-hazardous nature, MNMs have been widely utilized in the field of nanotechnology since their first report in 2000. Magnetic size and characteristics, like the length of MNMs, have resulted in interesting electrical and magnetic properties [7].

With unique magnetic properties like high magnetic susceptibility, high coercivity, superparamagnetic properties, and low Curie temperature, magnetic nanomaterials (MNMs) attract a lot of attention for a variety of uses (Fig. 9.1). Cobalt, nickel, iron, gadolinium, and the corresponding oxides and derivatives are among the magnetic metals that have been studied so far for a variety of uses [8].

In recent years, magnetic nanomaterials (MNMs) have gained popularity due to their wide range of applications (Fig. 9.2). However, just magnetic materials are not ideal for real-time applications due to certain limitations. To increase their practical use and efficiency, their physicochemical characteristics, such as surface area,

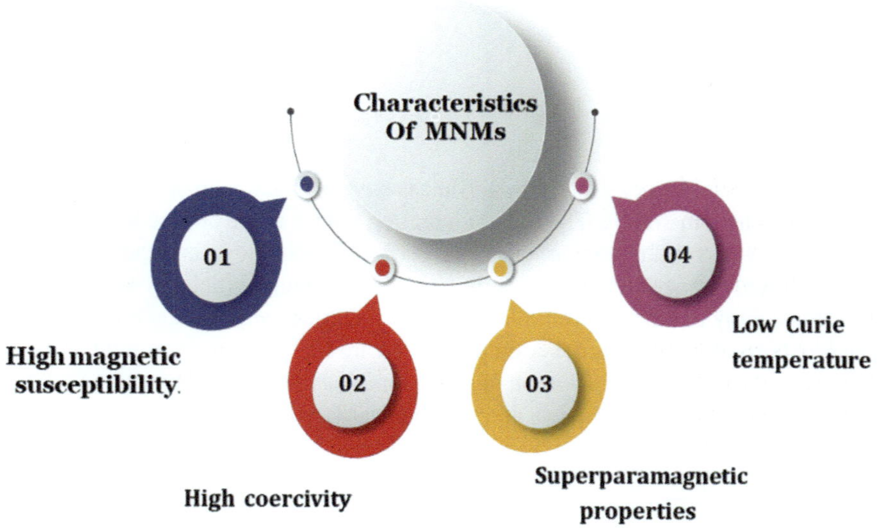

Fig. 9.1 Characteristics of MNMs

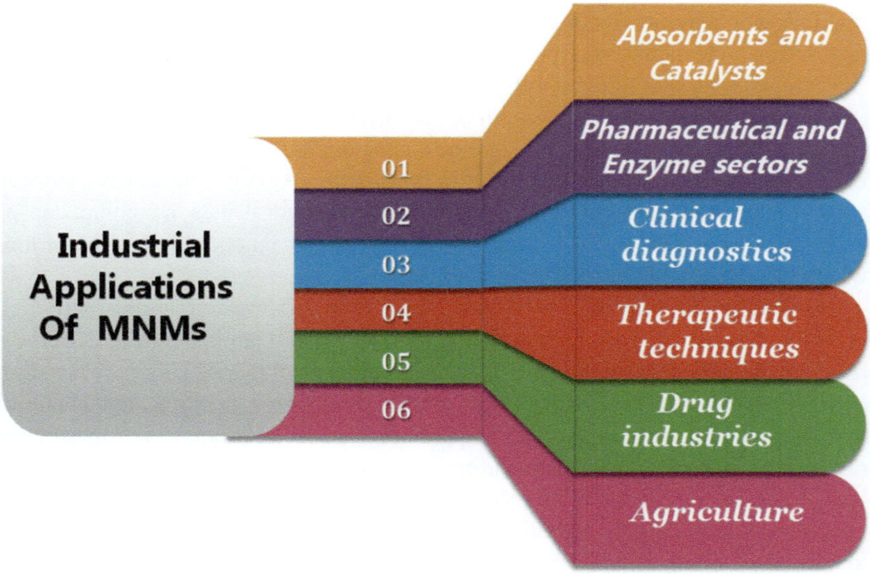

Fig. 9.2 Industrial applications of MNMs

charge transport characteristics, electronic and optical activity, and reactive sites on the surface, must be modified. One method for modifying the physicochemical characteristics of MNMs is functionalization. Polymer functionalization, surface modification, and bio-functionalization are some of the methods that have been reported for functionalizing magnetic materials. These methods have been found to be helpful in a variety of biomedical, energy, and environmental applications, including magnetic bioseparation and the detection of biological entities (protein, nucleic acid, cell, bacterial, enzyme, virus, etc.) [8].

MNMs are known for their potential usefulness in the catalysis industry. Because of the affordability, high removal efficiency, ease of use, quick kinetics, and design flexibility of MNMs, they have demonstrated a lot of potential in removing harmful contaminants from water bodies [9]. As an important class of nanoscale materials, MNMs are currently being used as next-generation drug carriers and in therapeutic approaches and clinical diagnostics. Nanocatalysts based on MNMs are resilient, affordable, easy to develop, magnetically recoverable, non-toxic, and recyclable.

MNMs are effective alternatives for the solid-supported heterogeneous catalysts that are currently in use. Because of their distinct physicochemical characteristics, MNMs have emerged as superior and more powerful than traditional nanomaterials [9]. The modified nanocomposites made by functionalizing active species onto magnetic nanomaterials (MNMs) show outstanding advantages in terms of catalyst recovery techniques as well as notable catalytic efficiency in completing organically significant conversions that were previously challenging to achieve.

MNM applications are not only in the catalytic domain but also in the pharmaceutical and enzyme industries. Iron, cobalt, nickel, and their oxides, such as cobalt ferrite ($FeCoO_4$), chromium oxide (CrO_4), maghemite (γ-Fe_2O_3), hematite, and magnetite (Fe_3O_4), are used for the preparation of certain NMs [10].

Because iron has strong paramagnetic characteristics and is reasonably abundant in the Earth's crust, MNMs that use iron in its many oxidation states have attracted a lot of interest [11]. Occupying octahedral and tetrahedral sites in a cubic, densely packed lattice and having an inverse spinel structure, magnetite (Fe_3O_4) is the most widely used magnetic nanomaterial [12].

Franklinite and other synthetic zinc MNMs have shown remarkable activity in breaking down diclofenac in water. In surface chemistry applications, MNMs have been created via physical, chemical, and biological techniques [11]. In the sector of agriculture, there has been a lot of interest in the creation of functional nanomaterials that are biocompatible, cost-effective, and biodegradable. Examples of these materials include nanofertilizers, nanoadditives, nanopesticides, herbicides, and nanosensors. These nanomaterials have shown potential in boosting crop yields, facilitating the effective application of fertilizers and agrochemicals, improving plant resilience to a range of stressors, and serving as nanosensors for the identification of different pollutants, plant illnesses, and inadequate plant nutrition [10].

In this chapter, we will be dealing with the general methodology of preparation of MNMs and their myriad applications. A brief discussion on the possible future prospects of MNMs is also given at the end of the chapter.

9.2 General Methodology

Magnetic nanomaterials possessing metallic and semiconducting properties are highly valuable across various domains, including energy conversion, energy storage, environmental protection, biomedical advancements, and agricultural applications. This is because of their charge carrier mobility, non-toxic nature, large surface area, capacity to recover and recycle, and optical band structure [8].

However, MNMs are not suitable for real-time applications due to their low stability, agglomeration, inadequate biocompatibility, rapid electron-hole recombination, and disintegration in acidic environments. A variety of techniques are used to overcome these issues, including surface functionalization, elemental doping, and nanocomposites. Functionalization is one of the most useful ways to change the size, morphology, electron mobility, and crystalline phase of MNMs. Functionalized MNMs have been widely studied for diverse applications, especially in biomedical and environmental fields, such as MRI, solar hydrogen production, drug delivery, purification of water and air, catalysis, and sensors [8].

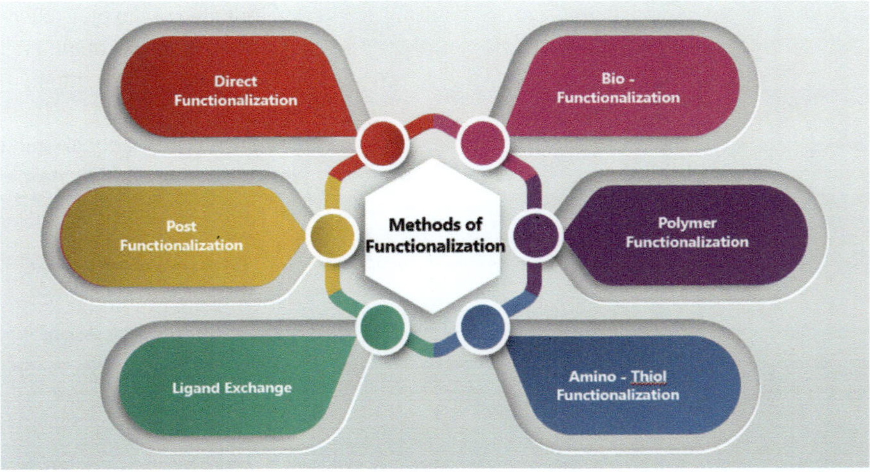

Fig. 9.3 Methods of functionalization of magnetic nanoparticles

Surface functionalization of magnetic nanoparticles should exhibit the following characteristics to be effectively utilized in biological applications:

- Should make MNMs more water compatible,
- Should improve their magnetic controllability,
- Should make the MNM surface more stable,
- MNMs should be non-hazardous and biocompatible for biological and medicinal applications.
- Should give accessible functional groups for biologically active materials to attach to for nano-bio applications [13].

The following are some methods for the functionalization of magnetic nanoparticles (Fig. 9.3).

(1) **Directfunctionalization**: The surface of the nanomaterial is altered by ligand molecules through this process. The ligand is supposed to have two functions: the complexing group that is used to bind the ligand to the nanomaterial's surface, and the active functional group. Benefits of this approach include the ability to employ a broad range of functional groups, better control over the amount of ligand integrated into the nanomaterial, and consistent surface coverage in one-pot synthesis.

(2) **Post-Functionalization**: It is a functionalization procedure, also commonly known as grafting, that often follows the synthesis of nanomaterials. In this technique, process ligands are attached to the surface of the MNMs. Bifunctional molecules are involved in this process, where the coupling site group may react with a bifunctional (binding-chelating) group to become the final functional group in the second phase.
The post-functionalization (Fig. 9.4) structure can be described as:

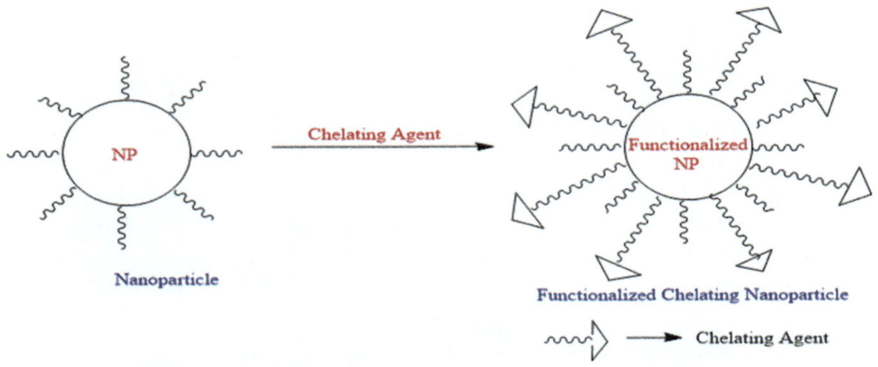

Fig. 9.4 Basic post-functionalization procedure

(3) **Polymer Functionalization**: A popular method that combines the qualities of polymers and nanomaterials. Because polymers are linked to the MNMs' surface, they increase interfacial interaction and enhance the MNMs' stability and dispersion as compared to bare MNMs. Polymers are favored over other small molecules due to their ability to stabilize nanoparticles and modify the chemical makeup and characteristics of a shell (homopolymer, hydrophilic/hydrophobic, or copolymer). Moreover, for multifunctional polymer-magnetic nanohybrids, a well-designed polymeric shell gives extra features like biocompatibility, biological activity, resistance to physiological conditions (such as enzymes or pH), or complexing capabilities.

(4) **Ligand Functionalization**: Ligand exchange is a post-functionalization process that includes the synthesis of nanoparticles, usually involving the introduction of ligands that control the composition, size, morphology, and dispersibility of the nanocrystals.

(5) **Biofunctionalization**: The three-dimensional structure and characteristic functions of proteins and peptides are lost when MNMs are prepared for biological applications. The biomolecules have to undergo step-by-step modifications to maintain their activity on MNMs. High surface area and magnetic field separability make MNMs functionalized with enzymes ideal for biological applications such as stability enhancement, enzyme recovery, specificity, selectivity, and inhibition reduction or purification.

(6) **Amino ThiolFunctionalization**: MNMs show increased metal selectivity when their surface is engineered with functional groups that contain donor atoms of nitrogen (-NH_2) and sulphur (-SH). It is based on the theory of hard and soft acids and bases (HSAB). The amino-thiol functionalization technique has several uses, including sensing and catalysis, but its main application is in environmental remediation [8].

9.3 Role of FNMs in Energy Storage and Conversion Technologies

Energy affects billions of people worldwide and is essential to both survival and economic advancement [14]. Researchers are focusing on creating safe energy sources for people because of the rising need for renewable resources [15]. To decrease the dependence on soon-to-be-extinct fossil fuels is one benefit of developing a biological and chemical method of producing efficient and environmentally friendly renewable energy. Because renewable energy sources emit fewer greenhouse gases than fossil fuels, they are becoming more and more important for a long-term, sustainable solution to climate change [16].

Nanotechnology offers a potentially helpful way to address the issues that lie in the use of non-renewable resources [5]. Iron-based magnetic nanoparticles, or Fe-MNMs, have garnered a lot of attention lately because of their significant applications in a variety of fields, including industrial, medicinal, agricultural, environmental, and energy. Since biodiesel is a clean, biodegradable, and environmentally friendly alternative to energy sources derived from fossil fuels, FNM-based systems have also demonstrated their potential in the fuel business [11].

Because of their internal surface area, high surface-to-volume ratio, high energy density, and two-dimensional (2D) material mobility, FNMs have shown their potential applications in supercapacitors and batteries [17].

Figure 9.5 shows us all the applications of FNMs that we have talked about in the rest of the chapter.

9.3.1 FMNanomaterials for Batteries

There have been numerous significant developments in the field of electronics and energy storage, which have contributed substantially to battery development[17, 18]. However, batteries are in greater demand due to evolving requirements and developments in modern society. Some ideal characteristics for modern batteries are longer life cycles and improved efficiency, i.e., high power and energy density. Recently, nanostructured materials have been suggested for the use of energy storage devices, specifically those with high charge and discharge current, such as lithium-ion batteries, which are often found in laptops and mobile phones [19].

Various researchers have been working on different applications of functionalized magnetic nanomaterials in batteries. Porous α-Fe_2O_3 nanorods varying in length-to-diameter ratios were prepared by Wu et al. [20]. They evaluated how well they worked as anode materials for lithium-ion batteries. They concluded that both the porosity and the length-to-diameter ratio of the α-Fe_2O_3 nanorods affected the batteries' electrochemical performance. They were able to shape-control the synthesis of hematite α-Fe_2O_3 nanostructures with gradients in diameters (from less than 20 nm to more than 300 nm) and surface areas (from 5.9 to 52.3 m^2/g) by using a high concentration

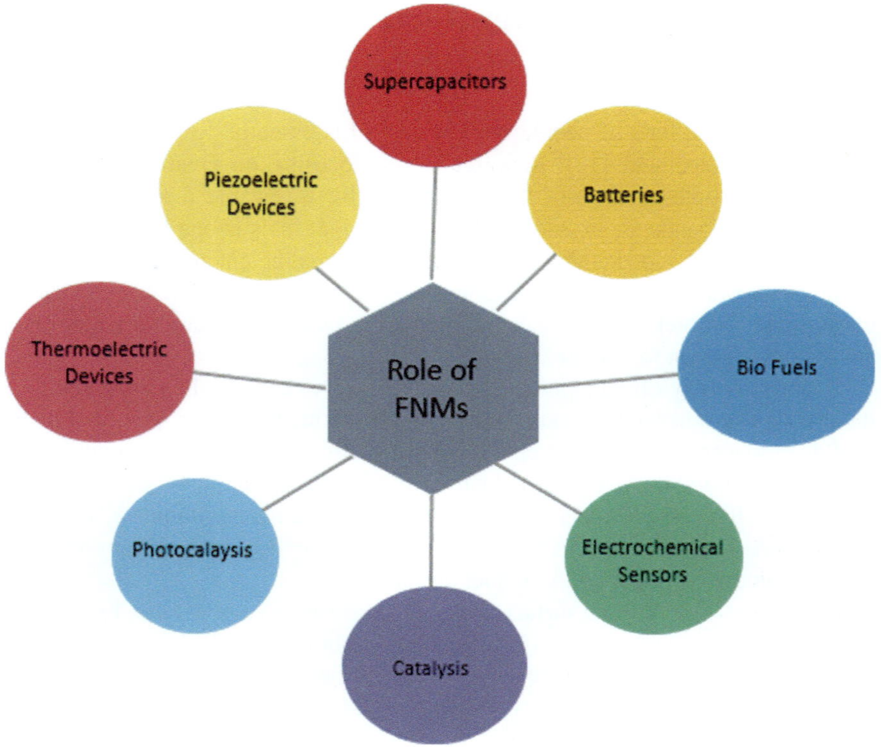

Fig. 9.5 Present and potential uses of FMNs in energy storage and conversion technologies

of inorganic salts and high temperatures during the synthesis process. The prepared FeOOH nanoparticles were heated to 520 °C with a ramping rate of 10 °C min^{-1} to create hematite nanostructures, which were then kept there for eight hours. After being decomposed in air, the obtained α-Fe_2O_3 products were washed with distilled water and vacuum-dried at 40 °C. A superconducting quantum interference device and a vibrating sample magnetometer were used to examine the magnetic characteristics of α-Fe_2O_3. A Teflon cell containing a lithium metal anode was used to test the α-Fe_2O_3's cathode performance. A mixture of â-FeOOH, acetylene black, and poly (vinylidene fluoride) with a weight ratio of 85/10/5 was used as the cathode. A glovebox containing highly pure argon gas (O_2 and H_2O values <5 ppm) was used to assemble the cell. At a current density of 0.2 mA cm^{-2}, a galvanostatic charge/discharge experiment was conducted between 3.0 and 0.5 V.

9.3.2 FMNanomaterials for Supercapacitors

Supercapacitors are multifunctional energy storage devices that bridge the gap between secondary batteries and traditional capacitors. They provide quick charge and discharge rates, a long lifespan, and a high energy density. Supercapacitors, in contrast to traditional batteries, store energy by ion adsorption or redox processes. The majority charge concentration is stored close to the electrode materials' surface. These capacitors are perfect for a variety of applications that require a high power density of a minimum 10 kW/kg because they have a magnitude of greater power density than lithium-ion batteries [18]. Supercapacitors are given a special place in electrochemical technology due to their ability to deliver stored energy faster than batteries and produce specific capacitance that is 6–9 orders of magnitude greater than that of conventional capacitors.

Nanomaterials increase supercapacitors' energy and power densities [21]. This development has created new opportunities for their use in fields like electric cars, grid stabilization, and portable electronics, where rapid energy bursts are crucial. A substantial contribution in overcoming the drawbacks of energy storage technologies can be made by focusing on creating and enhancing supercapacitors through nanomaterial modifications. This is essential for a sustainable and secure energy future in which renewable energy sources are smoothly woven into our energy systems. Various supercapacitors have been created by fusing nanomaterials and substrates, which is revolutionizing the energy storage sector. These components are a versatile option for energy harvesting applications because of their versatility and lightweight [16].

Electrochemical double-layer capacitors (EDLCs), hybrid supercapacitors, and pseudocapacitors are the three primary categories of supercapacitors. In EDLCs, energy is stored in electrodes made of porous carbon-based materials. At the electrode/electrolyte interface (EEI), these electrodes are attracted to one another electrostatically. Reversible Faradaic processes that take place at the EEI are used to store energy in pseudocapacitors, which utilize redox-active materials as electrodes, such as conducting polymers or metal oxides. Hybrid supercapacitors have combined features of both EDLCs and pseudocapacitors for achieving greater energy and power densities [22].

The following metal oxides are frequently used as electrode materials in supercapacitor research: cobalt oxide (Co_3O_4), manganese dioxide (MnO_2), nickel oxide (NiO), strontium oxide (SrO), titanium oxide (TiO), and copper cobalt oxide ($CuCo_2O_4$). Their large surface area and intrinsic porosity qualities make them suitable for the required purpose. Because of greater interaction between the electrolyte and electrodes at their interfaces, the mesoporous structure of these materials facilitates fast ion transport. The enhanced ion movement thus increases the capacitance. These oxides and their composite forms are commonly used in applications involving supercapacitors.

Nanostructured Co_3O_4 obtained through the sol-gel approach was reported by Devi et al. [23]. First, a three-electrode assembly was made for supercapattery applications, and the resulting sample was tested in 1 M KOH. Co_3O_4 was examined as a positive electrode and rGO as a negative electrode for supercapacitor applications. High stability has been verified by the constructed device's 100% Coulombic efficiency [16].

Similarly, Tarimo et al. used a hydrothermal technique to produce sulphur-rGO@cobalt oxide composites (rGO-S/Co_3O_4) with varying mass percentages of Co_3O_4. Three electrode configurations were used for analyzing the samples at a scan rate of 50 mV/s in Pw ranges of 0–0.5 V. After 10,000 cycles, the device retains 71.6% of its high capacitance and 99.5% of its Coulombic efficiency.

Chen et al. also used the hydrothermal approach and an annealing process to report the Co_3O_4 micro-bundle structure (MBs). These Co_3O_4 MBs have a huge surface area of 81.2 m^2/g, flakes that are 0.4 nm thick, and pores that are 6.14 nm in diameter. The electrode maintained 74.6% capacity during 4000 galvanostatic charge–discharge (GCD) cycles. Commercial AC was used to create the electrode material in a two-electrode assembly for the analysis of the hybrid device.

Iqbal et al. developed the ternary composite of Co_3O_4 NPs using multiwalled carbon nanotubes (MWCNTs) by adding AuNPs through a hydrothermal method. These NPs have been studied as electrode materials for practical supercapacitor device applications. After 3500 GCD cycles, the Co_3O_4/Au and MWCNTs//AC-based supercapattery demonstrated 91.90% capacitance retention with a coulombic efficiency of 85%.

Similarly, the hydrothermal process of aligning Co_3O_4 nanowires was used to examine a cage-like Co_3O_4. A cage-like structure with pores varying from 1 to 2 μm was created when the produced electrode was evenly distributed across the Ni-foam surface [24].

Electrical conductivity, charge carrier mobility, structural adaptability, and band gap optimization are the primary characteristics of the electrochemical energy storage mechanism of supercapacitors. It was found that by increasing the number of available free charge carriers, the electrochemical performance, as well as energy and power density, was enhanced in engineered nanomaterial supercapacitors [6].

Chandra & group [25] worked on the combination of dendrimer and iron oxide nanoparticles, which resulted in synergistic electrochemical behavior. Any electrode surface can be easily coated with the dendrimer-functionalized nanoparticles, which have excellent conductivity and stability. The manufactured system could be integrated into a supercapacitor because of its high surface area and mesoporous nature, which facilitates different kinds of electrolytes (aqueous, organic, and ionic) to reach the nanopores. This increased the energy density and facilitates quick and effective charge transfer. The resultant nanomaterial's low internal resistance increases power density and provides high energy capacitance. Fe_3O_4@D-NH_2 nanoparticles give enhanced current charging-discharging with increased retention time for up to thousands of cycles.

Since a lesser quantity of the material may demonstrate good performance and produce outcomes comparable to competitors, the nanomaterial and electrode fabrication approach also lowers the device's cost. This indicates that the created technology offers comparable performance to other gadgets at a lower cost.

The production of high-performance supercapacitors is ultimately intended to meet the increasing demand in a variety of applications. Supercapacitors are expected to become an essential component for energy storage in a wide range of applications, from renewable energy systems to electric automobiles, with continued research and development [16].

9.3.3 FMNanomaterials for Thermoelectric Devices

Research is driven by current environmental concerns to explore new ideas in renewable and sustainable energy production. The intrinsic capacity of materials to transform heat into electricity and vice versa is known as thermoelectricity. In thermal electric (TE) composites, integrating magnetic nanoparticles (NPs) into the carriers' transport pathway may result in an associated thermo-electro-magnetic effect.

Sani et al. in their research used FNMs in which a functional fluid composed of a stable colloidal solution of maghemite magnetic nanoparticles was suspended in water. This solution was examined from the perspectives of thermoelectrical and optical properties in order to assess its potential for producing electricity directly through the thermoelectric effect made possible by solar absorption. With a very small volume percentage, these nanoparticles were discovered to be very effective absorbers of solar radiation as well as a thermoelectric power output enhancer when the solution was heated from the top [26].

On nanoscale and mesoscopic scales, magnetic nanoparticles (NPs) implanted in a matrix generate micromagnetic fields (H_A), which can serve as carriers and phonon scattering centers. The majority of carriers (such as electrons) move from the hot side to the cold side with the higher (T_H) and lower temperature (L_T), respectively, due to thermal energy. If the drift velocities are not parallel to H_A, some of the electrons will be trapped or confined to helices or circular routes. Since magnetic nanoparticles are oriented randomly, TE composites contain at least three different types of electrons with varying velocities (v_1, v_2, and v_3). Since the velocities of electrons with v_1 are parallel to the magnetic field, they move linearly and are not influenced by H_A. However, due to the nonzero angle between H_A and the electron velocity, the electrons with velocities v_2 and v_3 are impacted. This leads to the restriction of electrons to circular pathways with v_2 perpendicular to H_A. Magnetic nanoparticles (NPs) trap electrons, which travel in a spiral manner when v_3 contains components that are parallel and perpendicular to H_A. Thus, in TE nanocomposites, the carrier concentration n will vary, due to which theα and σ values are affected. Furthermore, κ is decreased by phonon scattering caused by high magnetic NP. To some extent, the addition of magnetic nanoparticles to the TE matrix could serve as a catalyst for the decoupling of electron and phonon transports at the same spatial scale (the nanoscale and the mesoscopic scale). It most likely provides some new TE effects that are brought about by magnetic fields. As a result of this, magnetically enhanced thermoelectrics should attract quite a lot of interest in the TE community [27].

Sani et al. [26] explored the optical absorption and thermoelectric characteristics of diluted aqueous ferrofluids composed of Co-Polymer made of equimolar Acrylic and Maleic Acid Monomers (PAAMA) coated Fe_2O_3 magnetic nanoparticles. Thus, adding diluted ferrofluids was found to be beneficial for enhancing the thermocell's performance when the device's functioning requires the liquid to be heated from the top. Favorable optical absorption bands in the UV-visible wavelength range were shown by the studied ferrofluids, making them ideal for sunlight absorption. The effect of complete controllability of the optical absorption level and spatial distribution throughout the nanofluid volume can lead to the optimization of future sunlight-enabled thermocell geometries.

Functionalized Magnetic Thermoelectric nanoparticles excel in situations, particularly those where efficiency is of more importance than reliability and energy supply. They are exceptional and have widely contributed to a number of conceptual, structural, and performance advancements in the field of materials science.

9.3.4 FMNanomaterials for Piezoelectric Devices

The ability of piezoelectric materials to directly transform mechanical vibrations into electrical energy through the direct piezoelectric effect makes them particularly appealing for energy harvesting applications.

In non-centrosymmetric crystals, the piezoelectric effect can be defined as a linear electromechanical interaction between the mechanical and electrical states, which leads to an accumulation of electrical charge in response to applied mechanical stress. The direct piezoelectric effect can also be reversed through the application of electrical charge, in order to produce mechanical strain, which is known as the reverse piezoelectric effect [28].

Piezoelectric materials—a subset of ferroelectric materials—display localized charge separation known as electrical dipoles because of their non-centrosymmetric structure. These materials include both natural and manufactured materials like SiO_2, $NaKC_4H_4O_6 \cdot 4H_2O$, lead zirconium titanate, zinc oxide (ZnO), barium titanate ($BaTiO_3$), bismuth ferrite ($BiFeO_3$), Lead Zirconium Titanate {PZT ($Pb[Zr_xTi_{1-x}]O_3$} $0 < x < 1$), collagen, and so on.

The direct piezoelectric effect in synthetic bulk piezoelectric polymers is facilitated by the orientation and structure of the molecules (Fig. 9.6), which result from the naturally present molecular dipoles that are oriented through a poling process. The randomly oriented domains of a ferroelectric material cancel each other out, resulting in a weak or zero response. The application of an electric field at high temperatures aligns the randomly oriented dipoles along a particular trajectory in order to guarantee a strong piezoelectric response. It is known that composites' increased magnetic and piezoelectric qualities lead to high-performing energy harvesting devices [28].

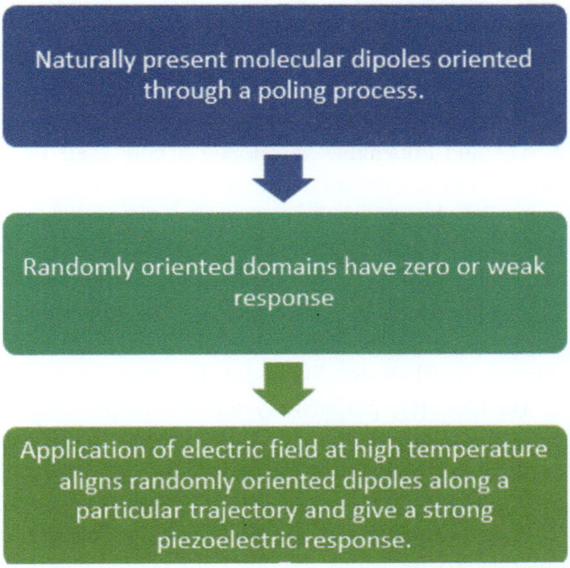

Fig. 9.6 Piezoelectric effect in synthetic bulk piezoelectric polymers

9.3.5 FMNanomaterials for Photocatalysis

One of the most difficult issues in ecological and environmental protection is the photocatalytic destruction of hazardous organic contaminants [29]. Photocatalysis provides possible solutions to this major issue. Photocatalysis uses the same concept as natural photosynthesis—that is, the use of light energy to produce chemical energy. However, instead of chlorophyll in photosynthesis, a semiconducting material known as a "photocatalyst" is used in photocatalysis. Photocatalyst compounds change the pace of a chemical process when exposed to light [5].

The interaction of light energy with nanomaterials is the basis for photocatalytic reactions of nanocatalysts[30]. Due to their ease of separation and prospective uses in the treatment of water and inorganic pollutants, magnetic photocatalysts have gained significant attention [31].

The photocatalyst material undergoes a photoexcitation state to generate a large number of electron-hole pairs (e^-/h^+) when light irradiations reach its surface. As catalytic materials, magnetic nanoparticles have been employed. Magnetic nanoparticles have the benefit of being easily retrieved after usage and used in a different reaction. In order to create a composite material that the magnet can recover, magnetic nanoparticles are often partially coated on the catalyst's surface. Following successful applications prove the enhancement of photocatalytic activity using different FNMs.

Fe_2O_3 was first used with TiO_2 for photocatalytic processes in 2011 by Liu et al. Using $Fe_2(SO_4)_3$ and $Ti(SO_4)_2$ as primary materials, they hydrothermally produced α-Fe_2O_3–TiO_2 composites. The findings demonstrated that up to 90% of auramine could be photodegraded by the α-Fe_2O_3–TiO_2 composites [32]. Additionally, the

$Fe_3O_4@TiO_2$ demonstrated strong photocatalytic activity in the photocatalytic breakdown of rhodamine B solution, indicating that it can also be a very effective recoverable photocatalyst [33]. The formation mechanism of $Fe_3O_4@TiO_2$ was proposed by Zhang and colleagues with respect to the core-shell structure [34].

Yin and colleagues developed a core-shell $Fe_3O_4/SiO_2/TiO_2$ nano-composite that exhibited exceptional photocatalytic activity [35]. The photo-degradation of acetaminophen in an aqueous solution was clearly affected by TiO_2/Fe_3O_4 and $TiO_2/SiO_2/Fe_3O_4$, and the stability of $TiO_2/SiO_2/Fe_3O_4$ is superior to that of nanoparticles of TiO_2/Fe_3O_4 due to the SiO_2 shell's capacity to inhibit the resolution of Fe_3O_4. A tiny magnet may readily separate and recover these catalysts due to their magnetism [34]. When exposed to sunlight, the produced magnetic nanoparticles showed excellent stability and activity for dye degradation [31].

The $CoFe_2O_4$-graphene system was prepared by Wang and colleagues to catalyze the degradation of Methylene Blue (MB) in the presence of visible light [36]. To catalyze the oxidation of carbamazepine, Zhou et al. produced $CoFe_2O_4$ as an effective magnetic photocatalyst. The strong photocatalytic activity was further enhanced by the easy transport of electrons from the $CoFe_2O_4$ conduction band to the reduced graphene oxide, which inhibited the recombination of electrons and holes.

Das et al. synthesized Pd and Au supported on magnetic chitosan-coated Fe_3O_4 (Fe_3O_4@Ch-MNMs) under UV light [37]. Fe_3O_4@Ch-AuNPs and Fe_3O_4@Ch-PdNMs have been tested for their catalytic ability to reduce MB to Leucomethylene Blue (LMB). The average diameters of Fe_3O_4@Ch-AuNPs and Fe_3O_4@Ch-PdNPs were shown in the low-resolution Field Emission Scanning Electron Microscopy (FESEM) pictures (4.21 ± 0.51 and 4.08 ± 0.4 μm, respectively). The homogeneous dispersion of Au and Pd nanoparticles on the surface of Fe_3O_4@Ch-MNMs was demonstrated by the high-resolution FESEM pictures. Under different reaction circumstances, Fe_3O_4@Ch-AuNPs showed more photocatalytic activity than Fe_3O_4@Ch-AuNPs [31].

Chen and colleagues created the magnetically separable hollow spherical Fe_3O_4/TiO_2 hybrid photocatalyst via a template approach [38]. The coprecipitation approach used to create the Fe_3O_4 monolayer as a template made of monodispersed poly(styrene-acrylic acid) (PSA) latex. The Fe_3O_4 monolayer's surface was then covered with the TiO_2 shell. Ultimately, a solvothermal treatment eliminated PSA. With a hollow sphere structure, the Fe_3O_4/TiO_2 catalyst exhibited remarkable activity in the photodegradation of RhB [31].

By reducing size, expanding surface area, integrating the photocatalyst with certain unique magnetic nanoparticles, and doping to boost hole concentration, many efforts have been made to improve photocatalytic performance. The interaction of light energy with magnetic nanomaterials is the basis for photocatalytic reactions of nanocatalysts [30]. Due to their ease of separation and prospective uses in the treatment of water and inorganic pollutants, magnetic photocatalysts have gained significant attention [31].

9.3.6 FMNanomaterials for Catalysis

As functionalized catalysts, magnetic nanoparticles have recently drawn a lot of attention for their synthesis and significance in nanocatalysis.

The straightforward recovery procedure increases the usefulness and durability of these non-toxic and versatile magnetic nanocatalysts, which is one of their benefits. For example, $MnFe_2O_4$@Silica-SO_3H's reusability was assessed by Nikmanesh et al. [39] while oxazolidine-2-one was being synthesized. A proper magnetic field makes it easy to extract the insoluble magnetic catalyst when the reaction is finished. Effective catalyst recovery and reuse are made feasible by this efficient magnetic separation method, which lowers expenses and waste. To get it prepared for more reactions, the isolated nanocatalyst was repeatedly cleaned with diethyl ether and dried. It was found that even after five reuse cycles, the catalytic activity of $MnFe_2O_4$@Silica-SO_3H remained relatively unchanged (92–97%). This outcome demonstrated the nanocatalyst's useful reusability.

Heterogeneous catalysis holds significant potential for magnetic nanoparticles, which have been extensively developed and used in several fields. FNMs provide a large surface area and promote active metal sites by enhancing activity. Hydrogenation reactions have made extensive use of Pd-based magnetic materials. Pd nanoclusters anchored on silica-coated Fe_2O_3 nanoparticles (Fe_2O_3@SiO_2) were reported by Ying's group Fe_2O_3@SiO_2 was first created using the microemulsion technique, and then two functional groups—mercapto and amine (HS^- and $-NH_2$)—were added to the SiO_2 shell. High activity and reusability were shown by the produced catalyst Pd/H_2N–SiO_2/Fe_2O_3 for the hydrogenation of nitrobenzene [31].

A Pd-based magnetic nanomaterial was synthesised by Yao et al. [40] and demonstrated extremely high activity (turnover frequency TOF = 3600 h^{-1}) for the hydrogenation of 2-methyl-3-buten-2-ol.

This catalyst utilizes FeO_x-coated Fe nanoparticles as the support to immobilize Pd, while Pd nanoparticles and Fe MNMs coated with the FeO_x shell form the catalyst. However, considering the catalyst's poor stability in water, there is a limitation in its practical applicability. Kim and colleagues [20, 31] employed Fe_2O_3 nanoparticles as the support to deposit Ag in order to overcome the low stability. Both bitruncated-octahedral and spherical Ag and α-Fe_2O_3 nanocrystals, as well as the functionalized MNMs (Ag/Fe_2O_3), have the ability to catalyze the hydrogenation of nitroarenes in water with significant activity and good recoverability.

The development of nanotechnology and nanosized catalytic systems has established new groundwork for sustainable catalysis and given rise to a new generation of catalytic systems. As intrinsic nanocatalysts or as nanosupports for the immobilization of transition metal complexes or metal nanoparticles with catalytic characteristics, ferrite MNMs have been breaking new ground in green catalysis. Magnetic nanocatalysts, such as heterogeneous catalysts, help in overcoming many of the drawbacks of homogeneous catalysts, including their low chemical and thermal stabilities and difficulty in separating from the reaction medium.

They can also increase the concentration of catalytically active sites in the reaction media and improve their accessibility to all reagents due to their nanosize, high surface area-to-volume ratio, and ease of dispersion in a variety of solvents. These features present various opportunities for enhanced catalytic performance since they can increase reaction rates and overcome the diffusion restrictions that arise with bulk heterogeneous materials. In this regard, FNMs facilitated the development of a new class of environmentally benign nanocatalysts, known as magnetically recyclable nanocatalysts [41].

9.3.7 FMNanomaterials for Electrochemical Sensors

A variety of analytical tools for assessing the strength of environmental contaminants have been developed. Their high sensitivity, efficiency, and selectivity make them unsuitable for setups with limited resources since their costs are very high. Moreover, the natural and physical conditions of the substance determine the analysis time.

Therefore, to fulfill the universal requirement for environmental pollution monitoring, basic, affordable, sensitive, selective, and portable sensors are crucial. An easy and affordable alternative to the other instruments could be electrochemical sensors [42].

It functions on the principles of electrochemistry, a highly significant electroanalytical approach with characteristics like high sensitivity, ease of miniaturization, portability, and instrument simplicity, cost-effective [43]. Nanomaterials are frequently used in environmental monitoring in two different ways when developing electrochemical sensors:

(1) in conjunction with an enzyme (enzymatic sensors) to enhance the enzyme's sensitivity and electron transfer rate, or
(2) as independent catalysts (nonenzymatic sensors) that promote the reactions and reduce the limits of detection (LODs) [42].

Magnetic nanoparticles are gaining more popularity due to their economical manufacture, magnetic characteristics, and nanosize, magnetic properties. It has been demonstrated that magnetic nanoparticles between 10 and 20 nm in size have a greater surface area, a high mass transfer capacity, remarkable supermagnetism, and a quicker reaction time. As a result, magnetic nanoparticles can be placed into the transducer materials so as to enhance their attraction to the biosensor's active sensing surface by an external magnetic field [43]. It was also found out that sensors composed of nanoscale metal oxide semiconductors, including SnO_2, In_2O_3, ZnO, TiO_2, WO_3, and NiO, are crucial for monitoring volatile organic compounds (VOCs) and explosive/toxic gases in the environment [42].

An example of a sensitive and useful electrochemical sensor was developed using $Fe_3O_4@SiO_2$ for simultaneously and individually detecting Pb^{2+} and Hg^{2+} ions. The proposed electrochemical sensor was found to have a wide linear range, good selectivity, and a low detection limit (DL) of 9.09 nmol/L and 6.06 nmol/L for Hg^{2+}

and Pb^{2+}, respectively. This electrochemical sensor was also used for detecting these heavy metal ions in milk, and the recoverability rate was found to be 95.1–114%. This easy-to-prepare and clean electrode with a fast pretreatment method has the potential to make more practical and versatile electrochemical sensors [44].

With the efforts of both industries and academics, the exploration of novel magnetic nanomaterials will develop into a creative source for developing environmental monitoring sensors, which could eventually have widespread applications.

9.3.8 FMNs in Biofuels

The use of nanomaterials in biodiesel and biofuel applications has grown significantly in the context of the current scientific and technological advancements in order to increase the efficiency of various bioprocesses and bioenergy [45].

Biodiesel is a mixture of fatty acid alkyl esters (FAAEs) mainly produced via transesterification reactions between short-chain alcohols and triglycerides catalyzed by chemical catalysts (e.g., KOH, NaOH). However, this method of chemical synthesis has many drawbacks: high energy consumption, expensive separation of the catalyst from the reaction mixture, and production of large amounts of wastewater during product separation and purification. However, the main drawback of this process is the high enzyme cost [46].

FNMs provide alternate ways to synthesize Biodiesel by overcoming the drawbacks of traditional reactions. A promising alternative is the immobilization of lipases onto magnetic nanoparticles, which allows the separation of the biocatalyst by simply applying an external magnetic field. Magnetic nanostructures benefit from both their nano and magnetic aspects and have several advantages for industrial applications (Fig. 9.7), such as low cost, biocompatibility, ease of synthesis, superparamagnetism, and magnetic separability, high loading capacity for the loading of different molecules, and standardized methods for their large-scale production [46]. Among MNMs, Fe_3O_4 (Magnetite) nanomaterials are most widely used for this purpose [47].

Sulphamic acid and sulfonic acid functionalized silica-coated crystalline Fe/Fe_3O_4 core/shell magnetic nanoparticles were produced by Wang et al. and showed high stability. To overcome the complexity of macromolecules in NMR analysis, oleic acid and glyceryl trioleate were chosen as model reagents for the synthesis of biodiesel [48].

Marcos et al. prepared biodiesel from soybean oil using FNMs in a research carried out in 2016. In the presence of methanol, Pseudomonas cepacia lipase was covalently bonded to magnetite nanoparticles covered in a thin polydopamine coating and used to enzymatically convert soybean oil into biodiesel. Without the use of particular coupling agents, this method examined the direct immobilization of the enzyme via Michael addition and aldolic condensation processes at the catechol rings [49].

The production of bioethanol was increased by using nickel oxide (NiO) nanoparticles (NPs) as a biocatalyst. High ethanol yields following fermentation are often

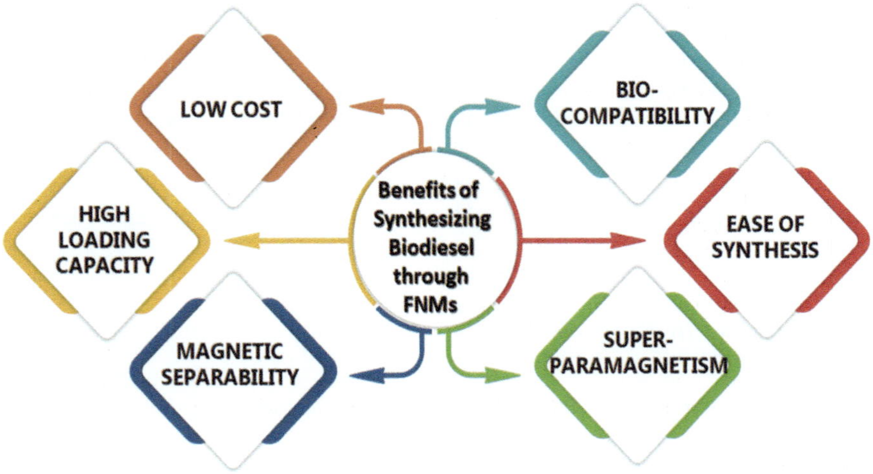

Fig. 9.7 Advantages of synthesizing biodiesel through FNMs over traditional methods

dependent on process factors such as temperature, pH, culture nutrients, and substrate availability [45].

Vasantha and co-workers studied the impact of Fe_3O_4 magnetic nanoparticles as additions in the production of biogas to find the ideal concentration of magnetic Fe_3O_4, which came out to be 20 mg/L, and the product yield was studied at regular intervals. When compared to a substrate treated with $FeCl_3$, the volume of biogas generated with Fe_3O_4 NPs treated substrate showed a threefold increase in yield. Because of their higher absorption, integration in metabolic intermediates, and role as a cofactor of important enzymes involved in hydrolysis, acidification, and methanation, NPs have an impact on improved biogas generation [50].

Vasantha et al. used Fe_3O_4 nanoparticles to produce methane from municipal solid waste at a concentration of 50–125 mg/L. The methane yield was higher than that of the control without NPs, and NPs enhanced the performance of AD [50].

Biodiesel has a lower sulphur and aromatic content, a higher combustion efficiency, and a significant reduction in greenhouse gas emissions and hazardous chemical emissions as compared to traditional fossil fuels. Blending biodiesel with petroleum diesel is another advantage of biodiesel. These blends can be used without modifying existing engines because of their high cetane values and flash points. The production of biodiesel from locally accessible natural resources boosts local economies [46].

Compared to other additives used in this process, cobalt nanoparticles have been shown to be the least toxic to humans, reduce greenhouse gas emissions, and cause the least amount of eutrophication and acidification, making them one of the most effective nanocatalysts for producing biogas from organic biomass. However, the usage of nickel nanoparticles causes the minimum depletion in the ozone layer.

Applications of FNMs are crucial in the biofuel sector due to these environmentally friendly factors.

9.4 Conclusion and Future Prospects

Nanotechnology produces clean, affordable energy, and it has a big impact on human life. As a result, it offers a notable advancement in a number of renewable energy devices for energy conversion and storage, as well as eco-friendly materials [19]. By transferring innovations at a broader and larger scale, nanotechnology offers a seamless platform for transforming and sustainably developing various areas (industry, R&D, medicine, homeland security, energy, safety, and environment) [1].

FNMs are receiving a lot of interest because of their high stability, distinctive functionalized surface, durability, non-toxicity, and magnetically recoverable nature. Because of their distinct physico-chemical characteristics, FNMs have emerged as a superior and more powerful support material than traditional supports [12].

As covered in this chapter, the research and applications of magnetic nanomaterials have great promise for the fields of agriculture, biosensing, biomedicine, catalysis, the environment, and energy [11]. The bar for sustainability and innovation has been raised by nanotechnology. The development of fresh, cutting-edge, and inventive products and technologies will have a significant impact on enhancing functionality, enhancing the environment, and reviving life [1].

The market for magnetite nanoparticles is expanding rapidly worldwide due to the growing demand for these particles in a variety of industries, including biomedical, wastewater treatment, electronics, and enzymes, among others. According to the report, the market for magnetite nanoparticles was estimated to be worth USD 43.9 million in 2018 and is projected to rise at a Compound Annual Growth Rate (CAGR) of 10.3% between 2019 and 2025 [12].

Despite the fact that magnetic properties are appealing for a variety of applications, their nanoscale size frequently results in issues with reduced stability and toxicity. While surface modification has an impact on MNM efficiency, it is essential in resolving these problems. To increase the synergy between functionalized MNMs' characteristics and applications, further study is needed [8].

The results of many lab-scale investigations should be translated to field applications in the future, where circumstances could change quickly. However, more study is required to completely comprehend the special characteristics of FNMs and pinpoint the underlying processes that control synthesis through innovative techniques. Assessing the possible toxicity of such nanomaterials to the environment and human health following exposure would help design commercially viable, large-scale, and sustainable engineered nanomaterial-driven systems [11].

We need more research, technologies, and applications of FNMs in fields like Magnetic energy storage, fuel cells, etc., to fulfill our aim to completely switch to Renewable energy sources for a sustainable future.

Abbreviations

CAGR	Compound Annual Growth Rate
DL	Detection Limit
EDLCs	Electrochemical double-layer capacitors
EEI	Electrode Electrolyte Interface
Fe_2O_3@SiO_2	Silica-Coated Fe_2O_3 Nanoparticles
Fe_3O_4@TiO_2	Fe_3O_4 coated with TiO_2
FESEM	Field Emission Scanning Electron Microscopy
FMNs	Functionalized Magnetic Nanomaterials
HSAB	Hard and Soft Acids and Bases
LMB	LeucoMethylene Blue
LOD	Limits of Detection
MB	Methylene Blue
MBs	Micro-Bundle Structure
MNMs	Magnetic Nanomaterials
MNPs	Magnetic Nanoparticles
MWCNTs	Multiwalled Carbon Nanotubes
NMR	Nuclear Magnetic Resonance
NP	NanoparticlesPAAMA: Co-Polymer made of equimolar Acrylic and Maleic Acid Monomers
PSA	Poly(Styreneacrylic) Acid
PZT	Lead Zirconium Titanate
rGO-S/Co_3O_4	Sulphur-rGO@cobalt oxide composites
TE	Thermoelectric
TOF	Turnover Frequency
VOC	Volatile Organic Compound

References

1. Meenakshi et al (2021) Impact of environmental changes on life and environmental health. In: Functionalized nanomaterials based devices for environmental applications. Elsevier, pp 1–12. https://doi.org/10.1016/B978-0-12-822245-4.00004-0
2. Dhahri S, Shall H, Thabet Mliki N (2024) Towards sustainable nanomaterials: exploring green synthesis methods and their impact on electrical properties. Inorg Chem Commun 168:112872. https://doi.org/10.1016/j.inoche.2024.112872
3. Wu S, Peng J, Lee SLJ, Niu X, Jiang Y, Lin S (2024) Let the two sides of the same coin meet—Environmental health and safety-oriented development of functional nanomaterials for environmental remediations. Eco-Environ Health 3(4):494–504. https://doi.org/10.1016/j.eehl.2024.06.001
4. Sourkouni G, Argirusis C, Argirusis N (2024) Recycling of surface-functionalized nanoparticles—A short review. Processes 12(11):2354. https://doi.org/10.3390/pr12112354

5. Hossen MdA, Ikreedeegh RR, Aziz AA, Zerga AY, Tahir M (2024) Carbon-based nanomaterials (CNMs) modified TiO2 nanotubes (TNTs) photo-driven catalysts for sustainable energy and environmental applications: a comprehensive review. J Environ Chem Eng 12(5):114088. https://doi.org/10.1016/j.jece.2024.114088
6. De S, Madhuri R (2020) Functionalized nanomaterials for electronics and electrical and energy industries. In: Handbook of functionalized nanomaterials for industrial applications. Elsevier, pp 269–296. https://doi.org/10.1016/B978-0-12-816787-8.00011-9
7. Chen Z, Wu C, Zhang Z, Wu W, Wang X, Yu Z (2018) Synthesis, functionalization, and nanomedical applications of functional magnetic nanoparticles. Chin Chem Lett 29(11):1601–1608. https://doi.org/10.1016/j.cclet.2018.08.007
8. Manoj GM, Shalini M, Thenmozhi K, Ponnusamy VK, Hari S (2024) Recent advancements in the surface modification and functionalization of magnetic nanomaterials. Appl Surf Sci Adv 21:100608. https://doi.org/10.1016/j.apsadv.2024.100608
9. You J, Wang L, Zhao Y, Bao W (2021) A review of amino-functionalized magnetic nanoparticles for water treatment: features and prospects. J Clean Prod 281:124668. https://doi.org/10.1016/j.jclepro.2020.124668
10. Anku WW et al (2021) Functionalized nanomagnetic materials for environmental applications. In: Functionalized nanomaterials based devices for environmental applications. Elsevier, pp 127–145. https://doi.org/10.1016/B978-0-12-822245-4.00001-5
11. Al-Anazi A (2022) Iron-based magnetic nanomaterials in environmental and energy applications: a short review. Curr Opin Chem Eng 36:100794. https://doi.org/10.1016/j.coche.2022.100794
12. Sharma S, Sharma H, Sharma R (2024) A review on functionalization and potential application spectrum of magnetic nanoparticles (MNPs) based systems. Chem Inorg Mater 2:100035. https://doi.org/10.1016/j.cinorg.2024.100035
13. Bohara RA, Thorat ND, Pawar SH (2016) Role of functionalization: strategies to explore potential nano-bio applications of magnetic nanoparticles. RSC Adv 6(50):43989–44012. https://doi.org/10.1039/C6RA02129H
14. Idisi DO, Meyer EL (2025) Engineering magnetic nanomaterials for enhanced organic solar cell performance: a review. Inorg Chem Commun 172:113742. https://doi.org/10.1016/j.inoche.2024.113742
15. Mohamed NA, Kiong TS, Ismail AF (2024) Pioneering sustainable energy solutions with rare-earth nanomaterials: exploring pathways for energy conversion and storage. Int J Hydrog Energy 93:607–649. https://doi.org/10.1016/j.ijhydene.2024.10.299
16. Shah SS et al (2024) Advanced strategies in electrode engineering and nanomaterial modifications for supercapacitor performance enhancement: a comprehensive review. J Energy Storage 79:110152. https://doi.org/10.1016/j.est.2023.110152
17. Awan HTA et al (2024) MXene-polymer hybrid composites for advanced energy storage: insights into supercapacitors and batteries. J Energy Storage 95:112449. https://doi.org/10.1016/j.est.2024.112449
18. Gohar O et al (2024) Nanomaterials for advanced energy applications: recent advancements and future trends. Mater Des 241:112930. https://doi.org/10.1016/j.matdes.2024.112930
19. Hala SH (2023) The state of the art of nanomaterials and its applications in energy saving. Bull Natl Res Cent 47(1):7. https://doi.org/10.1186/s42269-023-00984-4
20. Wu C, Yin P, Zhu X, OuYang C, Xie Y (2006) Synthesis of Hematite (α-Fe2O3) nanorods: diameter-size and shape effects on their applications in magnetism, lithium ion battery, and gas sensors. J Phys Chem B 110(36):17806–17812. https://doi.org/10.1021/jp0633906
21. Bello IT, Tsotetsi D, Shaku B, Adedokun O, Chen D, Dhlamini MS (2024) Advances in MoS2-based nanomaterials for supercapacitors, batteries and photovoltaics applications. J Energy Storage 103:114355. https://doi.org/10.1016/j.est.2024.114355
22. Sharma P, Bhatti TS (2010) A review on electrochemical double-layer capacitors. Energy Convers Manag 51(12):2901–2912. https://doi.org/10.1016/j.enconman.2010.06.031
23. Sankar Devi V, Athika M, Duraisamy E, Prasath A, Selva Sharma A, Elumalai P (2019) Facile sol-gel derived nanostructured spinel Co3O4 as electrode material for high-performance

supercapattery and lithium-ion storage. J Energy Storage 25:100815. https://doi.org/10.1016/j.est.2019.100815
24. Iqbal J et al (2019) Density functional theory simulation of cobalt oxide aggregation and facile synthesis of a cobalt oxide, gold and multiwalled carbon nanotube based ternary composite for a high performance supercapattery. New J Chem 43(33):13183–13195. https://doi.org/10.1039/C9NJ02473E
25. Chandra S, Patel MD, Lang H, Bahadur D (2015) Dendrimer-functionalized magnetic nanoparticles: a new electrode material for electrochemical energy storage devices. J Power Sour 280:217–226. https://doi.org/10.1016/j.jpowsour.2015.01.075
26. Sani E, Martina M, Salez T, Nakamae S, Dubois E, Peyre V (2021) Multifunctional magnetic nanocolloids for hybrid solar-thermoelectric energy harvesting. Nanomaterials 11(4):1031. https://doi.org/10.3390/nano11041031
27. Sun F-H et al (2021) Magnetically enhanced thermoelectrics: a comprehensive review. Rep Prog Phys 84(9):096501. https://doi.org/10.1088/1361-6633/ac105f
28. Soin N (2018) Magnetic nanoparticles—Piezoelectric polymer nanocomposites for energy harvesting. In: Magnetic nanostructured materials. Elsevier, pp 295–322. https://doi.org/10.1016/B978-0-12-813904-2.00010-3
29. Wu W, Changzhong Jiang CJ, Roy VAL (2015) Recent progress in magnetic iron oxide–semiconductor composite nanomaterials as promising photocatalysts. Nanoscale 7(1):38–58. https://doi.org/10.1039/C4NR04244A
30. Pathak VV, Rai A, Shukla SK, Jangra S, Tiwari S (2021) Prospects of iron oxide nanomaterial for remediation of wastewater. In: Functionalized nanomaterials based devices for environmental applications. Elsevier, pp 261–273. https://doi.org/10.1016/B978-0-12-822245-4.00012-X
31. Zhang Q, Yang X, Guan J (2019) Applications of magnetic nanomaterials in heterogeneous catalysis. ACS Appl Nano Mater 2(8):4681–4697. https://doi.org/10.1021/acsanm.9b00976
32. Liu H, Shon HK, Sun X, Vigneswaran S, Nan H (2011) Preparation and characterization of visible light responsive Fe2O3–TiO2 composites. Appl Surf Sci 257(13):5813–5819. https://doi.org/10.1016/j.apsusc.2011.01.110
33. Duan M, Shapter JG, Qi W, Yang S, Gao G (2018) Recent progress in magnetic nanoparticles: synthesis, properties, and applications. Nanotechnology 29(45):452001. https://doi.org/10.1088/1361-6528/aadcec
34. Xin T et al (2014) A facile approach for the synthesis of magnetic separable Fe3O4@TiO2, core-shell nanocomposites as highly recyclable photocatalysts. Appl Surf Sci 288:51–59. https://doi.org/10.1016/j.apsusc.2013.09.108
35. Ye M et al (2010) Magnetically recoverable core–shell nanocomposites with enhanced photocatalytic activity. Chem Eur J 16(21):6243–6250. https://doi.org/10.1002/chem.200903516
36. Fu Y, Chen H, Sun X, Wang X (2012) Combination of cobalt ferrite and graphene: high-performance and recyclable visible-light photocatalysis. Appl Catal B Environ 111–112:280–287. https://doi.org/10.1016/j.apcatb.2011.10.009
37. Parandhaman T, Pentela N, Ramalingam B, Samanta D, Das SK (2017) Metal nanoparticle loaded magnetic-chitosan microsphere: water dispersible and easily separable hybrid metal nano-biomaterial for catalytic applications. ACS Sustain Chem Eng 5(1):489–501. https://doi.org/10.1021/acssuschemeng.6b01862
38. Xuan S, Jiang W, Gong X, Hu Y, Chen Z (2009) Magnetically separable $Fe_3 O_4/TiO_2$ hollow spheres: fabrication and photocatalytic activity. J Phys Chem C 113(2):553–558. https://doi.org/10.1021/jp8073859
39. Nikmanesh S, Heidarizadeh F, Mahdavifar Z (2024) Synthesis and density functional theory study of functionalized Oxazolidine-2-ones using a novel MnFe2O4@SiO2–SiO3H magnetic nanocatalyst. J Saudi Chem Soc 28(3):101870. https://doi.org/10.1016/j.jscs.2024.101870
40. Yao Y, Rubino S, Gates BD, Scott RWJ, Hu Y (2017) In situ X-ray absorption spectroscopic studies of magnetic Fe@FexOy/Pd nanoparticle catalysts for hydrogenation reactions. Catal Today 291:180–186. https://doi.org/10.1016/j.cattod.2017.02.049

41. Cruz IF, Freire C, Araújo JP, Pereira C, Pereira AM (2018) Multifunctional ferrite nanoparticles: from current trends toward the future. In: Magnetic nanostructured materials. Elsevier, pp 59–116. https://doi.org/10.1016/B978-0-12-813904-2.00003-6
42. Maddiboyina B, Sunaapu O, Chandrashekharappa S, Sivaraman G (2022) Advantages and limitations of functionalized nanomaterials based electrochemical sensors environmental monitoring. In: Functionalized nanomaterial-based electrochemical sensors. Elsevier, pp 165–174. https://doi.org/10.1016/B978-0-12-823788-5.00016-8
43. Rajendrachari S, Ramakrishna D (2022) Functionalized nanomaterial-based electrochemical sensors: a sensitive sensor platform. In: Functionalized nanomaterial-based electrochemical sensors. Elsevier, pp 3–25. https://doi.org/10.1016/B978-0-12-823788-5.00010-7
44. Zhang M, Guo W (2023) A facile electrochemical sensor based on amino-functionalized magnetic nanoparticles for simultaneous detection of lead and mercuric ions. J Food Compos Anal 119:105232. https://doi.org/10.1016/j.jfca.2023.105232
45. Shahbaz A, Hussain N, Saleem MZ, Saeed MU, Bilal M, Iqbal HMN (2022) Nanoparticles as stimulants for efficient generation of biofuels and renewables. Fuel 319:123724. https://doi.org/10.1016/j.fuel.2022.123724
46. Hajareh Haghighi F, Binaymotlagh R, Palocci C, Chronopoulou L (2024) Magnetic iron oxide nanomaterials for lipase immobilization: promising industrial catalysts for biodiesel production. Catalysts 14(6):336. https://doi.org/10.3390/catal14060336
47. Krishnan S, Diagaradjane P, Cho SH (2010) Nanoparticle-mediated thermal therapy: evolving strategies for prostate cancer therapy. Int J Hyperthermia 26(8):775–789. https://doi.org/10.3109/02656736.2010.485593
48. Wang H, Covarrubias J, Prock H, Wu X, Wang D, Bossmann SH (2015) Acid-functionalized magnetic nanoparticle as heterogeneous catalysts for biodiesel synthesis. J Phys Chem C 119(46):26020–26028. https://doi.org/10.1021/acs.jpcc.5b08743
49. Andrade MFC, Parussulo ALA, Netto CGCM, Andrade LH, Toma HE (2016) Lipase immobilized on polydopamine-coated magnetite nanoparticles for biodiesel production from soybean oil. Biofuel Res J 3(2):403–409. https://doi.org/10.18331/BRJ2016.3.2.5
50. Vasantha VL, Sharvari S, Alfia NS, Praveen N (2021) Application of nanotechnology toward improved production of sustainable bioenergy. In: Nanomaterials. Elsevier, pp 445–479. https://doi.org/10.1016/B978-0-12-822401-4.00010-6

Chapter 10
Nanocomposites Incorporating Functionalized Magnetic Nanomaterials (FMNs)

Parul, Vivek Das, and Kriti Bhandari

Abstract The functionalized magnetic nanomaterials (FMNs) incorporating nanocomposite materials represent a significant advancement in the development of multifunctional materials with enhanced and robust properties tailored for various industrial and biomedical applications. By integrating FMNs such as graphene and reduced graphene oxide (rGO), carbon nanotubes, metal/metal-oxide nanoparticles, and polymer-based nanostructured materials with composite matrices, researchers have unlocked enhanced mechanical strength, durability, electrical conductivity, high thermal resistance capacity, and biocompatibility. This chapter delves into the methodologies employed to fabricate FMN-based nanocomposites by incorporating polymeric matrices and metallic and ceramic systems, highlighting the challenges and innovations in achieving uniform dispersion and interfacial bonding. The synergistic effects of FMNs within nanocomposites are examined, focusing on how they address specific industrial requirements due to their lightweight yet robust materials. The wide-scale applications of such materials are in food packaging industries, aerospace and automotive applications, as well as their role in creating responsive, biocompatible platforms for drug delivery, tissue engineering, diagnostic tools in biomedical fields, energy and environmental factors, catalysis, etc. Furthermore, emerging trends and future perspectives in the synthesis, characterization, and application of FNM-based nanocomposites are discussed, providing insights into their transformative potential across diverse sectors.

Parul and Vivek Das are authors contributed equally.

Parul (✉) · K. Bhandari
Department of Chemistry, Shaheed Rajguru College of Applied Sciences for Women, Vasundhara Enclave, University of Delhi, Delhi, India
e-mail: dr.parul809@gmail.com

K. Bhandari
e-mail: kritibhandari007@gmail.com

V. Das
Department of Applied Sciences (Chemistry), Lloyd Institute of Engineering & Technology, Greater Noida, Uttar Pradesh, India
e-mail: vivek.das@liet.com

S. Gulati (ed.), *Functionalized Magnetic Nanomaterials*, Nanostructure Science and Technology, https://doi.org/10.1007/978-3-031-97199-0_10

Keywords Nanomaterials · Nanocomposites · Functionalized magnetic nanomaterials (FMNs) · Polymer · Carbon nanotubes (CNTs) · Sol–gel synthesis · Chemical vapor deposition

10.1 Introduction

Nanocomposites and functionalized magnetic nanomaterials (FMNs) are bringing novel advancements in material science and offering transformative solutions across various biomedical fields and industries. Together, these materials promote innovation by enhancing functionality, conserving resources, and facilitating innovative solutions in a sustainable and efficient manner. FMNs have nanoscale dimensions with intrinsic magnetic properties and versatile functionalities. These are specially tailored at the nanometer scale, like nanoparticles, nanowires, thin films, etc., which impart them extraordinary size-dependent properties such as strong magnetic responsiveness, catalysis, tunable optical characteristics, high drug loading capacity, high conductivity, and exceptional catalytic efficiency, quantum effect, etc. Examples include iron oxide NPs, cobalt and nickel NPs, Ferrites, etc. Nanocomposites, on the other hand, are a host matrix (polymer, ceramic, or metal-based) to which nanoscale reinforcement, such as nanoparticles and nanotubes have been added to improve the properties of the material. The integration of FMNs into nanocomposites creates multifunctional materials with enhanced capabilities. These properties of nanocomposites have caused researchers and companies to consider using this material in several fields. For example, if an FMN is added to a polymer matrix, the properties of the composite, like physical, magnetic, electrical, and optical, show a drastic change in their magnitude. The incorporation of Fe_3O_4 into SiO_2-PEG-Dox matrix can result in targeted drug delivery with minimized side effects. Manganese ferrite-based graphite oxide-chitosan nanocomposite shows efficient dye removal from aqueous and textile wastewater under UV and sunlight irradiation [1]. In addition to the host matrix, the selection of the correct FMNs is also essential to ensure improvement in the desired properties. The key feature of the properties depends on the particle size and the concentration of the magnetic nano filler. The particle size <100 nm is considered the result oriented in these properties [2]. The characteristics of the particle nature also play a role in properties like electrical, magnetic, and optical. Although for mechanical and physical properties, the particle sizes are important [3]. This synergy between FMNs and nanocomposites addresses critical technological challenges, paving the way for advancements in energy, medicine, and sustainability. The combination of FMNs and nanocomposites allows the creation of materials that are not only more robust but also capable of performing specialized tasks, thereby expanding their potential applications across a wide range of industries. Due to its magnetic properties, high surface area-to-volume ratio, high chemical and thermal stability, durability, and high resistance, these nanocomposite materials are used in industries such as food packaging industries, aerospace and automotive industries, biomedicine, environmental protection, and energy. The chapter describes the fabrication techniques, challenges occurring during the usage of FMN-based nanocomposite materials, and applications in various industries.

10.2 Fabrication Techniques for FMN-Based Nanocomposites

10.2.1 Melt Intercalation

In this method polymer is melted and mixed with FMNs, which in turn allows the magnetic nanomaterial to disperse into the polymer matrix. This is a step in the overall communication process and widely used due to its simplicity and compatibility with existing polymer processing techniques [4, 5]. It is widely known that adding FMNs to polymer matrices enhances their hardness, magnetic responsiveness, resistivity, and fire resistance. The top-down process of intercalation necessitates altering the nanoplatelet surface to allow platelet-like magnetic nanofillers to diffuse in the polymer matrix. Diffusion of the polymer chains into the gallery space of the multilayer structure results in an extra strain. Both chemical and mechanical techniques can be used to distribute individual magnetic nanoplatelets. The chemical process involves dispersing the FMNs in the monomer, followed by a polymerization reaction, the dispersion of the magnetic nanoplatelets in the polymer, and then another polymerization step. Through the process of polymerization, magnetic nanoplatelets interact with the monomer solution to produce polymers between the formed sheets [6]. By mixing the solution, the polymer is directly injected into the magnetic nanoplatelets as part of the repair process. The coating dissolves the polymer, and the bits of platelets react with the solvent. The two solutions are combined, and the polymer chains from the solution enter the cavity of the magnetic nanoplatelets and extract the solvent. The working surfaces' relationship is crucial to the intercalation method's efficacy since it influences the intercalated object's performance and connection. The worth of a connection can be undermined by poor social contact, which can result in issues like temporary or unpleasant separation [7] (Fig. 10.1).

10.2.2 Layer by Layer (LBL) Assembly

This technique involves sequential alternating layering of FMNs and polymers, giving rise to a multilayered nanocomposite. Such nanocomposites with precise control over composition and shape show performance many times better than that of traditional nanocomposites obtained by batch-mixing or extrusion. In other words, this method achieves uniform dispersion and tailored properties of nanocomposites [8] (Fig. 10.2).

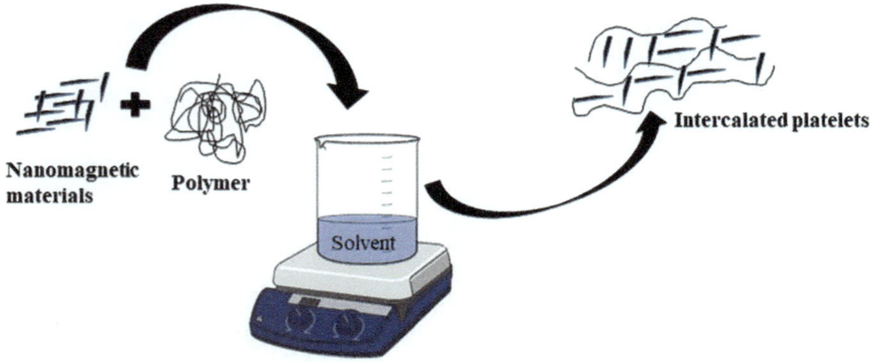

Fig. 10.1 Synthesis of FMN-based materials via the melt intercalation process

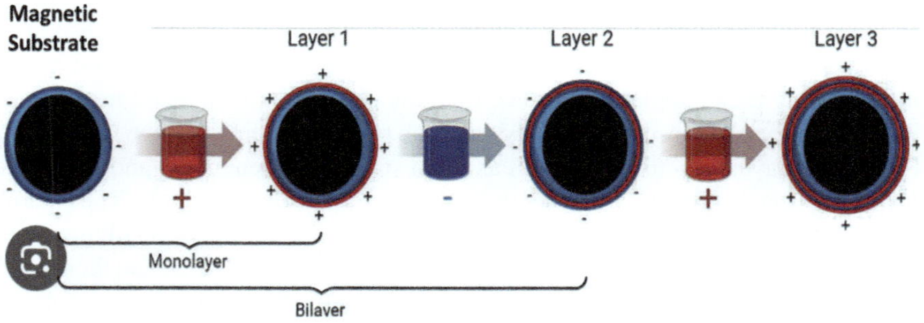

Fig. 10.2 LBL process for synthesis of FMN-based composite materials

10.2.3 In Situ Polymerization

The process involves the dispersion of FMNs in the monomers, followed by polymerization [9, 10]. The polymerization of monomer is carried out under the influence of heat, initiators, or ultraviolet radiation. Consequently, exfoliated or intercalated nanocomposites are formed. In situ template manufacturing is another technique, where the polymer matrix regulates the platelets' size and they move through a high fluid, such that the polymer chains that govern the platelets' nucleation go ahead and do them. However, this approach suffers from the limitation that it accelerates the breakdown of polymer chains at elevated temperatures [11] (Fig. 10.3).

10.2.4 Sol–Gel Method

This process includes two reaction phases: sol and gel. The solid is dispersed in a liquid, forming a colloidal suspension called "sol", whereas a "gel" is a cohesive

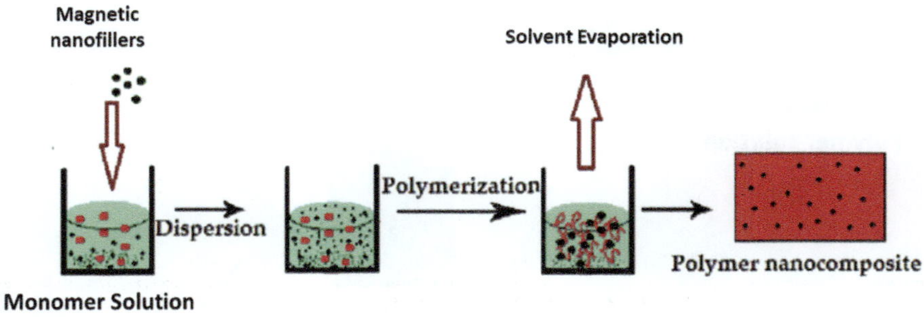

Fig. 10.3 In situ polymerization process for synthesis of FMN-based composite materials

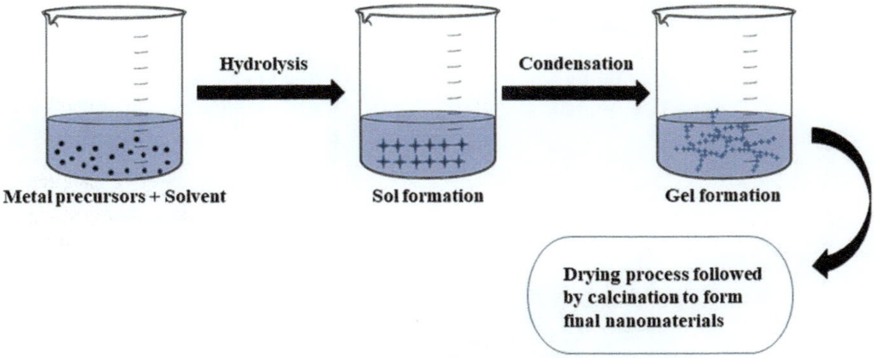

Fig. 10.4 Process of formation of FMNs using sol–gel technique

network that forms between these processes. This process includes dispersion of FMNs or magnetic fillers in a monomer solution (sol) that emerges from the layers of the particle, resulting in the formation of an interactive network of polymers and magnetic nanoparticulates [12, 13]. So, the contribution of the polymer is carried out in the nucleation and growth of FMNs (Fig. 10.4).

10.2.5 *Electrospinning*

This is the most common technique for synthesizing polymer nanocomposite fibres. It involves creating fibers from a polymer solution containing functionalised magnetic nanoparticles while it is exposed to an electric field [14]. The nanocomposites obtained from this method are of superior mechanical, electrical, and thermal properties by precisely controlling fiber diameter, shape, and content (Fig. 10.5).

These fibers are diversely used in filtration, energy storage, catalysis, and biomedical fields like magnetically targeted drug delivery, magnetic hyperthermia, and tissue engineering. This technique is useful for creating innovative materials with some

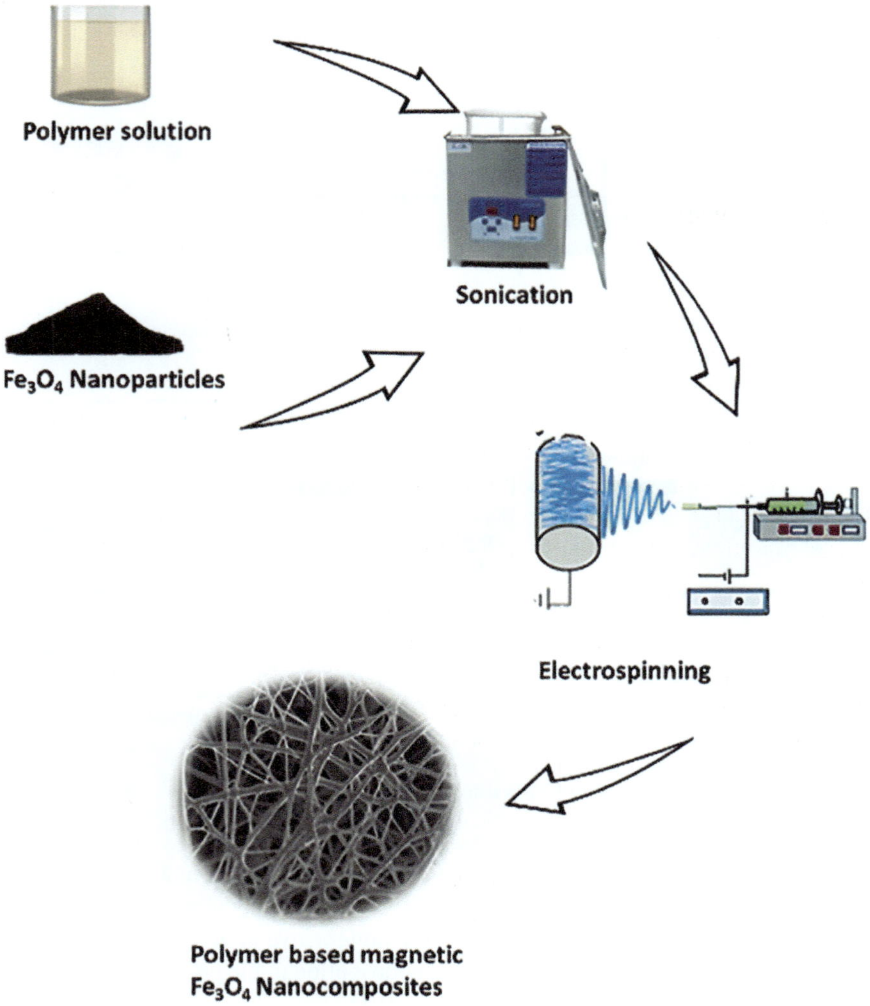

Fig. 10.5 Electrospinning process for synthesis of FMN-based composite materials

specific functionalities. The Polymer nanocomposites (PNC) synthesis methods and materials are listed in Table 10.1.

Table 10.1 Synthesis techniques for polymer nanocomposites

S. No	Polymers	Components	Precursors	Methods	References
1	Polymethyl methacrylate (PMMA)	PLA	AIBN (azophloxine)	Optical filtering and thermal polymerization	[15, 16]
2	Polyaniline (PANI)	TiO_2	APS	Ex-situ and in-situ chemical oxidation polymerization	[17]
3	Propylene	Al_2O_3, MoO_3 and Fe_2O_3	TIP	Chemical vapor deposition (CVD)	[18, 19]
4	Epoxy	Polyaniline	Ammonium persulfate (APS)	oxidative chemical polymerization	[20, 21]
5	Tetragonal zirconia polycrystals	ZrO_2	Zirconium oxychloride octahydrate ($ZrOCl_2$)	Sol gel preparation, hydrothermal and co-precipitation	[22]
6	Epoxy	SiO_2 and TiO_2	TEOS and TIP	Combination of sol–gel and in situ polymerization	[23]
7	Epoxy	SiO_2	HCl	Thermal treatment	[24]
8	Epoxy	TiO_2	$TiCl_4$	Direct precipitation	[25]
9	Polyacrylonitrile	TiO_2 and Ag	NaOH	Sol–gel solution	[26, 27]
10	Adenine-containing polyimide	Nitrogen (N)	Cellulose and lignin	In situ polymerization	[28, 29]
11	Resin	OMMT	Acetyltrimethylammonium chloride (MMT-CTAC)	Hydrothermal andsol-gel preparation	[30, 31]

10.3 Challenges and Limitations of FNM-Based Nanocomposites

The functionalized magnetic nanocomposite materials have gained considerable attention due to their enhanced properties, such as electrical, mechanical, chemical, magnetic responsiveness, thermal, durability, high resistance, and large surface area. However, several challenges hinder their usage. The limitations and potential risks of FNM-based nanocomposites must be considered. A few of the challenges are discussed as follows.

10.3.1 Challenges in Synthesis

Due to the complex functionalization process, chemical modifications, and the synthesis of FMN-based nanocomposites are often challenging and time-consuming. The variation in synthesis method may lead to non-reproducibility of the materials, which may prohibit the use of such materials in large quantities. Many synthetic approaches may limit the synthesis of nanocomposites in small-scale laboratories, which restricts to scale-up to an industrial level.

10.3.2 Limitation in Performance

Although the nanocomposites have enhanced properties, however, enhancing one property may lead to a degradation in other properties. For instance, while improving the mechanical property of the material may deteriorate its flexibility and electrical properties. Some functionalized nanocomposites may also degrade their performance in harsh environmental conditions such as high temperature and pressure.

10.3.3 Dispersion and Compatibility

The atoms on the surface of functionalized magnetic nanomaterials (FMNs) have a high surface energy, which is stabilized by binding amount atoms, increasing the surface area, i.e., they have a high tendency of agglomeration due to the presence of van der Waals forces [32]. The increase in surface area results in a change in many properties of magnetic nanomaterials. To address this issue, the particles are dispersed, which breaks the agglomerated particles into small fragments, which can be achieved by splitting the agglomerate by applying stress to it. Notably, some factors, such as the force of attraction or the length of magnetic nanoparticles, influence dispersion. The dispersion of agglomerates can be done by the following methods.

- Physical technique where the mechanical vibrations created by electric currents are allowed to travel through the solution of water containing magnetic nanomaterials. Homogeneous solutions of water and nanomaterials were not obtained.
- Chemical method, where surfactants or functional groups, were used to reduce the surface tension of water. This method had a stable homogeneous dispersion, but it also had a drawback that it was totally dependent on the concentration of the surfactant's dispersion capability [33].

Compatibility of the magnetic nanomaterials depends on the dispersion. For good dispersion, FMNs microstructures were improved, ensuring the uniform pore size increased the mechanical strength of the nanomaterials. Although in the case of techniques such as physical, where there was no formation of homogeneous dispersion, the mechanical strength was not improved.

10.3.4 Scalability and Cost-Effectiveness

With the extensive requirements of the FMNs, there is a major challenge, which is scalability. For the authenticity of the nanocomposite materials, the synthesized nanocomposites must be characterized to check their structural identity, which requires expensive analytical instruments. The cost of fabrication of nanocomposites is expensive, which limits the synthesis at a larger scale in industries. To meet the required amount, magnetic nanomaterials cannot be produced in a large quantity at once, as this increases the production cost and requirement of expensive materials [34]. All FMNs have special properties according to which they work; to ensure that these properties do not cease, the scale of production of the FMNs cannot be increased [35].

10.3.5 Long-Term Stability

Stability of the FMNs is also a major challenge; for example, magnetic nanocomposites, especially iron based, are very much prone to oxidation in a humid atmosphere, and even the functional coating of polymer or other organic molecules also degrades with time or with long-time exposure to UV radiation, causing instability of the nanocomposites. Though the long-term stability of nanomaterials can be increased via various techniques but at the same time, the cost of making them stable also increases [36].

10.3.6 Toxicity

Few FMNs, such as cobalt-based nanocomposites, may also pose a threat to human life due to their toxic nature (leaching of Co^{2+} ions), which eventually damages DNA and disrupts the cell membrane.

It is, therefore, crucial to address these challenges for the application of such nanocomposite materials.

10.4 Applications of FMN-Based Nanocomposites

The nanocomposite incorporating functionalized magnetic nanoparticles has tailored several properties and is widely used across various industries due to its unique properties, such as a large surface area and reactivity. Moreover, the magnetic polymer nanocomposite (MPNC) materials possess great advantages in imaging techniques due to the presence of magnetic properties such as superparamagnetic behavior. The flow chart describes the specific fields in which the nanocomposite material is implemented on a wide scale.

10.4.1 Food Packaging Industries

Food packaging plays a vital role in the preservation and extension of the shelf life of food products until consumption. The packaged food product prevents unfavourable changes due to microbial actions, moisture, temperature change, light, etc., thus maintaining the quality of the food products. The packaged materials create a proper condition that acts as a barrier for moisture, gases, and micro-organisms [37, 38]. The usage of new varieties of packaging materials is being explored, which work as a protective barrier with advanced functions. The novel technology for food packaging products involves the use of nanotechnology. The advantage of using nanotechnology for the production of materials is that the nano-dimensional materials possess a large surface area-to-volume ratio. As a result, the total surface energy, surface area, and reactivity increase [39]. Among all nanotechnology products, the polymeric magnetic nanocomposite nanomaterials have created widespread attention in the food packaging industry. The magnetic nanocomposites are integrated with sensors to detect any changes in the temperature and gas composition in food packaging products. Furthermore, these FMNs capture and remove the heavy metals and contaminants through magnetic separation techniques (Fig. 10.6).

The polymeric nanocomposite materials have potential applications in food packaging for processed food products such as bakery products, fruit juices, dairy products, and for the manufacture of carbonated beverages [40]. The food package consisting of polymeric nanocomposites exhibits a high potential in maintaining

Fig. 10.6 Schematic representation of applications of FMN materials

the quality of food products and extending the shelf life of products. During the manufacturing of packaging materials, it is essential to resist gases like oxygen and water vapor and prevent the migration of CO_2 outside the beverage bottles to preserve the shelf life of carbonated beverages. Polymer nanocomposites have been developed for improved gas and barrier properties. For improving the barrier properties, nylon and clay-based nanocomposites are often utilized as the combination of plastic and nylons provide strength and toughness to the structure. Nylon-MXD6 is used in foil-based packaging materials. The combination of nylon-MXD6 and clay nanocomposites possesses gas barrier properties and tear resistance of films [41]. The chitin nanofibril-based biodegradable polymers are also being used on a wide scale to increase the mechanical properties and reduce the permeability of vapor in nanocomposite films [42].

10.4.2 Aerospace and Automotive Industries

Nanocomposite materials have a wide range of applications in the aerospace and automotive industries. It is essential to have a coating of nanocomposites on the turbine blades of aircraft to provide durability and low grain growth at high temperatures [43].

The coating of magnetic conductive carbon nanotubes was found to have larger flexibility for de-icing applications in the aerospace industry [44]. Some of the compositions of magnetic nanoparticles with polymers include magnetite-polyaniline, cobalt ferrite-polypyrrole, etc. Magnetic nano-silica-based nanocomposites are thermally resistant material that helps in protecting the parts of spacecraft materials. Due to the good thermal expansion coefficient and low erosion rate, nanocomposite materials provide protection against lightning. The graphene-based nanocomposite materials are nowadays utilized for the detection of UV exposure in space missions. The alumina ceramic-graphene-based composites have been found to have high densities and are capable of withstanding high temperatures, which fulfills the requirement for specific components of an aircraft engine [45].

The automotive sector focuses on lightweight, durability, and efficiency. For the manufacturing of lightweight and high-strength materials, it is essential to use carbon nanotubes (CNTs) or graphene-reinforced polymer composites. The carbon fibre-based nanocomposites improve the crash resistance capacity of automobiles. In addition to crash resistance, to enhance wear resistance at high-temperature areas like the exhaust system or turbocharger, nanoceramic materials are used. These materials are also used in the making of engine components. Similarly, nanolubricant materials are used to reduce the friction and abrasion in engine components, which results in higher efficiency and longer lifespan. Nano-enhanced Li-ion batteries improve the battery life and increase the efficiency of electric vehicles, which are in high demand in the current scenario and in the future.

A large-scale use of FMNs is seen in aerospace and automotive industries due to lightweight, thermal stability, durability, high strength, and fuel efficiency. These materials will play a vital role in the future of transportation.

10.4.3 Biomedical Applications

Nanocomposites comprise multiphase materials, which are extensively used in the pharmaceutical industries. These nanocomposites are made of hydrogels that mainly consist of biopolymers having nano-metal oxides. These materials have been effectively utilized in skin treatment [46]. The inorganic nanocomposite materials have been used as anti-cancer drugs, anti-tumor treatments, etc. The usage of nanocomposite materials as anti-bacterial, anti-microbial is due to tunable porosity and a large surface area-to-volume ratio.

Schematic Fig. 10.1 describes the fundamental area in which FMN-based magnetic nanocomposites are applied. The most common devices used for diagnosing disease are bio-imaging techniques and biosensor devices. The nanomaterials, particularly multiphase composite materials, are also utilized for the manufacturing of sensing devices, which can analyze any outgrowth of cells, such as tumor/cancer cells, in the human body. The nanocomposites, such as graphene-based and gold-based functionalized materials, have been used as biosensors for detecting cancer. Along with the manufacturing of such devices, the functionalized nanocomposite

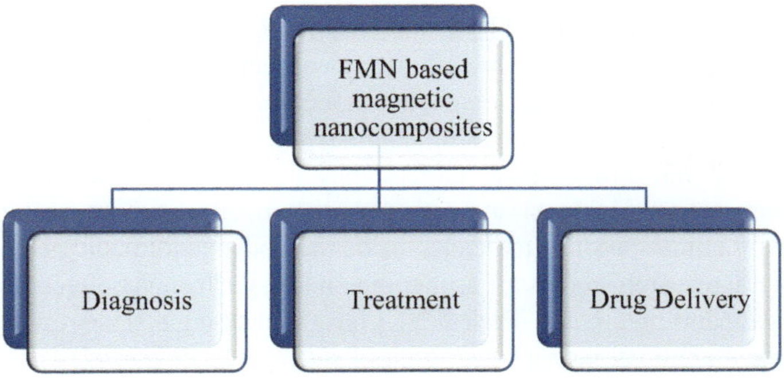

Fig. 10.7 FMN-based magnetic nanocomposites used in biomedical applications

materials have several applications in the treatment of diseases such as cancer . The functionalized nanocomposites can act as regenerative medicines, like nervous system regeneration and wound treatment, plastic surgery, [47] joint replacements, [48] etc. Tissue malfunctioning has been a widespread concern around the world and hence requires attention to solve the issues related to tissue and organ replacement. Nanomedicine plays a role in solving such complications. The nanocomposites used as nano-medicine act as a substrate for cell growth and proliferation, which further act as biological molecules for the growth of new tissue or organ [49] (Fig. 10.7).

During the regeneration technique, various biodegradable polymeric nanocomposites such as polylactic acid, polyglycolic acid, polycaprolactone, etc. are used for the cell growth and replacement of damaged cells and tissues [50, 51]. Moreover, nerve regeneration is one of the challenging operations, and nanocomposite materials show a promising physicochemical property for such regeneration.

Various inorganic materials are used for the regeneration of damaged nerves. The examples include metallic nanoparticles, magnetic nanoparticles, and quantum dots. In addition, polymeric nanoparticles, [52] dendrimers, [53] and carbon-based nanocomposite materials [54] also contribute to the regeneration of neural cells. The fact that the nanocomposite materials act better as regenerative medicine, as they can mimic the surface properties of damaged tissues and cells. Nanomaterials have been proven for various regeneration applications, such as in bone, [55] ligaments, [56] and vascular organs [57]. For instance, hydroxyapatite-functionalized nanocomposites help in bone regeneration by mimicking the structure of natural bone. Furthermore, the study on nanocomposites shows that the silver functionalized nanocomposites are much better at aiding in wound healing and preventing infections.

Due to the increased rate of cardiovascular diseases, the nanocomposites are employed for the treatment of heart valves and the cardiovascular system. Notably, the largest market for biomaterials is used in the treatment of cardiovascular diseases, followed by the orthopaedic treatment.

Drug and gene delivery is desired in numerous situations to achieve improved results, and hence, magnetic nanomaterials are preferred as drug carriers. For instance, a 600 nm polyglycolic acid nanomaterial stacked with DNA resulted in better delivery in rodents. The FMN, specifically polymeric-based composites, has been shown to improve therapeutic efficiency without having any side effects.

Although the functionalized magnetic nanocomposites have shown promise as a promising material to act as nano-medicines for the diagnosis of diseases, therapeutic agents, and drug delivery targets, they face several challenges as well. A few of them include high cost, expensive characterization instrumental techniques required, sometimes produce toxicants inside the body, genotoxicity, etc. [58]. As a result, several research studies are still going on for synthesizing effective nanocomposite-based nano-medicines. The MPNCs have gained wide-scale research interest in biomedical applications by integration of magnetic nanoparticles such as ferric oxide and cobalt ferrite with polymeric matrices. These MPNCs are utilized as Magnetic Resonance Imaging (MRI) contrast agents to detect neural stem cells and for in-vivo imaging. The MPNCs can act as the drug delivery targets within the body, as these materials can be directed to their desired location by applying an external magnetic field. In tissue engineering, the magnetic properties of MPNCs allow control of scaffold architecture, which helps in the formation of tissues [59].

10.4.4 Energy and Environment

Due to the remarkable mechanical, electrical, and chemical properties and large surface area of functionalized magnetic nanocomposites, these are utilized in both the energy and environmental sectors. Currently, widespread research is going on in the field of energy storage devices such as batteries, supercapacitors, and fuel cells.

The graphene oxide (GO) based functionalized nanocomposites have revolutionized the areas of sustainable energy and environmental technologies. These materials possess highly efficient heat transfer, a high surface-to-volume ratio, and electron conductivity. The fabrication of such materials requires functionalization of graphene with organic molecules, proteins, metal nanoparticles, metal oxides, etc. [60]. Nowadays, society demands the usage and storage of energy on a large scale, which requires the generation of supercapacitors. The graphene functionalized with RuO_2, TiO_2, and Fe_3O_4 nanoparticles has been shown to act as a high-performance supercapacitor. These materials enhance the charge storage capacity, and the specific capacitance sustains up to 85% at higher voltage sweep for each nanocomposite [61]. In case of solar harvesting energy, nanocomposites help in improving the performance and power conversion efficiency of photovoltaic cells, [62] and perovskite solar cells [63, 64]. Quantum dot sensitized solar cells (QDSSCs) have also gained much interest in generating development towards future solar cells. The synthesis of QDSSCs is very similar to dye-sensitized solar cells (DSSCs); however, the power conversion efficiency of QDSSCs is much higher [65]. The efficiency of QDSSCs can also be improved by introducing carbon nanotubes (CNTs) [66].

The functionalized nanocomposites are also used for removing pollutants from water. For the water purification process, nanocomposite materials functionalized with chelating agents assist in removing heavy metals such as Pb, Hg, and As from wastewater. The TiO_2-based nanocomposites also help in degrading organic pollutants and dyes in the presence of UV–visible light [67]. The cerium-doped TiO_2 facilitates the degradation of volatile organic pollutants and greenhouse gases [68].

Moreover, the nanocomposites contribute to climate change by carbon capturing and storage, i.e., CO_2 adsorption and catalytic CO_2 conversion. The amine-functionalized graphene nanocomposites exhibit a high adsorption capacity of CO_2. For this process, a new, improved mesoporous nanocomposite CO_2 adsorbent is synthesized by the mixture of improved mesoporous silica nanotubes (IMSiNTs) and polyethyleneimine (PEI). These adsorbent materials have high thermal stability and good stability towards adsorption or desorption cycles [69].

The magnetic iron oxide nanoparticles (MIONPs) mixed at a certain ratio induce the required electromagnetic properties within the materials, such as metal oxides, clay, and polymers. An efficient separation technique, wastewater treatment, storage devices, and sensors involve the usage of high gradient magnetic separation (HGMS) in which magnetic iron oxide nanocomposites (MIONCs) are utilized. The size of the MIONC depends on the synthesis procedure, in which the most economical approach is to synthesize via the hydrothermal procedure. The electromagnetic characteristics within the target material depend on the size of MIONC [70].

The FNM materials have transformed energy and environmental technologies by providing high efficiency and sustainability. The applications range from energy storage devices and treatment of wastewater, air pollution prevention, along with carbon capture for improved climatic conditions. Further research is being carried out to explore new nanocomposite materials for addressing global energy and environmental challenges.

10.5 Conclusion

The integration of functionalized magnetic nanomaterials with polymer composites due to synergistic interactions between their constituent materials allows a widespread application in industries from electronics, packaging industries, biomedicine, to energy and environmental remediation. High surface area, tunable reactivity, biocompatibility, high thermal, electrical, and chemical resistance, and magnetic properties make the use of such materials at a larger scale. The widely used magnetic nanocomposite materials include cobalt and iron oxides. The fabrication techniques involve polymerization processes, sol–gel techniques, etc., which can be carried out easily. The future perspective of FMNs-based magnetic nanocomposite materials will focus on scalability, economic and environmental impact during the fabrication process, with a more diverse application.

Abbreviations

APS	Ammonium Persulfate
CNTs	Carbon Nanotubes
CVD	Chemical Vapor Deposition
DSSCs	Dye-sensitized Solar Cells
FMNs	Functionalized Nanomaterials
LBL	Layer by Layer
GO	Graphene Oxide
HGMS	High Gradient Magnetic Separation
IMSiNTs	Improved Mesoporous Silica Nanotubes
MIONCs	Magnetic Iron Oxide Nanocomposites
MPNC	Magnetic Polymer Nanocomposite
PANI	Polyaniline
PEI	Polyethyleneimine
PHMB	Polyhexamethylene Biguanide Hydrochloride
PMMA	Polymethyl Methacrylate
PNC	Polymer Nanocomposites
QDSSCs	Quantum Dot Sensitized Solar Cells
rGO	Reduced Graphene Oxide

References

1. Hosseini NM, Sheshmani S, Shahvelayati AS (2025) Manganese ferrite-graphite oxide-chitosan nanocomposite for efficient dye removal from aqueous and textile wastewater under UV and sunlight irradiation. Nature 15:866
2. Bharadwaj RK, Mehrabi AR, Hamilton C, Trujillo C, Murga M, Fan R, Chavira A, Thompson AK (2002) Structure–property relationships in cross-linked polyester–clay nanocomposites. Polymer 43(13):3699–3705
3. Ghuman BS (2020) Nanocomposites—a review. J Chem Chem Sci 5:506–510
4. Bokov D, Turki Jalil A, Chupradit S, Suksatan W, Javed Ansari M, Shewael IH, Valiev GH, Kianfar E (2021) Nanomaterial by sol-gel method: synthesis and application. Adv Mater Sci Eng 2021(1):5102014
5. Mahto TK, Chowdhuri AR, Sahoo B, Sahu SK (2014) Polyaniline-functionalized magnetic mesoporous nanocomposite: a smart material for the immobilization of lipase. Polym Compos 37(4):1152–1160
6. Hasan M, Liu Q, Kanwal A, Tariq T, Mustafa G, Batool S, Ghorbanpour M (2024) A comparative study on green synthesis and characterization of Mn doped ZnO nanocomposite for antibacterial and photocatalytic applications. Sci Rep 14(1):7528
7. Gobena ST, Woldeyonnes AD (2024) A review of synthesis methods, and characterization techniques of polymer nanocomposites for diverse applications. Discov Mater 4(1):52
8. Kotov NA, Mamedov AA, Guldi DM, Prato M, Wicksted J, Hirsch A (2002) Nanocomposites, layer-by-layer assembly. In: Encyclopedia of polymer science and technology
9. Rashid IA, Tariq A, Shakir HF, Afzal A, Ali F, Abuzar M, Haider T (2022) Electrically conductive epoxy/polyaniline composite fabrication and characterization for electronic applications. J Reinf Plast Compos 41(1–2):34–45

10. Tanasa E, Zaharia C, Radu IC, Surdu VA, Vasil BS, Damian CM, Andronescu E (2019) Novel nanocomposites based on functionalized magnetic nanoparticles and polyacrylamide: preparation and complex characterization. Nanomaterials 9(10):1384
11. Kumar A, Kumar S, Pathak AK, Ansari AA, Rai RN, Lee Y, Kim SY, Van Le Q, Singh L (2023) Recent progress in nanocomposite-oriented triboelectric and piezoelectric energy generators: an overview. Nano-Struct Nano-Objects 36:101046
12. Batra RC, Pydah A (2020) Impact analysis of PEEK/ceramic/gelatin composite for finding behind the armor trauma. Compos Struct 237:111863
13. Sy DT, Nguyen T, Thanh NT, Thinh PV (2020) Preparation and characterization of magnetic $NiFe_2O_4$ decorated exfoliated graphite via sol-gel method. Asian J Chem 32(9):2233–2236
14. Venmathi Maran BA, Jeyachandran S, Kimura M (2024) A review on the electrospinning of polymer nanofibers and its biomedical applications. J Compos Sci 8(1):32
15. Muller O, Hege C, Guerchoux M, Merlat L (2022) Synthesis, characterization and nonlinear optical properties of polylactide and PMMA based azophloxine nanocomposites for optical limiting applications. Mater Sci Eng, B 276:115524
16. Ahlawat A, Sahdev RK, Gupta RK, Chhabra D (2021) 3D FDM printable polymer composites and polymer nanocomposites: state of the art. J Nano Electr Phys 13(2)
17. Kalaiarasi J, Balakrishnan D, Al-Keridis LA, Al-mekhlafi FA, Farrag MA, Kanisha CC, Murugan M, Pragathiswaran C (2022) Sensing and antimicrobial activity of polyaniline doped with TiO_2 nanocomposite synthesis and characterization. J King Saud Univ Sci 34(3):101824
18. Dorbani T, Bouleklab MC, Settar A, Chetehouna K, Naoui Y, Revo S, Hamamda S (2022) Synthesis and characterization of Al+ 0.25% MWCNTs nanocomposite with reduced thermal expansion coefficient. J Market Res 19:1484–1492
19. Mittal V (2012) Thermal characterization of fillers and polymer nanocomposites. In: Characterization techniques for polymer nanocomposites, pp 13–32
20. Teijido R, Ruiz-Rubio L, Echaide AG, Vilas-Vilela JL, Lanceros-Mendez S, Zhang Q (2022) State of the art and current trends on layered inorganic-polymer nanocomposite coatings for anticorrosion and multi-functional applications. Prog Org Coat 163:106684
21. Panwar LS, Panwar V, Anoop G, Park S (2022) Carbon nanofiber-polyelectrolyte triggered piezoelectric polymer-based hydrophilic nanocomposite for high sensing voltage generation. J Mater Res Technol 17:3246–3261
22. Hu C, Sun J, Long C, Wu L, Zhou C, Zhang X (2019) Synthesis of nano zirconium oxide and its application in dentistry. Nanotechnol Rev 8(1):396–404
23. Adnan MM, Dalod AR, Balci MH, Glaum J, Einarsrud MA (2018) In situ synthesis of hybrid inorganic–polymer nanocomposites. Polymers 10(10):1129
24. Moosa AA, Saddam BF (2017) Synthesis and characterization of nanosilica from rice husk with applications to polymer composites. Am J Mater Sci 7(6):223–231
25. Wang M, Abbineni G, Clevenger A, Mao C, Xu S (2011) Upconversion nanoparticles: synthesis, surface modification and biological applications. Nanomed Nanotechnol Biol Med 7(6):710–729
26. Yerli-Soylu N, Akturk A, Kabak Ö, Erol-Taygun M, Karbancioglu-Guler F, Küçükbayrak S (2022) TiO_2 nanocomposite ceramics doped with silver nanoparticles for the photocatalytic degradation of methylene blue and antibacterial activity against Escherichia coli. Eng Sci Technol Int J 35:101175
27. Gebisa AW, Lemu HG (2017) A case study on topology optimized design for additive manufacturing. IOP Conf Ser Mater Sci Eng 276(1):012026. IOP Publishing
28. Liu C, Wang L, Liu S, Tong L, Liu X (2020) Fabrication strategies of polymer-based electromagnetic interference shielding materials. Adv Ind Eng Polym Res 3(4):149–159
29. Su X, Wang R, Li X, Araby S, Kuan HC, Naeem M, Ma J (2022) A comparative study of polymer nanocomposites containing multi-walled carbon nanotubes and graphene nanoplatelets. Nano Mater Sci 4(3):185–204
30. Sreenivasalu PKP, Dora CP, Swami R, Jasthi VC, Shiroorkar PN, Nagaraja S, Asdaq SMB, Anwer MK (2022) Nanomaterials in dentistry: current applications and future scope. Nanomaterials 12(10):1676

31. Roshanali M, Nodehi A, Atai M (2020) Synthesis and characterization of core-shell nanoparticles and their application in dental resins. J Mech Behav Biomed Mater 110:103926
32. Büyüktiryaki S, Keçili R, Hussain CM (2020) Functionalized nanomaterials in dispersive solid phase extraction: advances & prospects. TrAC, Trends Anal Chem 127:115893
33. Parveen S, Rana S, Fangueiro R (2013) A review on nanomaterial dispersion, microstructure, and mechanical properties of carbon nanotube and nanofiber reinforced cementitious composites. J Nanomater 2013(1):710175
34. Rizvi M, Gerengi H, Gupta P (2022) Functionalization of nanomaterials: synthesis and characterization. In: Functionalized nanomaterials for corrosion mitigation: synthesis, characterization, and applications. American Chemical Society, pp 1–26
35. Xu L, Ma W, Wang L, Xu C, Kuang H, Kotov NA (2013) Nanoparticle assemblies: dimensional transformation of nanomaterials and scalability. Chem Soc Rev 42(7):3114–3126
36. Hoang VT, Mai M, Thi Tam L, Vu NP, Tien Khi N, Dinh Tam P, Quang Huy T, Le AT, Xuan Dinh N, Tran VH (2020) Functionalized-AgNPs for long-term stability and its applicability in the detection of manganese ions. Adv Polym Technol 2020(1):9437108
37. Marsh K, Bugusu B (2007) Food packaging—roles, materials, and environmental issues. J Food Sci 72(3):R39–R55
38. Lee SJ, Rahman AM (2014) Intelligent packaging for food products. In: Innovations in food packaging. Academic Press, pp 171–209
39. Mariano M, El Kissi N, Dufresne A (2014) Cellulose nanocrystals and related nanocomposites: review of some properties and challenges. J Polym Sci Part B: Polym Phys 52(12):791–806
40. Smolander M, Chaudhry Q (2010) Nanotechnologies in food packaging, pp 86–101
41. Dasari A, Njuguna J (2016) Functional and physical properties of polymer nanocomposites
42. Shankar S, Reddy JP, Rhim JW, Kim HY (2015) Preparation, characterization, and antimicrobial activity of chitin nanofibrils reinforced carrageenan nanocomposite films. Carbohyd Polym 117:468–475
43. Zhao W, Li M, Peng HX (2010) Functionalized MWNT-doped thermoplastic polyurethane nanocomposites for aerospace coating applications. Macromol Mater Eng 295(9):838–845
44. Bhat A, Budholiya S, Aravind Raj S, Sultan MTH, Hui D, Md Shah AU, Safri SNA (2021) Review on nanocomposites based on aerospace applications. Nanotechnol Rev 10(1):237–253
45. Gao X, Fu Y, Jiang D, Wang D, Yang J, Weng L, Hu M, Sun J (2018) Structural, mechanical, and tribological properties of WS 2-Al nanocomposite film for space application. Tribol Lett 66:1–12
46. Mabrouk M, Das DB, Salem ZA, Beherei HH (2021) Nanomaterials for biomedical applications: production, characterisations, recent trends and difficulties. Molecules 26(4):1077
47. Biggs M, Zeugolis D, Pandit A (2017) Foreword to special issue on two-dimensional biomaterials in regenerative medicine. Nanomed Nanotechnol Biol Med 14(7):2351–2353
48. Dzedzickis A, Bucinskas V, Viržonis D, Sesok N, Ulcinas A, Iljin I, Sutinys E, Petkevicius S, Gargasas J, Morkvenaite-Vilkonciene I (2018) Modification of the AFM sensor by a precisely regulated air stream to increase imaging speed and accuracy in the contact mode. Sensors 18(8):2694
49. Siddique S, Chow JC (2020) Gold nanoparticles for drug delivery and cancer therapy. Appl Sci 10(11):3824
50. Nance E (2019) Careers in nanomedicine and drug delivery. Adv Drug Deliv Rev 144:180–189
51. Castro AG, Diba M, Kersten M, Jansen JA, van den Beucken JJ, Yang F (2018) Development of a PCL-silica nanoparticles composite membrane for guided bone regeneration. Mater Sci Eng, C 85:154–161
52. Kumar R, Aadil KR, Ranjan S, Kumar VB (2020) Advances in nanotechnology and nanomaterials based strategies for neural tissue engineering. J Drug Deliv Sci Technol 57:101617
53. Shah S, Solanki A, Sasmal PK, Lee KB (2013) Single vehicular delivery of siRNA and small molecules to control stem cell differentiation. J Am Chem Soc 135(42):15682–15685
54. Péan JM, Venier-Julienne MC, Filmon R, Sergent M, Phan-Tan-Luu R, Benoit JP (1998) Optimization of HSA and NGF encapsulation yields in PLGA microparticles. Int J Pharm 166(1):105–115

55. Wen CJ, Zhang LW, Al-Suwayeh SA, Yen TC, Fang JY (2012) Theranostic liposomes loaded with quantum dots and apomorphine for brain targeting and bioimaging. Int J Nanomed:1599–1611
56. Wang D, Imae T, Miki M (2007) Reprint of "fluorescence emission from PAMAM and PPI dendrimers [J Colloid Interface Sci 306:222–227]. J Colloid Interface Sci 312(1):8–13
57. Yin T, Wang P, Li J, Zheng R, Zheng B, Cheng D, Li R, Lai J, Shuai X (2013) Ultrasound-sensitive siRNA-loaded nanobubbles formed by hetero-assembly of polymeric micelles and liposomes and their therapeutic effect in gliomas. Biomaterials 34(18):4532–4543
58. Sun Q, Zhou Z, Qiu N, Shen Y (2017) Rational design of cancer nanomedicine: nanoproperty integration and synchronization. Adv Mater 29(14):1606628
59. Vijayasri K, Tiwari A (2023) A review on magnetic polymeric nanocomposite materials: emerging applications in biomedical field. Inorg Nano-Metal Chem:1–25
60. Hassan T, Salam A, Khan A, Khan SU, Khanzada H, Wasim M, Khan MQ, Kim IS (2021) Functional nanocomposites and their potential applications: a review. J Polym Res 28(2):36
61. Mishra AK, Ramaprabhu S (2011) Functionalized graphene-based nanocomposites for supercapacitor application. J Phys Chem C 115(29):14006–14013
62. Socol M, Preda N (2021) Hybrid nanocomposite thin films for photovoltaic applications: a review. Nanomaterials 11(5):1117
63. Alasmari A, Shariq M, Alhazmi NE, Alzahrani HS, Bouzgarrou SM, Alkhayri F, Alzahrani A, Alharbi SH, Azeez NA (2024) Enhancing perovskite solar cells with graphene-based nanocomposites for sustainable energy: a comprehensive review. Diam Relat Mater:111517
64. AlZaidy GA, Alanazi HT (2024) Recent advancements in perovskites solar cell materials and the investigation of transition metal oxide-based nanocomposites for usage in perovskites solar cells. ECS J Solid State Sci Technol 13(5):055006
65. Zolfaghari-Isavandi Z, Shariatinia Z (2018) Fabrication of CdS quantum dot sensitized solar cells using nitrogen functionalized CNTs/TiO_2 nanocomposites. Diam Relat Mater 81:1–15
66. Davari M, Karimi S, Tavasoli A, Karimi A (2014) Enhancement of activity, selectivity and stability of CNTs-supported cobalt catalyst in Fischer-Tropsch via CNTs functionalization. Appl Catal A 485:133–142
67. Gao F (2019) An overview of surface-functionalized magnetic nanoparticles: preparation and application for wastewater treatment. ChemistrySelect 4(22):6805–6811
68. Zhu CZ, Tian QH, Wang BH, Xu MT, Jin QJ, Zhang ZY, Le SK, Wu Y, Wei YC, Xu HT (2024) Application of modified cerium dioxide for photocatalytic air pollution purification. Rare Met 43(11):5473–5486
69. Taheri FS, Ghaemi A, Maleki A, Shahhosseini S (2019) High CO_2 adsorption on amine-functionalized improved mesoporous silica nanotube as an eco-friendly nanocomposite. Energy Fuels 33(6):5384–5397
70. Sarfraz N, Ashraf M, Ali S, Khan I (2023) Magnetism-driven iron oxide nanocomposites for energy and environmental solutions: harnessing magnetism. Mater Today Sustain 24:100589

Chapter 11
Quantum Dots and Functionalized Magnetic Nanomaterials (FMNs): Synergistic Effects in Advanced Technologies

Roopali Sharma, Sakshi Singh, and Jaya Tuteja

Abstract Quantum dots (QDs) and functionalized magnetic nanomaterials (FMNs) have emerged as significant materials in the development of advanced technologies, owing to their unique properties and wide range of applications. This chapter explores the synergistic effects of combining QDs and FMNs, highlighting their enhanced functionalities when used together in various fields, including electronics, energy, medicine, and environmental management. The chapter begins with an overview of nanomaterials and their evolution in advanced technologies, emphasizing the demand for multifunctional materials. Quantum dots, due to their size-dependent electronic and optical properties, are discussed in detail, including their structure, synthesis, and diverse applications such as in optoelectronics and biological imaging. Furthermore, functionalized magnetic nanomaterials are examined, with a focus on their types, functionalization methods, and applications in catalysis, drug delivery, and environmental remediation. The integration of FMNs with QDs enhances their magnetic, catalytic, and optoelectronic properties, leading to more efficient and sustainable technologies. The chapter concludes by exploring future directions for the development of QD-FMN hybrid materials, which hold the potential for transformative advancements in energy storage, medical diagnostics, and environmental solutions.

Keywords Quantum dots · Functionalized magnetic nanomaterials · Hybrid materials · Advanced technologies · Synergic effects

R. Sharma · S. Singh · J. Tuteja (✉)
School of Basic Sciences, Galgotias University, Greater Noida, U.P., India
e-mail: jayatuteja89@gmail.com

S. Gulati (ed.), *Functionalized Magnetic Nanomaterials*, Nanostructure Science and Technology, https://doi.org/10.1007/978-3-031-97199-0_11

11.1 Introduction

11.1.1 Overview of Nanomaterials in Advanced Technologies

Nanomaterials have found a place in novel research for decades now, and have been utilized for a variety of applications in various fields from defence to medicine. The recent approach for developing multifunctional materials is the need of the hour. We need new and improved materials, particularly multifunctional nanomaterials with more than one function, to progress beyond current technology. For example, structural batteries integrate energy storage into the body of the car, and there is no requirement for massive battery packs in the boot or under the floor. By contrast, theranostic (therapeutic and diagnostic) nanoparticles enable targeted drug delivery in the body through the integration of therapeutic and diagnostic functions. These nanoparticles may also possess other functions, such as the ability to convert light to heat to destroy cancer cells or to fluoresce for tracking [1]. The development of multifunctional nanomaterials paves the way for revolutionary advances in a range of fields, including medicine. Creating new and novel features that the bulk materials with typical properties cannot achieve is most frequently the goal. Two kinds of nanomaterials are merged to achieve hybrid materials in case one lacks all desired features or multifunctional properties. Mechanically robust yet electrically conducting graphene or nanotubes, for instance, can be able to host catalytic metal or oxide particles. Conducting particles, on the other hand, can be dispersed into an insulating dielectric polymer host to achieve nanocomposites [2]. There is a wide array of materials to pick from, and unlimited applications where multifunctional materials can be used. Fullerenes, ceramic and metal nanopowders, and colloidal ceramic, metal, and organic particles such as proteins, DNA, carbon, and other nanotubes, nanofibers, and nanowires can all be used as zero-dimensional nanoparticles to prepare multifunctional nanomaterials. Two-dimensional (2D) materials are one of the fastest-developing areas, the most renowned of which is graphene. This is supplemented by many recently discovered 2D compounds such as silicene, germanene, borophene, and MXenes (2D transition metal carbides and nitrides). Conventional materials also fall into this category that are being rediscovered, i.e., boron nitride, transition metal dichalcogenides, and oxides/hydroxides, such as clays. 2D materials are preferred due to their property of interconnecting or interlocking into sheets and layers. The rising need for printable and solution-processable nanoparticles is on the rise due to upcoming technologies like the Internet of Things and flexible and wearable electronics, which can further be utilized for the development of new flexible, multifunctional nanomaterials [3].

11.1.2 Quantum Dots (QDs): Structure, Properties, and Applications

Quantum dots have emerged as be recent material of interest in various sectors of nanomaterials for a wide range of applications. A quantum dot is defined as a tiny artificial structure that can hold single particles of electrons or thousands of them. The majority of these "charge droplets" are just a few nanometers or a couple of microns within the size range. However, advanced skills in nano-fabrication enable precise control of the shape, size, or interactions of an object. "False atoms" is one of the ways through which one can define a quantum dot. Quantum dots are typically considered to be equivalent to atoms that are formed only from electron-devoid holes that have a positive charge since they lack an electron in the valence band. Holes are commonly referred to as dots due to their small diameters. At the nanoscale level, electrons begin orbiting around the holes. The absence of protons and neutrons from the core of quantum dots results in quantum dots possessing a noticeably lighter mass relative to traditional atoms. Because quantum dots have a mass that is far less than an atom's, they are, as a result, able to have larger orbits than those of traditional atoms. This weaker nuclear force further allows electrons to be more freely bound within the orbital levels [4]. The simplest way to describe finite quantum systems with interacting particles is to assume that the interactions among the particles and any external constraining forces give rise to a so-called "mean field." Because this field can often be mimicked using an effective potential, we can consider the particles to be moving independently [5]. Quantum dots behave like semiconductor materials owing to their charge confinement properties, and due to the close confinement of charge carriers, only a few self-assembled quantum dots with only a few discrete energy levels—three to five—exist. Due to this confinement, quantum dots possess limited space for only a few electrons and holes, and thus, they are ideal for the construction of devices that need few charge carriers to function. A structure that encapsulates charge carriers in one dimension is known as a quantum well. A small layer with a reduced band gap is placed within layers of a material with a greater band gap, and it is formed using layer-by-layer growth of two or more disparate materials. Semiconductor heterostructures known as self-assembled quantum dots have charge carriers in all three dimensions. The structures are fabricated using various fabrication methods. By slicing quantum wells into slender pillars or bars, microfabrication techniques such as lithography and etching allow the generation of quantum wires or quantum dots with only a small portion of the original well remaining. For confinement of charge carriers in a reduced volume, a repulsive potential is introduced following the formation of electrical contacts on the surface of a quantum well. Moreover, a spontaneous self-assembly process is a natural process that enables ensembles of quantum dots to spontaneously self-assemble. The natural process has been found to be one of the best methods of preparing high-quality 0D semiconductor structures [6]. Quantum dots find a variety of applications in different fields owing to their excellent electronic and optical properties. Over the past ten years, quantum-dot-based light-emitting diodes, or QLEDs, have attracted a lot of attention from researchers and industry. Down conversion, or the transformation of high-energy light into lower energy light, is a common use for quantum dots (Qdots). Compared to conventional inorganic phosphors and organic dyes, quantum dots have a number of benefits, such as high quantum yield (QY), a wide spectrum of emission, more stability, more color saturation than traditional phosphors. The simultaneous stimulation of quantum dots of varying sizes is made possible by a broad absorption window that stretches from the band gap to the UV. QD-based photonic devices like room-temperature lasers and optical amplifiers are still being researched. QDs, on the other hand, are being used in biology for a number of reasons, including their higher extinction coefficients, higher QYs, reduced photobleaching, size-tunable absorbance and emissions, generally broad excitation windows but narrow emission bands, capacity to be used in a wide range of assays with minimal interference, potentially lower toxicity than traditional organic dyes, and capacity to be functionalized with a range of bioactive molecules [7].

11.1.3 Functionalized Nanomaterials (FMNs): Types and Functionalization Methods

With its widespread uses in industry ranging from medical research, electronics, energy, and environmental technologies, nanomaterials have had profound effects on the material science discipline and are in rising demand worldwide. The size-dependent peculiar properties of these materials render them highly suitable for novel applications, with usually varying sizes in the range between 1 and 100 nm. The quantum confinement effect of electrons significantly enhances a material's electrical, structural, magnetic, optical, and catalytic properties as the material is reduced to the nanoregime. Moreover, small materials' large surface area-to-volume ratios can be utilized in a way to influence reaction rates with fault or surface area dependence. In all, nanomaterials offer permanent solutions to crucial challenges in energy, medicine, and the environment. Control over their dispersion in different matrices and understanding the behavior of NPs with other molecules are important in order to realize the full potential of the NPs in other fields like biomedicine, nanocomposites, imaging, sensing, energy, and electronics. Through the proper functionalization of the NPs, where the surfaces are engineered accordingly to enhance their behavior with the surrounding environment, it can be achieved. Dispersion, assembly, and colloidal stability of NPs are all controlled by functionalization [8]. Functionalized magnetic nanomaterials can be of different types, including carbon-based, metal and metal oxide-based, polymeric or silica-based, ceramic-based, composite-based based or quantum dot-based. Fullerenes, carbon nanotubes, graphene and graphene derivatives, graphene oxide, nanodiamonds, and quantum dots based on carbon are all materials that come under the ambit of carbon-based nanomaterials. On account of their enhanced mechanical, electrical, thermal, optical, and chemical characteristics and distinctive structural dimensions, such materials have been of great interest in a wide variety of sectors, including biomedicine [9]. Polymer chains can be introduced into soft polymer nanoparticles to stabilize them and modify their response to various solvents. Their applications across all industries have promising new opportunities because of this modification [10]. The self-assembly fabrication, scalability, and distinct optical signatures of colloidal photoluminescent quantum dots (QDs) are well established. They currently form a new class of optical probes widely utilized for biomedicine research after successful functionalization with engineered surfaces [11]. Functionalized nanomaterials can be synthesized by various methods, viz. Direct Functionalization, postsynthetic functionalization, grafting functionalization, functionalization by a noncovalent bond via physisorption, and functionalization by a covalent bond via chemisorption. Functionalized nanomaterials can also be synthesized using intrinsic surface engineering, which can be done by defect engineering.

11.1.4 Synergistic Effects of QDs and FMNs

Synergistic activity of nanomaterials is critical in the design of new, low-toxicity intracellular and subcellular drug delivery systems. Nanoparticle translocation, outer encapsulation, embedding, and internal cellular anchoring are all suitable to explain the underlying interaction that is a result of this synergy at the bio/nano interface. Shape, size, surface area, ligand chemistry, and charge of the nanoparticles are critical parameters that affect translocation. Quantum dots (QDs) and functionalized nanomaterials complement each other's unique properties and operate more efficiently in various applications, including energy harvesting, sensing, imaging, and catalysis. These materials interact chemically, electrically, or optically to produce these effects. Functionalized nanomaterials like metal nanoparticles and graphene can act as energy donors or acceptors, enabling Förster Resonance Energy Transfer (FRET) [12] and enhancing the efficiency of QDs in light emission and absorption. Additionally, through plasmonic effects, metal nanomaterials such as silver (Ag) and gold (Au) can enhance the photoluminescence (PL) intensity of QDs, resulting in improved photostability and brightness. When conductive nanomaterials such as graphene or carbon nanotubes (CNTs) are combined, they enhance the electrical characteristics of compounds and enable faster charge transfer. This enhancement leads to improved photocatalysis and increased efficiency in solar cells. Moreover, this combination reduces electron–hole recombination and facilitates more effective charge separation, both crucial for enhancing performance in energy-related applications. Additionally, functionalized nanomaterials (FMNs) protect quantum dots (QDs) from oxidation and degradation, providing defence. Consequently, QDs have a prolonged lifespan in biological and environmental settings, making them more beneficial for specific purposes such as medication delivery and biological imaging [13]. As a result of this process, the boosting or quenching of fluorescence is enhanced, making quantum dots (QDs) very useful in environmental monitoring, chemical sensing, and biosensing. Studies have shown that the engagement of QDs and functionalized nanomaterials (FMNs), particularly ceramic-based FMNs, results in enhanced catalytic and photocatalytic activities [14]. These improvements are achieved by increasing the absorption of light and separating charges, thus optimizing CO_2 capture, pollution destruction, and photocatalytic water splitting. Furthermore, this synergy facilitates multi-step catalytic reactions, which increases the overall efficiency of organic synthesis and hydrogen fuel production.

11.2 Quantum Dots: Fundamentals and Innovations

QDs are small semiconductor nanoparticles with dimensions of a few nanometers (ranging from a few to tens of nanometers) [15] that possess one of the most important properties of quantum confinement [16]. QD has different properties from the characteristics of bulk materials, such as size, because the size of the QD is small. The high surface/volume ratio increases the bulk material, but it is very responsive (or active). At this scale, the electronic and optical properties of the material start to deviate from those of bulk materials, such as sharp absorption and emission spectra. In addition to this, QDs have a remarkable ability to tune the band gap, which can be effectively controlled by tuning the size of the QDs [17, 18]. Bulk materials lack this characteristic. When these properties are carefully regulated, QDs prove highly desirable for a variety of uses in fields including drug delivery, sensing, and light-based therapies [19].

11.2.1 Synthesis Methods of Quantum Dots

Numerous pathways have been established for the synthesis of QDs with regulated dimensions, size, shape, and structure; nevertheless, commonly employed methods can be divided into two categories: top-down and bottom-up [20]. Biosynthesis using microorganisms is a green alternative to chemical methods [21]. Only when nanostructured materials are made available with the required size, shape, morphology, crystal structure, and chemical composition will new physical attributes and uses of nanoparticles become feasible [22].

11.2.1.1 Top-Down Approach

The top-down method divides the main material into particles or structures that are nanoscale in size. Physical procedures are the foundation of this strategy. These are merely a continuation of those techniques, which are utilized to create micro-sized particles. Since top-down methods rely on either the removal or division of bulk material or the miniaturization of any bulk production process to create the necessary nanostructures, they are comparatively simpler. However, this method has a flaw in that the surface structure that is created is not ideal. This method produces nanostructured materials that are not smooth, have surface structural flaws, and are contaminated with various contaminants. High-energy wet ball milling, laser ablation, electron beam lithography, liquid phase exfoliation, atomic force manipulation, gas-phase condensation, and aerosol spray are a few examples of top-down approaches [23, 24].

11.2.1.2 Bottom-Up Approach

The bottom-up method is another strategy for this [21, 25, 26]. The majority of these techniques rely on chemicals. These are also cost-effective. With this method, a material is constructed from the bottom up, atom by atom, molecule by molecule, cluster by cluster, etc. Examples include template-assisted sol–gel, hydrothermal synthesis, electrodeposition, reverse-micelle synthesis, organometallic chemical synthesis, and colloidal precipitation. QDs can be created using a variety of approaches, such as electrochemical methods, hydrothermal and solvothermal synthesis, colloidal synthesis, and molecular beam epitaxy (MBE).

11.2.1.3 Electrochemical Methods

Electrochemical synthesis is another method in which metal ions are reduced in an electrochemical cell to produce QDs that form nanoparticles on the electrode surface, often involving electrochemical exfoliation [27]. This method allows for fine adjustment of QD properties by adjusting parameters such as applied voltage, electrolyte composition, and reaction time [28]. For the fabrication of QDs in thin-film or patterned formats for electrical and sensing applications, electrochemical synthesis is especially helpful. Another method used is electrodeposition, which uses an electrical potential to provide controlled deposition of nanomaterials onto conductive surfaces.

11.2.1.4 Hydrothermal and Solvothermal Synthesis

Another important synthesis method for QDs is hydrothermal and solvothermal synthesis, which uses a sealed autoclave with high temperatures and pressures. The solvothermal approach uses organic solvents or a combination of solvents, whereas the hydrothermal approach uses water as the solvent. This process involves the gradual crystallization of precursors and is especially helpful for creating QDs with intricate compositions, such as doped and transition metal-based QDs[29]. Because of its environmentally friendly nature and mild reaction conditions, hydrothermal synthesis has become a popular substitute for conventional chemical processes [30]. In their study, Zhu et al. demonstrated a simple hydrothermal process for creating homogeneous, monodispersed SnO_2 QDs with a small diameter range (2.3 to 3.1 nm).

In this method, the hydrazine hydrate, or $N_2H_4H_2O$, is essential to the hydrothermal process explained here since it functions as a ligand and an alkali forms a complex cluster by coordinating with the Sn ions [31]. In another study by Wei-hai Yang et al. [32], the hydrothermal approach provides a simple means of producing high-quality cysteamine-stabilized CdTe QDs, with a quantum yield (QY) of up to 19.7% under optimal circumstances. Compared to the aqueous technique, which involved growing the CdTe QDs at 220 °C for 20 min, the growth rate is faster. The absorbance peak moved from 490 to 640 nm throughout this period, while the mean particle sizes went from 2 to 4 nm.

Hydrothermal synthesis enables the production of quantum dots, including nitrogen-doped carbon quantum dots (CQDs), creating NCQDs. This has effectively demonstrated its value as a useful imaging agent that can be advantageous for specific applications like bioimaging. Due to its simple operating conditions, like minimal energy with minimal requirements and low-cost equipment, hydrothermal synthesis is considered the most accessible method. Moreover, hydrothermal synthesis has few drawbacks. Obtaining very uniform and monodisperse QDs might be tough due to the high-pressure and temperature conditions that are necessary [33].

11.2.1.5 Colloidal Synthesis

Colloidal synthesis is a commonly utilized technique with the high-temperature variant for creating QDs efficiently [34]. In 1993, Murray, Norris, and Bawendi initially suggested this chemical method for generating QDs. The procedure entails dissolving metal ions or organometallic compounds, referred to as precursors, in a liquid medium, and subsequently, exposing them to different chemical or physical processes to produce the intended QDs. The benefit of this synthesis technique is its capability to easily create QDs of different sizes, rendering it highly adjustable [35, 36]. A well-known technique for colloidal synthesis is the 'Hot Injection Method.' In this process, a solution with the necessary components is swiftly added to a hot, high-temperature reaction mixture. Consequently, the precursors swiftly react to produce QDs, which then rapidly cool and solidify. By modifying the type, the amount of the precursor, the time of reaction, and the temperature during the synthesis, the dimensions and makeup of Colloidal QDs can be regulated [37]. This method enables the quick and effective creation of premium-quality QDs with a uniform size distribution available in large amounts.

Irshad Ahmed et al. [38] showcased the production of CdSeQD using the 'Hot Injection Technique' compared to the 'Ambient Temperature Technique. It was observed that the smallest QD formed by employing the HI method measured 2.5 nm in size. The RT method could only produce QDs with a size of 3.3 nm. This approach to synthesis provides numerous benefits compared to alternative methods, encompassing exact regulation of the dimensions and form of the QDs, superior purity, and minimal defect density.

11.2.1.6 Molecular Beam Epitaxy

A more advanced technology for precisely fabricating quantum dot structures is molecular beam epitaxy (MBE), a physical vapor deposition method. Using this technique, ultra-pure elemental sources are deposited layer by layer onto a substrate after being evaporated in a high vacuum atmosphere. Lattice mismatches between the deposited and substrate materials cause the spontaneous generation of quantum dots, which is how QDs grow through strain-induced self-assembly. Because MBE offers superior control over the homogeneity of quantum dots, it is well suited for use in sophisticated semiconductor devices, lasers, and quantum computing.

11.3 Surface Modification and Passivation

Modifying and passivating the surfaces of quantum dots (QDs) is crucial for enhancing their stability, optical characteristics, and suitability for different applications. Quantum dots, as nanoscale substances, possess a high surface-to-volume ratio, making their surface chemistry essential for influencing their effectiveness. Quantum dots that are non-passivated frequently experience surface defects, oxidation, and non-radiative recombination, resulting in decreased photoluminescence efficiency and low stability [39]·. To address these challenges, numerous strategies have been created for surface modification and passivation. Passivation with ligands is a well-established method to reduce surface trap states and prevent non-radiative processes.

Surface modification typically entails the chemical treatment of QDs to add functional groups or ligands that can adjust their solubility, biocompatibility, and binding affinity. Passivation, conversely, entails forming a protective layer or coating around the QD core to reduce surface flaws and avoid deterioration. Ligand exchange is a key surface modification technique that involves swapping out the original surface ligands with new ones in order to accomplish particular functionality. For example, halogenation can first passivate Carbon quantum dots (CQD) surfaces using a cascade surface modification (CSM) technique [39]. This is followed by reprogramming with functional ligands to adjust the solubility and doping character. Both n-type and p-type CQD inks that are miscible and retain surface passivation can be produced using this technique. By producing stable colloids using compounds like 1-thioglycerol (TG), 2-mercaptoethanol (ME), and cysteamine (CTA), the CSM approach expands the variety of molecules that can be used on CQD surfaces. Figure 11.1 depicts the surface modifications of quantum dots.

Electronic states within the gap caused by surface flaws on QDs can negatively impact charge carrier dynamics and overall performance. By minimizing these flaws, passivation methods seek to improve photoluminescence and quantum yield. There are several ways to accomplish effective passivation, such as applying inorganic shells, like $Cd1 - xZnxSe$, on the QDs or employing Lewis acidic (Z-type) ligands to eliminate surface traps. For instance, by passivating unsaturated indium atoms on

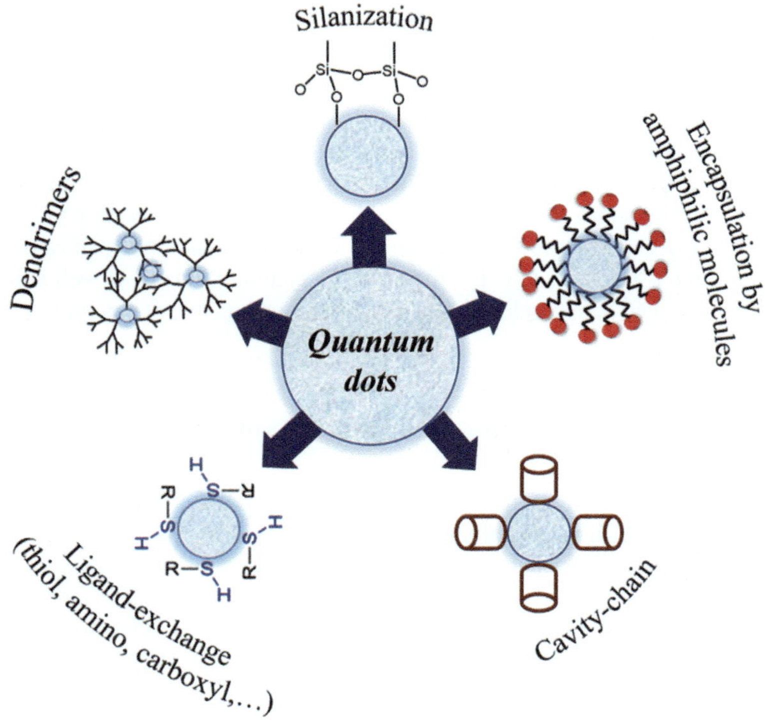

Fig. 11.1 Surface modification strategies for quantum dots [42]. (Open access)

the QD surface, BZA can specifically eliminate electron trap states in InP QDs [39]. A typical method for passivation involves developing a core–shell structure, in which a thin layer of a broader bandgap material, like ZnS, is applied around a core material such as CdSe or InP [40]. This core–shell design minimizes surface imperfections and restricts charge carriers to the core, resulting in improved quantum yield and stability. Moreover, inorganic passivation using substances like silica (SiO_2) or metal oxides offers chemical strength and resistance to environmental deterioration [41].

Passivation and surface engineering facilitate the incorporation of quantum dots into hybrid and composite materials, including perovskite-QD heterostructures or QD-polymer films for optoelectronic uses. The inorganic passivation layers are built on the surface of a perovskite as shown in Fig. 11.2. They enhance charge transport characteristics and boost device efficiency in uses like light-emitting diodes (LEDs), solar panels, and sensors. In general, strategies for surface modification and passivation are essential for maximizing the potential of quantum dots in different fields. Ongoing improvements in these methods are resulting in more stable, efficient, and eco-friendly quantum dots, thereby broadening their range of applications in innovative technologies.

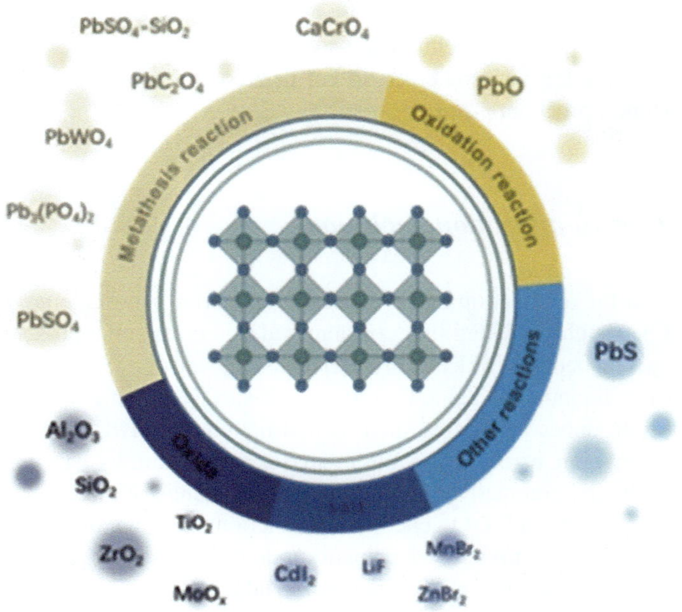

Fig. 11.2 Classification of inorganic passivation layers built on the surface of a perovskite [42] (Open access)

11.4 Functionalized Magnetic Nanomaterials: Fundamentals and Types

Among the many drawbacks of bare nanomaterials are their increased aggregation, low surface reactivity, and poor mobility. Functionalization can eliminate these restrictions, which aids in the creation of environmentally acceptable engineered NMs with enhanced surface reactivity, stability, and mobility, as well as a narrow bandgap. The functionalized magnetic nanomaterials (FMNs) are nanomaterials that have been modified after treatment to have unique properties that are not found in their volumetric analogues [43]. Nanomaterials, in principle, are chemicals or materials with a unit size of 1 to 100 nm in at least one dimension. Functionalization performed using chemical or physical methods includes the assembly of various organic and inorganic materials on a surface of size with covalent bonds [44].

To fully investigate the potential of magnetic nanomaterials in the fields of biomedicine, nanocomposite technology, imaging, sensing, energy, electronics, etc. It is essential to regulate how FMNs interact with other molecules and how they disperse in various matrices. This is only possible if the FMNs are properly modified to functionalize their surface, which improves their interaction with the environment [45, 46]. The overall functionalization of NPs affects regulated assembly, dispersion, and colloidal stability. For instance, surface functionalization of biomolecules in medical applications can offer overall biocompatibility, high cellular internalization ability, effective intracellular transport, and stealth characteristics. When it comes to customizing nanomaterials to get desired characteristics and functions, functionalization procedures are essential.

11.4.1 Functionalization Techniques

Depending on the application and kind of nanomaterial being altered, these techniques can be roughly divided into covalent and non-covalent approaches, each of which has special benefits [47]. Nanomaterials work using direct or indirect methods, such as organic ligands, inorganic fragments, and surface polymerization. Changing the surface introduces properties such as hydrophobicity, hydrophilicity, conductivity, and corrosion resistance. Simple surface modifications can easily induce the hydrophobicity, conductivity, and corrosion resistance of nanomaterials. TiO_2, Al_2O_3, Fe_2O_3, ZnO, and ZrO_2 chemically functionalized nanomaterials improve performance and reduce toxicity when incorporated into organic/epoxy coatings [8]). Direct or indirect functionalization of nanomaterials can be achieved by surface polymerization, inorganic moieties, and organic ligands. Properties including conductivity, corrosion resistance, hydrophilicity, and hydrophobicity can be added by surface modification. Nanomaterials' hydrophobicity, hydrophilia, conductivity, and corrosion resistance can all be readily induced by simple surface alteration [48].

11.4.1.1 Covalent Functionalization

Strong chemical bonds are created between functional groups and the surface of the magnetic nanomaterial during covalent functionalization [49]. One popular method is surface modification by silanization, in which reactive groups like amines, thiols, or carboxyls are added to nanoparticles by silane coupling agents. Organofunctional alkoxysilanes like 3-mercaptopropyl)-trimethoxysilane (MPTMS), 3-aminopropyl)-triethoxysilane (APTES), and 3-glycidoxypropyl)-dimethyl-ethoxysilane (GPMES) are frequently used in covalent functionalization techniques. N-hydroxysuccinimide (NHS), glutaraldehyde (GA), and 1-ethyl-3-(3-dimethylaminopropyl)carbodiimide (EDC) chemistry are also frequently used [8, 50]. These procedures can be used to functionalize proteins, peptides, oligonucleotides, and other NPs. This technique is frequently used to improve the stability and dispersibility of metal oxide and silica nanoparticles in a variety of liquids.

11.4.1.2 Non-covalent Functionalization

Noncovalent functionalization, on the other hand, is based on low interactions such as hydrogen bonding, π-π stacks, van der Waals forces, and electrostatic interactions.

This technique is particularly advantageous for preserving the inherent properties of nanomaterials while transmitting new features. Non-coded functionalization is based on weak interactions with the receptor system. Researchers add surface functionality through in situ synthesis and post-intake modifications. Various ligands, especially active surface materials, small molecules, polymers, dendrites, biomolecules, and inorganic materials, are used to functionalize NPs through a variety of methods and strategies.

11.4.1.3 Grafting Techniques

Another notable method is to vaccinate polymer chains for nanomaterials by polymerization of free radicals or radical polymerization (ATRP), improving their biocompatibility and solubility. In addition, the functionalization of carbon-based nanomaterials, such as graphene and carbon nanotubes (CNTs), can be obtained thanks to oxidation processes that introduce hydroxyl, carboxyl, and epoxy groups. These functional groups allow for other chemical modifications and allow for applications in drug administration, biological detection, and catalysis [45]. Polymer graft functionalization techniques include grafting and implantation methods.

11.4.2 Types of Functionalized Magnetic Nanomaterials

Functionalized magnetic nanomaterials encompass those that are derived from iron, iron oxide, cobalt, and nickel ferrite nanoparticles. These substances are studied for use in agriculture due to their distinctive magnetic characteristics and adaptability in functionalization. Magnetic nanoparticles (MNPs) typically comprise a magnetic material element, like iron, nickel, or cobalt, along with another functional unit that has a chemical origin.

Core Magnetic Nanoparticles

Magnetic nanoparticles may be fabricated with different shells. Numerous techniques are available for synthesizing MNPs, such as micro emulsion, thermal decomposition, co-precipitation, solvo-thermal, wave-based (sound, microwave), chemical vapor deposition, and combustion. The design approach incorporates size-based synthesis, the discovery of biocompatible linkers and ligands, tailoring size in relation to magnetism, and the determination of the most effective magnetic detection method for functional validation [51].

Materials forFunctionalization

Dextran, chitosan, polyethylene glycol (PEG), polyvinyl alcohol (PVA), oleic acid, polyoxamers, and polyamines are the usual coating materials. These can be bonded using chemical reagents or cross-linkers onto the surface of cobalt, nickel, manganese, or iron oxide MNP cores. Iron (Fe)-oxide-based magnetic nanoparticles are frequently used because they are biocompatible [51].

The classification of functionalized nanomaterials is based on their composition and the methods used to enhance their properties for specific applications. The main types encompass carbon-based nanomaterials, metal and metal oxide nanoparticles, polymeric nanoparticles, and hybrid nanomaterials.

Carbon-based nanomaterials: Due to their exceptional electrical, mechanical, and thermal properties, carbon-based nanomaterials, including graphene, carbon nanotubes (CNTs), and fullerenes, are currently undergoing extensive research. These nanomaterials can be modified through processes such as oxidation, polymer grafting, or biomolecule conjugation, making them suitable for various applications like energy storage, medication delivery, and biosensors [52]. For instance, modified CNTs allow for targeted medication administration and bioimaging, while functionalized graphene enhances the efficiency of supercapacitors.

Metal and metal oxide nanoparticles: Metal and metal oxide nanoparticles, such as gold (Au), silver (Ag), iron oxide (Fe_3O_4), titanium dioxide (TiO_2), and zinc oxide (ZnO), are widely employed in catalysis, medical imaging, and environmental remediation (Singh et al. 2018). Stability, dispersibility, and biocompatibility are enhanced through functionalization. For instance, gold nanoparticles with thiol-based ligands are utilized in biosensors and cancer treatments, whereas polymer-coated iron oxide nanoparticles act as contrast agents in magnetic resonance imaging (MRI) [53].

Polymeric nanoparticles: To enhance drug delivery, gene therapy, and controlled release systems, polymeric nanoparticles—such as dendrimers, micelles, and nanogels—are functionalized through surface modifications or encapsulating procedures. These materials are perfect for pharmaceutical applications because of their exceptional biocompatibility and adjustable qualities [54].

Hybrid nanomaterials: They integrate several components, such as metal–organic frameworks (MOF) or nanoparticles of nuclear-shelling, to achieve increased functionality. Functionalization allows them to be used in modern applications, such as target therapy, high-performance sensors, and energy-transforming technologies [55].

Table 11.1 summarizes different types of functionalized nanomaterials, their examples, functionalization methods, and key applications.

11.4.3 Challenges in Functionalization

Achieving uniform and regulated functionalization is one of the main obstacles. It can be challenging to guarantee constant attachment of functional groups due to differences in nanoparticle size, shape, and surface chemistry. This can result in batch-to-batch variability and decreased reproducibility in applications like drug administration and catalysis. Another major challenge is the retention of the intrinsic properties of nanomaterials with new functionalities. Covalent functionalization, though it gives strong and stable modifications, could affect electrical conductivity,

Table 11.1 Here's a tabular representation of the types of functionalized magnetic nanomaterials

S. no	Type of magnetic nanomaterial	Core material	Functionalization	Applications
1	Metallic magnetic nanoparticles	Fe, Co, Ni	Polymer coating, Silica shells	Catalysis, Biomedical imaging
2	Metal oxide nanoparticles	Fe^3O_4, γ-Fe_2O_3	Organic ligands, Surfactants	Drug delivery, MRI contrast agents
3	Core–shell nanoparticles	Fe^3O_4@SiO_2, Fe^3O_4@Au	Biomolecules, Polymers	Targeted drug delivery, Biosensors
4	Doped magnetic nanoparticles	Mn-doped Fe^3O_4, Co-doped Fe_3O_4	Ligands, Polyelectrolytes	Magnetic hyperthermia, Bioseparation
5	Hybrid magnetic nanomaterials	Magnetic core with carbon, gold, or polymer shell	Antibodies, DNA, Enzymes	Immunoassays, Targeted therapy
6	Superparamagnetic nanoparticles	Fe^3O_4 (magnetite), γ-Fe^2O^3	PEGylation, Silanization	MRI contrast, Hyperthermia treatment

optical properties, or mechanical strength, thereby reducing their performance in applications such as nanoelectronics and sensing [49]. Non-covalent functionalization, while retaining these properties, typically yields weaker interactions, making the functional groups less stable in harsh environments.

Biocompatibility and toxicity pose another obstacle, in particular for biomedical applications. The introduction of functional groups or surface coverings can modify the interaction of nanomaterials with biological systems [49], which potentially leads to undesirable effects. Scalability and cost-effectiveness are also significant concerns. Many functionalization techniques involve complex chemical processes, expensive reagents, and time-consuming synthesis steps, making large-scale production difficult.

11.5 Synergistic Effect in FMNs Integrated with QDs and Their Applications

11.5.1 Synergistic Effects in FMNs Integrated with QDs

Recently, hybrid nanostructures comprising functionalized magnetic nanomaterials integrated with quantum dots have illustrated exceptionally tunable desired magnetic, optical, and electronic properties. These heterostructures display immense potential for advancements in the technological and scientific fields. To address the challenges faced in the widespread commercialization of quantum dot technology related to its thermal, mechanical, and chemical instability, researchers have proposed a promising solution of integrating the quantum dots with the functionalized magnetic nanomaterials. Herein, the quantum dots are embedded in the protective, stable matrix of FMNs to improve the chemical, mechanical, magnetic, and thermal stability of hybrid nanocomposites of QDs and FMNs for a broader adoption by numerous industries. In a report, solvothermal deposition and electrochemical deposition were employed in a two-stage process to form composites consisting of a carbon nanotube (CNT) film with an Fe_3O_4/MnO_2 coating. The CNT film had a nanoscale core–shell structure formed on the surface by the magnetite nanoparticles and the MnO_2 sheets. Due to the high specific surface area and interfacial polarization, the composites' unique core–shell and three-dimensional network structure is believed to be beneficial for their ability to absorb electromagnetic waves [56]. N-MWCNTs with Co@Au core–shell NPs at their tips were made by Terrones and colleagues. Due to its enormous surface area and biocompatibility, the carbon nanotube (CNT) was used as a drug delivery unit in this composite material, which could also be used as a magnetoplasmonic system. Ferromagnetic and superparamagnetic-like behavior of these core–shell NPs rendered them attractive to be used in nanomedicine, where they can be used to target or destroy cancerous cells. Equipment used to synthesize the said composite material comprised a high-vacuum deposition chamber [57]. There has been research on the liquid-phase filling of carbon nanotubes (CNTs) with uniformly shaped magnetic Co-based nanoparticles reported by Baaziz et al. Such composites find applications in wastewater treatment, batteries, electrochemical devices, magnets, and catalysis [58]. $CoFe_2O_4$-CdS core–shell nanostructures were recently synthesized with a chemical precipitation synthesis route, which is a novel method. It was examined how the rice husk ash nanocomposites and microparticles influenced the mechanical properties of epoxy resin. With the introduction of both types of particles, the tensile strength was increased, which was attributed to the interactions among the epoxy fibers, the NPs, and the rice husk ash particles. The capacity of the epoxy fibers to interact with each other was increased [59]. A low-cost, facile modified solvothermal synthesis was employed to efficiently synthesize a novel magnetic PECQDs/$MnFe_2O_4$ nanocomposite. Due to the abundance of amide groups, the obtained PECQDs/$MnFe_2O_4$ nanocomposites possessed high U(VI) adsorption capacity and preserved the fluorescent and magnetic properties of $MnFe_2O_4$. The present study provided PECQDs/$MnFe_2O_4$ at a low price for the adsorption of wastewater containing U (VI) and shows potential for practical application [60]. Synergistic engineering of hybrid heterogeneous magnetic nanostructures has opened advantageous possibilities to fabricate the desired material with sustainable design, tunable properties, environmentally benign, biocompatible, stable nanomaterials for myriad technologically advanced fields. The applications of these tunable nanostructures in biomedical, electrical, and optical fields are discussed in the next section of this chapter.

11.5.2 Biomedical, Environmental, and Optoelectronic Applications of FMNs Integrated with QDs

As the QDs, due to their unique and tunable optical, magnetic, mechanical, chemical, and electrical characteristics, offer a replacement for the conventional fluorescent materials, their integration with various functionalized nanomaterials has opened various opportunities in the biomedical field. The synergistic effect due to the combination of distinct optical properties of quantum dots along with targeting capabilities of functionalized magnetic nanomaterials facilitates the efficient delivery of imaging agents and drugs to the target site inside the body. The integrated FNM-QDs hybrid nanostructures can yield significantly improved functionality in comparison to the individual FMNs and QDs alone. This makes the combined hybrid nanostructures more promising in the field of therapeutic techniques, target drug/gene delivery, optimized diagnosis, photodynamic therapy, biosensing, and bioimaging. Earlier, Shibu et al. conjugated superparamagnetic nanoparticles of iron oxide (Fe_2O_3 NPs) with quantum dots (CdSe/ZnS QDs) to design a photouncaging Fe_2O_3@CdSe/ZnS QDs magnetic probe having both magnetic and fluorescent characteristics. Upon photoactivation, the uncaging of the particles in a systematic manner was reported with continuous variation in the NMR, mass, and absorption spectra of the ligands. The biocompatibility of these conjugated magnetic NPs and QDs shows cell viability over 90%, which was evaluated with the help of MTT cytotoxicity assays. These site-specific probes were used for magnetic resonance imaging (MRI) of mouse melanoma cells and human lung epithelial adenocarcinoma cells, in vivo and in vitro studies with fluorescence images showing renal clearance after the desired targeted application [61]. He et al. reported designing of DNA-templated Au NPs-QD disassembly for the catalytic molecular imaging of microRNAs in living cells (miRNA), which are related to cancer, with the capacity of significantly amplifying the photoluminescence (PL) of QDs. The report shows that a single miRNA molecule may catalytically dismantle several QDs with the GNP, resulting in a significant enhancement of QD's PL for miRNA imaging. The limit of detection via the catalytic amplification method is three orders of magnitude higher in comparison with the imaging probe used, which is non-catalytic in nature. This sensitive detection method makes the utilization of miRNA imaging in living cells to accurately and easily differentiate between the non-cancerous and cancerous cells in situ and in vitro, creating new possibilities for the ultra-sensitive detection of low-abundant miRNA molecules in living cells, providing a significant contribution to clinical diagnostics [62]. In another report, Yao et al. developed a multifunctional probe fabricated using magnetic mesoporous silica nanoparticles (MMSN) and graphene quantum dots (GQDs) for synergistic therapeutic application for controlled drug delivery, photothermal therapy, and magnetic hyperthermia. The MMSN for carrying drugs and thermoseeds, and GQDs as caps and photothermal generators locally. The researchers studied this probe for magnetic hyperthermia performance, drug release kinetics, structural properties, photothermal traits, and complete therapeutic efficacy synergistically. As per the report, the 100 nm monodisperse MMSN/GQDs could load doxorubicin (DOX), followed by its release at lower pH. Additionally, the nanoparticles could generate heat, achieving the hyperthermic temperatures in the presence of an alternating magnetic field/near-infrared irradiation. Using breast cancer 4T1 cells as an experimental model system, the findings show that the combination of chemo-magnetic hyperthermia therapy or chemo-photothermal therapy, utilizing the DOX-loaded MMSN/GQDs nanosystem, offers a highly enhanced synergistic effect over chemotherapy or magnetic hyperthermia or photothermal therapy alone, resulting in increased efficacy in cancer cell destruction. Hence, the MMSN/GQDs provided a multifunctional platform with significant potential for the development of cancer therapy and enhanced therapeutic efficacy [63].

For FL/CT/MRI triple-modal bioimaging, Li et al. have presented the synthesis of multifunctional C-Fe_3O_4 QDs nanoprobes by a new and facile one-pot hydrothermal process. The resultant C-Fe_3O_4 QDs exhibit excellent water dispersibility, FL performance (optimal FL quantum yield of approximately 21.6%), photostability, superparamagnetic behavior, and biocompatibility in the mentioned concentration range due to this synthesis process. The as-prepared C-Fe_3O_4 QDs have a high r_2 value of approximately 154.10 mm^{-1} s^{-1} based on T_2 relaxometry experiments. Most importantly, the X-ray attenuation measurements indicate that the C-Fe_3O_4 QDs, as synthesized, distinctly demonstrate an X-ray attenuation effect. The synthesized C-Fe_3O_4 QDs can be used as a potential contrast agent for triple-modal FL/MR/CT imaging of tumors in biomedical applications, based on the in vivo FL, MR, and CT bioimaging studies (Liu, et al., 2016). Recently, Zhang et al. proposed a dual-functional nanoplatform of Fe_3O_4@CQDs coated PEGylated SWCNTs that can serve as a drug loading carrier for cancer combination therapy and imaging, as well as a photothermal heater, an ROS generator under NIR, and an imaging probe for multimodal imaging at the same time. The modification with the sgc8c aptamer enhances the material's ability to be absorbed by cells, and the prepared SWCNTs-PEG-Fe_3O_4@CQDs appear to be of good biocompatibility. In addition, various studies on the chemotherapeutic activity have been performed on HeLa tumor-bearing animals that were administered intravenous injections of SWCNTs-PEG-Fe_3O_4@CQDs/DOX-Apt over the years. Compared to the control group, the total inhibition of tumor growth by the aliquots of SWCNTs-PEG-Fe_3O_4@CQDs/DOX-Apt proves that they can specifically accumulate in cancerous tumor issues. Hence, nanocarriers fabricated from SWCNTs-PEG-Fe_3O_4@CQDs are a promising drug delivery technology for the therapy of cervical cancer and other diseases that require targeted drug delivery [64].

Recently, the synergistic effects obtained by conjugating the QDs and functionalized magnetic nanomaterials have highlighted new developments in the field of environmental remediation by providing environmentally-benign nanoprobes for the detection and eradication of toxic pollutants. These fabricated nanoprobes have been successfully used for designing the electrochemical, opto-chemical, magnetic, and chemical sensors for detecting and quantifying the pesticides, toxic gases, volatile organic compounds, heavy metals, toxic dyes, and chemicals, etc. Recently,

Ma et al. demonstrated a simple and effective one-pot synthesis route to simultaneously encapsulate two functional material types (dual-emissive RhB-CDs and Fe_3O_4 MNPs) within the MOF cavities. This strategy unites the numerous properties of encapsulated functional materials with MOFs' strong adsorption ability. The as-prepared MOF-based nanoprobe performed satisfactorily in the detection of HClO and shared the advantages of magnetic separation, stable MOF adsorption, and ratiometric sensors. Based on the experimental results, this could be a simple and efficient way of developing new functional MOF-based materials [65]. Huang et al. reported the fabrication of biocompatible photocatalyst carbon quantum dots and $ZnFe_2O_4$ for purifying the air and removing NO. Due to the synergistic effect, the CQDs/$ZnFe_2O_4$ (15 vol%) illustrated better photocatalytic activity (8 times higher) in comparison with pristine $ZnFe_2O_4$ for the removal of NO_x gas and greater selectivity for the formation of nitrate under the radiation of visible light. The CQDs have proven to act as a reservoir of electrons and transporter along with a highly powerful energy-transfer component for the photocatalytic conversions using CQDs/$ZnFe_2O_4$, indicating a promising probe for the photocatalysis of NO_x in order to have purified air [60]. In another report, Naghshbandi et al. synthesized several new magnetic nanocomposites of the MFe_2O_4/GQD type (M = Ni, Co) via a straightforward co-preparation route. Subsequently, the new nanocomposites' catalytic performance for the reduction of p-nitrophenol was investigated. $NiFe_2O_4$/GQD > $CoFe_2O_4$/GQD > $NiFe_2O_4$ > $CoFe_2O_4$ was the sequence of catalytic activity. Notably, among all the materials tested, the nanocomposite $NiFe_2O_4$/GQD exhibited the highest catalytic activity. For this material, the time of completion of reduction to p-aminophenol was 60 s, and the observed rate constant was 3.56 min^{-1}. These results suggest that the $NiFe_2O_4$/GQD nanocomposite is a promising catalyst for the reduction of nitro aromatic compounds and other environmental processes [66]. In another report by Arvand et al., a glassy carbon electrode modified with Fe_3O_4 nanoparticles, GQDs, and functionalized multi–walled carbon nanotubes (f–MWCNTs) was synthesized, and its electrocatalytic activity towards the oxidation of progesterone (P4) was investigated. Based on electrochemical studies, the prepared sensor possesses some advantages such as a high effective surface area, increased reactive sites, and better electrochemical catalytic activity towards P4 oxidation. The maximum currents for P4 were also discovered to vary linearly with its concentrations within the wide range of 0.01–0.5 and 0.5–3.0 μM after the optimization of the analytical conditions. The estimated sensitivity and detection limit of the electrode were 16.84 $\mu A \mu M^{-1}$ and 2.18 nM, respectively. Good stability, selectivity, sensitivity, and reproducibility were exhibited by this sensor, which can be applied to determine P4 in pharmaceutical products and human blood samples with good recovery and without interference from other present chemicals [67]. In a recent study, as a fluoride ion receptor and fluorophores, respectively, highly luminescent carbon quantum dots (CD) and magnetically separable nickel ethylenediaminetetraacetic acid (EDTA) complex-bound silica-coated magnetite nanoparticles ($Fe_3O_4@SiO_2$–EDTA–Ni) have been used. The exchange process between the CD and F^-, which persuades fluoride to bind to a magnetic receptor, is the basis of the test. For fluoride ions in water solution, this method is extremely sensitive, fast, and selective. Fluoride exhibited a linear range of response ranging from 1–20 μM with a detection limit of 0.06 μM (R_2 = 0.992). Removal and precise measurement of fluoride ions become possible due to the receptor's high magnetic properties and superparamagnetic characteristics. Based on tap water, the practicability of the method is comprehensively evaluated [68].

Considerable recent research efforts have reflected the role of the synergistic effect of functionalized nanomaterials conjugated with quantum dots in optoelectronic and energy-related applications, emphasizing photovoltaics, photocatalysis, light-emitting diodes, supercapacitors, and batteries. Energy storage systems have emphasized the development of robust, low-cost electrode materials with excellent electrochemical properties. For supercapacitor usage, N-doped graphene quantum dots (N-GQDs) on Fe_3O_4-halloysite nanotubes (Fe_3O_4-HNTs) have been synthesized as active anode materials. Following coprecipitation-synthesized Fe_3O_4 nanoparticles being in-situ deposited on HNT surfaces, 3-aminopropyl)-triexthoxysilane was deposited on top to bond 4–10 nm N-GQDs via the formation of an amide bond. In neutral electrolyte solutions, the N-GQD@Fe_3O_4-HNTs have a good rate capability and a high specific capacitance of 418 F g^{-1}. Moreover, with energy and power densities of 10.4–29 W h kg^{-1} and 0.25–5.2 kW kg^{-1}, respectively, the anode materials have excellent electrochemical performance. While N-GQDs provide accessible electroactive sites for rapid transfer of electrons and electrolyte ions and enhance the electrical conductivity of Fe_3O_4-HNTs, Fe_3O_4-HNTs provide a one-dimensional matrix to reduce the path of electron diffusion and electrolyte ions, alleviate mechanical stress on cycling, and offer redundant sites for charge storage. Since the future generation of energy storage devices with high energy and power densities requires high-performance anode materials, the findings of this work conclusively demonstrate that metal oxide-HNTs can be a viable support for anchoring N-GQDs nanomaterials [69]. In another report, Kumar et al. demonstrated the synergistic impact of graphene quantum dots (GQDs) and barium titanate nanoparticles ($BaTiO_3$) on the conductivity and dielectric characteristics of poly(vinylidenefluoride-co-hexafluoroethylene) films. For the same, the researchers developed films of PVDF-HFP/$BaTiO_3$/GQDs polymer nanocomposite at low frequency with high dielectric constant (ϵ) and excellent flexibility. The PVDF-HFP matrix contained different compositions of GQDs, a conductive nanofiller, and $BaTiO_3$, an insulating ceramic nanofiller, as a co-filler. Dielectric properties of nanocomposite films, such as AC conductivity, dielectric loss ($\tan\delta$), and dielectric constant (ϵ), were also observed with respect to temperature and frequency. For all polymer nanocomposite films, the maximum ϵ with relatively low $\tan\delta$ was observed at low frequency at the highest temperature. The frequency increases from lower (100 Hz) to higher (up to 100 kHz) and then drops to zero for the next increase in frequency. Based on these results, PVDF-HFP/$BaTiO_3$/GQDs nanocomposites illustrated promising results to be used for energy storage applications [70]. Co-assembling semiconducting quantum dots (QDs) and magnetic nanoparticles (MNPs) in a nematic liquid crystalline (LC) matrix resulted in composite structures for which a small applied magnetic field (<100 mT) systematically and reversibly manipulated the emission intensity of the quantum dots. This field-induced brightening, which enhanced peak intensity by a factor of two to three, is an authentic cooperative effect: the co-assemblies were formed through the LC phase transition, reclustering of MNPs lead to LC re-orientation at an anomalously low external field, and the clusters were compressed due to this re-arrangement, which resulted in detection of enhanced QD emission. These results presented a reversible, all-optical, and synergistic means of magnetometer detection of magnetic fields. Additionally, since the clusters were self-assembled in a fluid environment, they offered the potential for a variety of fluid-based applications for such sensors in optoelectronics [71].

11.6 Challenges and Limitations

11.6.1 Toxicity and Biocompatibility Issues

Functionalized nanomaterials show great potential in several areas of biomedicine. These include delivering drugs to specific targets, enhancing medical imaging, and improving diagnostics. They also help to study theranostics, which keeps track of medicines and disease processes at the same time. However, before we can use these in practice, we need to check how toxic and compatible these nanoparticles are with living systems. The main focus of this check should be to understand which physical and chemical properties cause unwanted reactions in the body. The same is true for the biodistribution of nanomaterials within the living system. Understanding how the body takes in and spreads nanomaterials is key to making them better for future medical uses and spotting possible risks. Learning more about these steps helps us create safer and more effective nanoparticles for medical purposes. How nanoparticles break down in the body is another key thing to think about. If the body can't get rid of them or break them down well enough, they might build up in organs and tissues. This could lead to problems down the road. To understand these risks, we need long-term studies with living things. These studies would help us keep an eye on how these dangers pile up and what they do over time. There is much yet to be learned about the biocompatibility, biodistribution, and biodegradation of functionalized nanomaterials if we are to wisely leverage their remarkable capacity to improve healthcare. Important features like nanoparticle size and charge are a root cause determinant of how they behave, with important implications both for risk and for therapeutic effect. Without proper assessment and control, it is an uncomfortable likelihood that those same nanoscale features that are increasing their valuable potential for electronics and medicine will ultimately make unanticipated contributions on their own accord. In spite of actual progress in the creation of nanobiotechnology, very real, serious concerns are yet outstanding [72].

A majority of the research revolves around the interaction between blood and foreign materials, and it has been proven that compatibility is highly reliant on the surface characteristics of the materials. The reaction of a material to blood contact depends on a range of factors such as surface area, charge, and hydrophobic or hydrophilic character of the surface. Of high significance in determining the way functionalized nanomaterials (FMNs) interact with blood are their shape, size, and surface features. New data on the possible toxicity of functionalized nanomaterials (FMNs) are appearing as researchers continue to investigate the materials. We still have yet to find their full impact, though. Researchers are trying to address this problem in a variety of different ways. Individual nanoparticle types are being investigated, dose–response curves are being analyzed, and in vitro (test tube) and in vivo (living organism) tests are being run. They are even trying to develop a complete database that specifies the toxic and safe concentrations of compounds at the nanoscale. To find out more about the biological activity of FMNs, it is important to investigate their interaction with various types of cells [73].

11.6.2 Stability and Scalability

Quantum dots and functionalized nanomaterials have gained research interest in various fields owing to a wide range of properties exhibited by them at the nano scale. But, this combination, facilitating the amalgamation of various properties of nanomaterials, comes with its own challenges mainly in terms of stability and scalability of FMNs. There are a number of problems in the stability of functionalization, which ensures that the chemical or biological modification is intact and functional in the long run. The reasons for such problems are environmental conditions, interaction with biological systems, and intrinsic properties of the nanomaterials themselves. Besides this, there is a unique set of problems in functionalizing nanomaterials (FMNs) on a large scale from lab synthesis. Though FMNs are full of promise for applications in electronics, energy, medicine, and environmental remediation, consistency, efficiency, stability, and cost-effectiveness must be addressed with the utmost care during mass production. Serious challenges that metal-based functionalized nanomaterials (FMNs) have to face can weaken their efficacy in several applications. Their surface properties can be significantly altered by oxidation, hydrolysis, agglomeration, and coagulation. Their stability in high-temperature environments is further impaired by exposure to ultraviolet (UV) light or high temperatures, making them less reliable. Moreover, enzymes can weaken the desired efficacy of functional coatings over time by degrading them, particularly those organic or polymer-based. Substitution of FMNs in a solution after they have been dried can make them lose their initial dispersion properties, which would further influence how effective they are.

When scaling up from small laboratory-scale production to manufacturing, it is much more difficult to keep all the nanoparticles in a batch equally functionalized. Variability in performance can be induced by non-uniform coatings, which is especially undesirable for drug delivery and catalysis applications. Moreover, large-scale synthesis is more likely to lead to contamination with solvents, reagents, or byproducts, all of which have an impact on nanomaterials' properties. Lab functionalization usually works, but scaling up can generate more waste, more consumption of solvent, and lower reaction efficiency, making the process less green. Because of their complexity, high expense, and labor-intensive nature, many advanced functionalization techniques, such as plasma treatment, layer-by-layer assembly, or self-assembled monolayers, are very difficult to scale. There are massive economic hurdles to mass production since many of the functionalization processes entail costly or rare chemicals, i.e., specialty ligands, polymers, or surfactants. Apart from being energy-consuming, most of the processes demand high temperature, vacuum, or large amounts of organic solvents, which are environmentally toxic. Functionalized nanomaterials also need compatibility with subsequent processes, i.e., coating, printing, or encapsulating, which are critical in applications such as coatings, electronics, and pharmaceuticals. Mass production must meet safety, environmental, and quality control specifications, which generally means stringent testing and adjustments to production processes. Nanomaterials' efficiency and sustainability market development depend on these factors being addressed.

11.6.3 Integration into Commercial Products

FMNs have found a place in various industrial and commercial applications. Humans specifically engineer and design nanomaterials to have specific properties. Functionalized nanomaterials are of two kinds: visionary nanomaterials and legacy nanomaterials. Legacy nanomaterials, including titanium dioxide and carbon black, were commercially produced prior to contemporary nanotechnology and are only incremental improvements over existing colloidal materials. Incidental nanoparticles, on the other hand, occur unintentionally as by-products of mechanical and industrial processes. Car exhaust, welding smoke, combustion residues, and domestic heating and cooking emissions are some of the accidental nanoparticle sources. Generally, ultrafine particles, incident atmospheric nanoparticles, can potentially pollute the atmosphere. In addition, harmless and naturally occurring nanomaterials are frequently found in living organisms. In the numerous chemical environments of the Earth's crust, crystal formation naturally produces inorganic nanomaterials. Food applications are increasingly employing nanomaterials with specific functions. Through the development of high-yielding crop varieties, chemical fertilizers, and pesticides, the "Green Revolution" technology and engineering have significantly boosted agricultural yields. Packaging along the supply chain is critical to preserving food quality and protecting it from environmental factors. Intelligent packaging, which includes internal or external indicators, delivers real-time information about food freshness. Advanced packaging technologies are getting more efficient and cost-effective with the help of innovations such as nanosensors and intelligent materials [74]. Nanotechnology is driving rapid advances in the energy and environmental sectors. It is utilized by fossil fuels, nuclear energy, windmills, fuel cells, and solar power, among other sources of energy [9]. It is enhancing the efficiency and longevity of conventional energy production, making existing methods more effective and paving the way for a transition to renewable and low-emission technologies. For example, it is making conventional solar panels, windmills, nuclear reactors, and fossil fuels more efficient. Third-generation solar cells that employ nanomaterials to achieve continuously improving efficiency levels are one of the most hopeful areas of study. Likewise, nanotechnology plays a critical role in fuel cell technology, particularly when it involves achieving the greatest possible utilization of costly catalyst materials, making such sources of energy more viable and sustainable [75].

11.7 Conclusion and Future Perspectives

Their technological and engineering applications, particularly designed and functionalized nanomaterials, remain largely untapped. To keep pace with scientific progress and the increasing demand for sustainable engineering solutions, the field of environmental engineering must be re-examined and redefined as environmental issues evolve. How effectively functionalized magnetic nanomaterials (FMNs) are utilized and how well nanotechnology is attained will decide how much responsibility for the development of sustainable and environmentally friendly technologies rests with them. Scientific imagination, vision, and determination—key indicators of modern global scientific innovation—are what drive actual progress in this field. A comprehensive overview of nanoscience and nanotechnology is provided in this discussion, as well as a focus on how they could be utilized to address the most urgent environmental issues. One of the principal issues discussed is the global water crisis, which has become extreme in nature. To further the scientific knowledge and engineering solutions, the limits of sustainability science are also explored. Redoing technology validation is imperative in this rapidly changing world to ensure that green sustainability projects remain effective and relevant [9].

Current advances in nanotechnology show its vast potential for a multitude of applications. Characterisation, classification, and synthesis of nanomaterials, along with surface modification and functionalization techniques, are the primary subjects of this discussion. Aside from the examination of the advantages and limitations of various synthesis methods, it explores functionalization methods such as defect engineering, covalent and noncovalent bonding, and heteroatom insertion. New models, concepts, and nanomaterials are constantly developed as research advances in order to better understand experimental findings. Nevertheless, due to their size and their possible influence on natural systems, the functionalization of nanomaterials raises concerns both for health and environmental risks. To overcome such hurdles, there is a need to develop synthesis methods that are safe, effective, and ecologically benign. Ultimately, the successful incorporation of nanomaterials into widespread commercial applications will rely on the implementation of smart and sustainable nanomanufacturing methods, which are paramount to the future of nanotechnology [8].

Abbreviations

2D	Two-dimensional
APTES	3-Aminopropyl-triethoxysilane
CNTs	Carbon nanotubes
CQDs	Carbon quantum dots
CSM	Cascade surface modification
EDTA	Ethylenediaminetetraacetic acid
FMNs	Functionalized nanomaterials

FRET Förster resonance energy transfer

GPMES	3-Glycidoxypropyl-dimethyl-ethoxysilane
LEDs	Light-emitting diodes
MBE	Molecular beam epitaxy
ME	2-Mercaptoethanol
MNPs	Magnetic nanoparticles
MOF	Metal–organic frameworks
MRI	Magnetic resonance imaging
MPTMS	3-Mercaptopropyl-trimethoxysilane
NHS	N-hydroxysuccinimide

PEG Polyethylene glycol

PL	Photoluminescence
PVA	Polyvinyl alcohol
QDs	Quantum dots
QY	Quantum yield
TG	Thioglycerol

References

1. Muhammed M, Tsakalakos T (2003) Nanostructured materials and nanotechnology : overview. J Korean Ceram Soc 40(11):1027–1046. https://doi.org/10.4191/kcers.2003.40.11.1027
2. Schiavo L, Cammarano A, Carotenuto G, Longo A, Palomba M, Nicolais L (2023) An overview of the advanced nanomaterials science. Inorg Chim Acta 559:121802. https://doi.org/10.1016/j.ica.2023.121802

3. Gohar O, Khan MZ, Bibi I, Bashir N, Tariq U, Bakhtiar M, Karim MRA, Ali F, Hanif MB, Motola M (2024) Nanomaterials for advanced energy applications: recent advancements and future trends. Mater Des 241:112930. https://doi.org/10.1016/j.matdes.2024.112930
4. Yadav M, Chaudhary A (2014). Quantum dots: an introduction. Int J Res Advent Technol (E-ISSN: 2321-9637) Special issue national conference "IAEISDISE 2014"
5. Reimann SM, Manninen M (2002) Electronic structure of quantum dots. Rev Mod Phys 74(4):1283–1342. https://doi.org/10.1103/revmodphys.74.1283
6. Riel BJ (2008) An introduction to self-assembled quantum dots. Am J Phys 76(8):750–757. https://doi.org/10.1119/1.2907856
7. Bera D, Qian L, Tseng T, Holloway PH (2010) Quantum dots and their multimodal applications: a review. Materials 3(4):2260–2345. https://doi.org/10.3390/ma3042260
8. Rizvi M, Gerengi H, Gupta P (2022) Functionalization of nanomaterials: synthesis and characterization. In: ACS symposium series, pp 1–26. https://doi.org/10.1021/bk-2022-1418.ch001
9. Palit S, Hussain CM (2018) Engineered nanomaterial for industrial use. Elsevier eBooks, pp 3–12. https://doi.org/10.1016/b978-0-12-813351-4.00001-8
10. Wichaita W, Kim Y, Tangboriboonrat P, Thérien-Aubin H (2020) Polymer-functionalized polymer nanoparticles and their behaviour in suspensions. Polym Chem 11(12):2119–2128. https://doi.org/10.1039/c9py01558b
11. Wang Y, Hu R, Lin G, Roy I, Yong K (2013) Functionalized quantum dots for biosensing and bioimaging and concerns on toxicity. ACS Appl Mater Interfaces 5(8):2786–2799. https://doi.org/10.1021/am302030a
12. Prabakaran G, Velmurugan K, David CI, Nandhakumar R (2022) Role of Förster resonance energy transfer in graphene-based nanomaterials for Sensing. Appl Sci 12(14):6844. https://doi.org/10.3390/app12146844
13. Hong G, Diao S, Antaris AL, Dai H (2015) Carbon nanomaterials for biological imaging and nanomedicinal therapy. Chem Rev 115(19):10816–10906. https://doi.org/10.1021/acs.chemrev.5b00008
14. Sampaio M, Bacsa R, Benyounes A, Axet R, Serp P, Silva C, Silva A, Faria J (2015) Synergistic effect between carbon nanomaterials and ZnO for photocatalytic water decontamination. J Catal 331:172–180. https://doi.org/10.1016/j.jcat.2015.08.011
15. Bangal M, Ashtaputer S, Marathe S, Ethiraj A, Hebalkar N, Gosavi SW, Urban J, Kulkarni SK (2005) Semiconductor nanoparticles. In: Proceedings of the international workshop on nanomaterials, magnetic ions and magnetic semiconductors studied mostly by hyperfine interactions (IWNMS 2004) held in Baroda, India, 10–14 Feb 2004 (Berlin)
16. Kuchibhatla SVNT, Karakoti A, Debasis B, Sudipta S (2007) One dimensional nanostructured materials. Prog Mater Sci 52:699–913. https://doi.org/10.1016/j.pmatsci.2006.08.001
17. Koole R, Groeneveld E, Vanmaekelbergh D, Meijerink A, De Mello Donegá C (2014) Size effects on semiconductor nanoparticles. Springer eBooks, pp 13–51. https://doi.org/10.1007/978-3-662-44823-6_2
18. Jang E, Jun S, Pu L (2003) High quality CdSeS nanocrystals synthesized by facile single injection process and their electroluminescence. Chem Commun 24:2964. https://doi.org/10.1039/b310853h
19. Steckel JS, Snee P, Coe-Sullivan S, Zimmer JP, Halpert JE, Anikeeva P, Kim L, Bulovic V, Bawendi MG (2006) Color-saturated green-emitting QD-LEDs. Angew Chem Int Ed 45(35):5796–5799. https://doi.org/10.1002/anie.200600317
20. Hussain S, Packirisamy G, Misra K, Tariq M, Sk MP (2022) Editorial: quantum dots for biological applications. Front Bioeng Biotechnol 10. https://doi.org/10.3389/fbioe.2022.930213
21. Joglekar PV, Mandalkar DJ, Nikam MA, Pande NS, Dubal A (2019) Review article on quantum dots: synthesis, properties and application. Int J Res Advent Technol 7(1):510–515
22. Agarwal K, Rai H, Mondal S (2023) Quantum dots: an overview of synthesis, properties, and applications. Mater Res Express 10(6):062001. https://doi.org/10.1088/2053-1591/acda17

23. Ahirwar S, Mallick S, Bahadur D (2017) Electrochemical method to prepare graphene quantum dots and graphene oxide quantum dots. ACS Omega 2(11):8343–8353. https://doi.org/10.1021/acsomega.7b01539
24. Zorab MM, Mohammadjani N, Ashengroph M, Alavi M (2022) Biosynthesis of quantum dots and their therapeutic applications in the diagnosis and treatment of cancer and SARS-COV-2. Adv Pharm Bull 13(3):411–422. https://doi.org/10.34172/apb.2023.065
25. Kaiser MS (2020b) Are bottom-up approaches in development more effective than top-down approaches? J Asian Soc Sci Res 2(1):91–109. https://doi.org/10.15575/jassr.v2i1.20
26. Jääskeläinen A, Uusi-Rauva E (2011) Bottom-up approach for productivity measurement in large public organizations. Int J Product Perform Manag 60(3):252–267. https://doi.org/10.1108/17410401111111989
27. Facure MHM, Schneider R, Lima JBS, Mercante LA, Correa DS (2021) Graphene quantum dots-based nanocomposites applied in electrochemical sensors: a recent survey. Electrochem 2:490–519. https://doi.org/10.3390/electrochem2030032
28. Lakshmanakumar M, Nesakumar N, Sethuraman S, Rajan KS, Krishnan UM, Rayappan JBB (2022) Development of an electrodeposited graphene quantum dot electrode for the electrochemical detection of C-Reactive protein (CRP) biomarker. ChemistrySelect 7(14). https://doi.org/10.1002/slct.202104511
29. Li J, Wu Q, Wu J (2015) Synthesis of nanoparticles via solvothermal and hydrothermal methods. Springer eBooks, pp 295–328. https://doi.org/10.1007/978-3-319-15338-4_17
30. Shabbir H, Tokarski T, Ungor D, Wojnicki M (2021) Eco friendly synthesis of carbon dot by hydrothermal method for metal ions salt identification. Materials 14(24):7604. https://doi.org/10.3390/ma14247604
31. Shouhua F, Li G (2011) Chapter 4—hydrothermal and solvothermal syntheses. In: Xu R, Pang W, Huo Q (eds) Modern inorganic synthetic chemistry. Elsevier, pp 63–95
32. Zhu H, Yang D, Yu G, Zhang H, Yao K (2006) A simple hydrothermal route for synthesizing SnO_2 quantum dots. Nanotechnology 17(9):2386–2389. https://doi.org/10.1088/0957-4484/17/9/052
33. Prakash A, Yadav S, Yadav U, Saxena PS, Srivastava A (2023) Recent advances on nitrogen-doped carbon quantum dots and their applications in bioimaging: a review. Bull Mater Sci 46(1). https://doi.org/10.1007/s12034-022-02846-7
34. Wang W, Wang S, Lv J, Zhao M, Zhang M, He G, Fang C, Li L, Sun Z (2018) Enhanced photoresponse and photocatalytic activities of graphene quantum dots sensitized Ag/TiO_2 thin film. J Am Chem Soc 101:5469–5476. https://doi.org/10.1111/jace.15795
35. Murray CB, Kagan CR, Bawendi MG (1995) Self-organization of CdSe nanocrystallites into three-dimensional quantum dot superlattices. Science 270(5240):1335–1338. https://doi.org/10.1126/science.270.5240.1335
36. Samokhvalov P, Artemyev M, Nabiev I (2013) Basic principles and current trends in colloidal synthesis of highly luminescent semiconductor nanocrystals. Chem Eur J 19(5):1534–1546. https://doi.org/10.1002/chem.201202860
37. Mir IA, Das K, Rawat K, Bohidar H (2016) Hot injection versus room temperature synthesis of CdSe quantum dots: a differential spectroscopic and bioanalyte sensing efficacy evaluation. Colloids Surf, A 494:162–169. https://doi.org/10.1016/j.colsurfa.2016.01.002
38. Martins CSM, LaGrow AP, Prior JAV (2022) Quantum dots for cancer-related mIRNA monitoring. ACS Sensors 7(5):1269–1299. https://doi.org/10.1021/acssensors.2c00149
39. Liu B, Altintas Y, Wang L, Shendre S, Sharma M, Sun H, Mutlugun E, Demir HV (2019) Record high external quantum efficiency of 19.2% achieved in light-emitting diodes of colloidal quantum wells enabled by hot-injection shell growth. Adv Mater 32(8). https://doi.org/10.1002/adma.201905824
40. Dabbousi BO, Rodriguez-Viejo J, Mikulec FV, Heine JR, Mattoussi H, Ober R, Jensen KF, Bawendi MG (1997) (CDSE)ZNS core−shell quantum dots: synthesis and characterization of a size series of highly luminescent nanocrystallites. J Phys Chem B 101(46):9463–9475. https://doi.org/10.1021/jp971091y

41. Choi M, De Arquer FPG, Proppe AH, Seifitokaldani A, Choi J, Kim J, Baek S, Liu M, Sun B, Biondi M, Scheffel B, Walters G, Nam D, Jo JW, Ouellette O, Voznyy O, Hoogland S, Kelley SO, Jung YS, Sargent EH (2020) Cascade surface modification of colloidal quantum dot inks enables efficient bulk homojunction photovoltaics. Nat Commun 11(1). https://doi.org/10.1038/s41467-019-13437-2
42. Gui Y, Shen J, Zhou Y, Wang R, Yang Y, Xue J (2024) Inorganic surface passivation strategies of metal halide perovskites. Inf Funct Mater 1(2):207–219. https://doi.org/10.1002/ifm2.20
43. Spanos A, Athanasiou K, Ioannou A, Fotopoulos V, Krasia-Christoforou T (2021) Functionalized magnetic nanomaterials in agricultural applications. Nanomaterials 11(11):3106. https://doi.org/10.3390/nano11113106
44. Lakshmi D, Diana MI, Helen PA, Selvin PC (2023) Functionalized nanomaterials, Classification, properties, and functionalization techniques. In: Materials horizons, pp 65–92. https://doi.org/10.1007/978-981-99-3021-0_3
45. Manoj GM, Shalini M, Thenmozhi K, Ponnusamy VK, Hari S (2024) Recent advancements in the surface modification and functionalization of magnetic nanomaterials. Appl Surf Sci Adv 21:100608. https://doi.org/10.1016/j.apsadv.2024.100608
46. Sperling RA, Parak WJ (2010) Surface modification, functionalization and bioconjugation of colloidal inorganic nanoparticles. Philos Trans R Soc Math Phys Eng Sci 368(1915):1333–1383. https://doi.org/10.1098/rsta.2009.0273
47. Mout R, Moyano DF, Rana S, Rotello VM (2012) Surface functionalization of nanoparticles for nanomedicine. Chem Soc Rev 41(7):2539. https://doi.org/10.1039/c2cs15294k
48. Mahajan G, Kaur M, Gupta R (2021) Green functionalized nanomaterials: fundamentals and future opportunities. Elsevier eBooks, pp 21–41. https://doi.org/10.1016/b978-0-12-823137-1.00003-8
49. Wieszczycka K, Staszak K, Woźniak-Budych MJ, Litowczenko J, Maciejewska BM, Jurga S (2021) Surface functionalization—the way for advanced applications of smart materials. Coord Chem Rev 436:213846. https://doi.org/10.1016/j.ccr.2021.213846
50. Biehl P, Schacher FH (2020) Surface functionalization of magnetic nanoparticles using a thiol-based grafting-through approach. Surfaces 3(1):116–131. https://doi.org/10.3390/surfaces3010011
51. Dash S, Das T, Patel P, Panda PK, Suar M, Verma SK (2022) Emerging trends in the nanomedicine applications of functionalized magnetic nanoparticles as novel therapies for acute and chronic diseases. J Nanobiotechnol 20(1). https://doi.org/10.1186/s12951-022-01595-3
52. Jayaprakash N, Elumalai K, Manickam S, Bakthavatchalam G, Tamilselvan P (2024) Carbon nanomaterials: revolutionizing biomedical applications with promising potential. Nano Mater Sci. https://doi.org/10.1016/j.nanoms.2024.11.004
53. Delille F, Pu Y, Lequeux N, Pons T (2022) Designing the surface chemistry of inorganic nanocrystals for cancer imaging and therapy. Cancers 14(10):2456. https://doi.org/10.3390/cancers14102456
54. Begines B, Ortiz T, Pérez-Aranda M, Martínez G, Merinero M, Argüelles-Arias F, Alcudia A (2020) Polymeric nanoparticles for drug delivery: recent developments and future prospects. Nanomaterials 10(7):1403. https://doi.org/10.3390/nano10071403
55. Wang Z, Zhao M, Huang X, Wang Y, Li W, Qiao J, Yang X (2024) Therapeutic types and advantages of functionalized nanoparticles in inducing ferroptosis in cancer therapy. Ann Med 56(1). https://doi.org/10.1080/07853890.2024.2396568
56. Shao Y, Lu W, Chen H, Xiao JQ, Qiu Y, Chou T (2018) Flexible ultra-thin Fe_3O_4/MnO_2 core-shell decorated CNT composite with enhanced electromagnetic wave absorption performance. Compos B Eng 144:111–117. https://doi.org/10.1016/j.compositesb.2018.02.015
57. Munoz-Sandoval E, Perea-Lopez N, Lima-Juarez R, Labrada-Delgado GJ, Rivera-Escoto BA, Zamudio A, Silva-Pereyra HG, Robles-Avila E, Terrones M (2014) Synthesis, characterization and magnetic properties of Co@Au core-shell nanoparticles encapsulated by nitrogen-doped multiwall carbon nanotubes. Carbon 77:722–737. https://doi.org/10.1016/j.carbon.2014.05.077

58. Baaziz W, Begin-Colin S, Pichon BP, Florea I, Ersen O, Zafeiratos S, Barbosa R, Begin D, Pham-Huu C (2012) High-density monodispersed cobalt nanoparticles filled into multiwalled carbon nanotubes. Chem Mater 24(9):1549–1551. https://doi.org/10.1021/cm300293b
59. Mozafari H, Hamidinezhad H (2019) The effect of adding $CoFe_2O_4$–CdS nanoparticles on the mechanical properties of rice husk ash epoxy composite: an experimental approach. Appl Phys A 125(5). https://doi.org/10.1007/s00339-019-2632-7
60. Huang S, Jiang S, Pang H, Wen T, Asiri AM, Alamry KA, Alsaedi A, Wang X, Wang S (2019) Dual functional nanocomposites of magnetic $MnFe_2O_4$ and fluorescent carbon dots for efficient U(VI) removal. Chem Eng J 368:941–950. https://doi.org/10.1016/j.cej.2019.03.015
61. Shibu ES, Ono K, Sugino S, Nishioka A, Yasuda A, Shigeri Y, Wakida S, Sawada M, Biju V (2013) Photouncaging nanoparticles for MRI and fluorescence imaging in vitro and in vivo. ACS Nano 7:9851–9859. https://doi.org/10.1021/nn4043699
62. He X, Zeng T, Li Z, Wang G, Ma N (2015) Catalytic molecular imaging of MicroRNA in living cells by DNA-programmed nanoparticle disassembly. Angew Chem 128:3125–3128. https://doi.org/10.1002/ange.201509726
63. Yao X, Niu X, Ma K, Huang P, Grothe J, Kaskel S, Zhu Y (2016) Graphene quantum dots-capped magnetic mesoporous silica nanoparticles as a multifunctional platform for controlled drug delivery, magnetic hyperthermia, and photothermal therapy. Small 13. https://doi.org/10.1002/smll.201602225
64. Zhang M, Wang W, Cui Y, Chu X, Sun B, Zhou N, Shen J (2018) Magnetofluorescent Fe_3O_4/carbon quantum dots coated single-walled carbon nanotubes as dual-modal targeted imaging and chemo/photodynamic/photothermal triple-modal therapeutic agents. Chem Eng J 338:526–538. https://doi.org/10.1016/j.cej.2018.01.081
65. Ma Y, Xu G, Wei F, Cen Y, Xu X, Shi M, Cheng X, Chai Y, Sohail M, Hu Q (2018) One-pot synthesis of a magnetic, ratiometric fluorescent nanoprobe by encapsulating Fe_3O_4 magnetic nanoparticles and dual-emissive rhodamine B modified carbon dots in metal-organic framework for enhanced HCLO sensing. ACS Appl Mater Interfaces 10(24):20801–20805. https://doi.org/10.1021/acsami.8b05643
66. Naghshbandi Z, Arsalani N, Zakerhamidi MS, Geckeler KE (2018) A novel synthesis of magnetic and photoluminescent graphene quantum dots/MFe_2O_4 (M = Ni, Co) nanocomposites for catalytic application. Appl Surf Sci 443:484–491. https://doi.org/10.1016/j.apsusc.2018.02.283
67. Arvand M, Hemmati S (2016) Magnetic nanoparticles embedded with graphene quantum dots and multiwalled carbon nanotubes as a sensing platform for electrochemical detection of progesterone. Sens Actuators, B Chem 238:346–356. https://doi.org/10.1016/j.snb.2016.07.066
68. Mohapatra S, Sahu S, Nayak S, Ghosh SK (2015) Design of FE_3O_4@SIO_2@carbon quantum dot based nanostructure for fluorescence sensing, magnetic separation, and live cell imaging of fluoride ion. Langmuir 31(29):8111–8120. https://doi.org/10.1021/acs.langmuir.5b01513
69. Ganganboina AB, Chowdhury AD, Doong R (2017) Nano assembly of N-doped graphene quantum dots anchored Fe_3O_4/halloysite nanotubes for high performance supercapacitor. Electrochim Acta 245:912–923. https://doi.org/10.1016/j.electacta.2017.06.002
70. Kumar YR, Pasha SK (2021) Synergistic effect of barium titanate nanoparticles and graphene quantum dots on the dielectric properties and conductivity of poly(vinylidenefluoride-co-hexafluoroethylene) films. Environ Res 204:112297. https://doi.org/10.1016/j.envres.2021.112297
71. Amaral JJ, Wan J, Rodarte AL, Ferri C, Quint MT, Pandolfi RJ, Scheibner M, Hirst LS, Ghosh S (2014) Magnetic field induced quantum dot brightening in liquid crystal synergized magnetic and semiconducting nanoparticle composite assemblies. Soft Matter 11(2):255–260. https://doi.org/10.1039/c4sm02015d
72. Kunzmann A, Andersson B, Thurnherr T, Krug H, Scheynius A, Fadeel B (2010) Toxicology of engineered nanomaterials: focus on biocompatibility, biodistribution and biodegradation. Biochimica et Biophysica Acta (BBA) Gen Subj 1810(3):361–373. https://doi.org/10.1016/j.bbagen.2010.04.007

73. Li X, Wang L, Fan Y, Feng Q, Cui F (2012) Biocompatibility and toxicity of nanoparticles and nanotubes. J Nanomater (1). https://doi.org/10.1155/2012/548389
74. Palit S, Hussain CM (2020) Functionalization of nanomaterials for industrial applications: recent and future perspectives. Elsevier eBooks, pp 3–14. https://doi.org/10.1016/b978-0-12-816787-8.00001-6
75. Vázquez CI, Iglesias RA (2018) Engineered nanomaterials in energy production industry. Elsevier eBooks, pp 713–723. https://doi.org/10.1016/b978-0-12-813351-4.00039-0

Chapter 12
Regulatory and Safety Aspects of Functionalized Magnetic Nanomaterials (FMNs)

Sini Kuriakose and Yukti Monga

Abstract The rapid expansion of functionalized magnetic nanoparticles (FMNs) in various scientific and industrial applications has led to increased demand among chemists and researchers. However, their widespread use raises critical concerns regarding safety, health risks, and environmental impact. This chapter emphasizes the significance of safety and regulatory considerations associated with FMNs, addressing potential toxicity, bioaccumulation, and environmental persistence. It provides a comprehensive review of global safety assessment guidelines and regulatory frameworks governing the responsible use, handling, and disposal of FMNs. Additionally, this chapter explores risk mitigation strategies, exposure limits, and recent advancements in toxicity evaluations to ensure their safe integration into biomedical, environmental, and industrial applications. By highlighting key safety aspects, this chapter aims to support the development of sustainable and regulatory-compliant FMNs for future technologies.

Keywords Magnetic nanoparticles · Environment · Regulatory issues · Safety issues

12.1 Introduction

The field of nanotechnology is one that is growing quickly. Researchers are learning that atomic particles and molecules behave differently at the nanoscale, and engineers and scientists are successfully creating materials at the nanoscale to benefit from improved properties, such as greater electrical conductivity, chemical reactivity, strength, and weight reduction, in comparison to their larger scale counterparts [1]. In many facets of study, from basic nanoscience to biological and environmental

S. Kuriakose
Department of Chemistry, Bangalore University, Bangalore, Karnataka, India

Y. Monga (✉)
Department of Chemistry, Shyamlal College, University of Delhi, Shahdara, Delhi, India
e-mail: yuktichem@shyamlal.du.ac.in

S. Gulati (ed.), *Functionalized Magnetic Nanomaterials*, Nanostructure Science and Technology, https://doi.org/10.1007/978-3-031-97199-0_12

applications, the ongoing separation of nanoparticles is essential. However, it is challenging to attain high separation performance, efficiency, or even high throughput because of the tiny size of nanoparticles. As a result, it is still difficult to separate nanoparticles with high purity and high resolution [2]. To overcome these problems, magnetic nanoparticles appeared as a viable solution. Compared to other separation techniques, magnetic nanoparticle separation is generally regarded as "easy" because it only requires applying a magnetic field to a solution containing the nanoparticles, which causes them to quickly clump together and separate from the liquid. This is especially true when using a permanent magnet because of the easily accessible magnetic force [3–5].

Nanomaterials consisting of magnetic elements like iron, cobalt, chromium, nickel, gadolinium, and manganese, and their chemical compounds are called magnetic nanoparticles. Due to their size, which is in the nanoscale, these are superparamagnetic and so offer great potential in various fields of applications [6]. These magnetic nanoparticles, which can be manipulated using a magnetic field, can be used with a surface coating with any functional group according to the specific use and may be transported to a target location under the influence of an external magnetic field provided by an electromagnet or a permanent magnet. These nanoparticles that combine the magnetic properties of the core with the specific surface modification are called functionalized magnetic nanoparticles (MNPs) [7]. The surface modification enhances the properties and is used in various biomedical applications such as drug delivery, magnetic resonance imaging (MRI), cancer therapy, gene delivery, analytical applications, and environmental applications (Fig. 12.1) [8–10].

However, a major area of concern while using MNPs for any application, be it biomedical, analytical, or environmental, toxicity is to be taken into consideration. Toxicity studies include the adverse effects on the physical, chemical, and biological condition of the animal or environment under consideration.

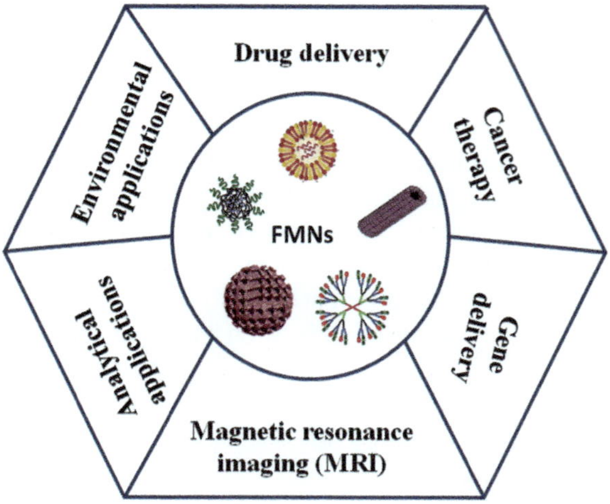

Fig. 12.1 Applications of FMNs

12.2 Importance of Regulatory Frameworks and Safety Measures with FMNs

Governments enact laws and regulations known as regulatory frameworks to guarantee adherence to the law and safeguard the rights of both persons and corporations. Ensuring compliance, safeguarding rights, encouraging fair competition, and improving public safety are the main goals of a regulatory framework [10]. Understanding the growing significance of handling nanomaterials safely, the Indian government's Nano Mission established a Nano-regulatory Task Force, of produced a thorough work on "Guidelines and Best Practices for Safe Handling of Nanomaterials in Research Laboratories and Industries" by a group of distinguished specialists [11].

Centre for Knowledge Management of Nanoscience & Technology (CKMNT) has released this extensive paper under the capable direction of the Nano Mission and Nano-regulatory Task Force, drawing from publicly available reports as well as published reports by regulatory organizations such as ISO, OECD, NIOSH, and OSHA [12].

Most jurisdictions, including the EU, consider magnetic nanoparticles (FMNs) to be nanomaterials and regulate them under existing nanomaterial regulations like REACH, which require thorough safety assessments before market introduction. The regulatory framework for MNPs is largely based on existing chemical regulations, with a focus on assessing potential toxicity based on factors like particle size, surface chemistry, and concentration. When considering their application, particularly in biomedical fields, special attention is paid to their unique characteristics, like agglomeration potential and biodistribution [13, 14].

12.3 Potent Health and Environmental Risks with FMNs

Over the last few decades, magnetic nanoparticles have been found to have great applications in various fields, especially biomedical applications [15–17]. This is because of their peculiar properties, such as high surface area, superparamagnetism, tunability, and easy functionalization of these nanoparticles [18–21]. Recent advances in synthesis and application have made it possible to engineer and manipulate magnetic nanoparticles with various functional groups for multiple uses [22, 23]. Therefore, it becomes the need of the hour to understand the toxicity of these functionalized MNPs. As such, the toxicity is found to depend on various factors like size, shape, functionalization group, pharmacokinetics, concentration, solubility, dosage, immunogenicity, biodegradation, metabolism, and bioavailability [24] (Fig. 12.1). In biomedical applications, since these MNPs have a large surface-to-volume ratio, they may interact with various organs in the body [25]. It is also necessary to check the biodegradation rate, as resistance to degradation will result in increased absorption and the possibility of increased toxicity. Toxicity studies should also take into consideration various factors such as chronic and long-term toxicity, stimulation, and inflammatory reaction (Fig. 12.2) [26].

Some of the key points that lead to the toxicity of the functionalized FMNs are listed [27–32]:

1. **Toxic Ligands or Coatings**: Molecules used for functionalization themselves may induce cytotoxicity, such as certain polymers or organic molecules.
2. **Unstable functionalization**: The coating may be unstable and may degrade and leak toxic substances into the environment.
3. **Increased reactive oxygen species (ROS)**: FMNs that are made using iron oxide core increase the production of ROS, which in turn causes damage to DNA, proteins, lipids, etc., leading to cytotoxicity.
4. **Metal Ion release**: Degradation of FMNs in the environment may lead to excess deposits of heavy metals and iron compounds in the environment.
5. **Nanoparticle size**: The smaller the size, the larger the surface-to-volume ratio and hence enhanced interaction with the molecules. This may lead to disruption in the cellular activities.
6. **Agglomeration**: Nanoparticles may agglomerate, leading to larger particles that may obstruct the blood flow in the capillaries.

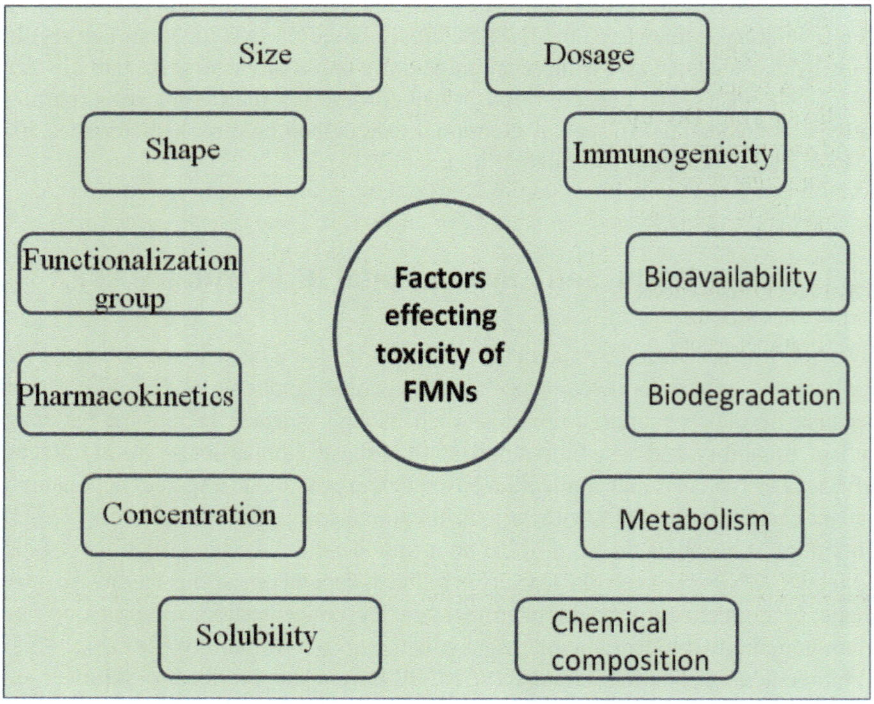

Fig. 12.2 Factors affecting the toxicity of FMNs

7. **Surface charge on FMNs**: A high positive surface charge on the MNPs strongly interacts with the cell membrane, which is negatively charged, and causes disruption and cell death.
8. **The environment of the surroundings**: Depending on the pH of the surrounding biological medium, the stability and toxicity of the FMNs may vary.

MNPs, which are toxic in nature, can reduce the therapeutic efficiency of the particle and may also result in inflammation or an immune response due to their interaction inside the body. The toxicity of the particle can produce an adverse effect on the nuclear process once it enters the cells. This can also cause the cell membrane to leak or its blocked, leading to cell proliferation.

Some of the health risks associated with FMNs are (Fig. 12.3) [33–39].

1. **Toxicity to cells and organs**: ROS generated by the FMNs can lead to DNA and cell damage. MNPs can also interfere with nerve functioning, leading to nerve damage or degeneration.
2. **Bioaccumulation**: If the FMNs are not metabolized effectively, they may accumulate inside the organs like the liver, kidney, spleen, etc., and lead to long-term toxicity.
3. **Inflammatory response**: The body's immune system may consider the functionalized MNPs as foreign bodies and may trigger inflammation in the body.
4. **Genotoxicity**: FMNs may directly interact with the DNA and disrupt normal DNA sequencing.
5. **Endocrine Disruption**: Some functionalizing agents may act as endocrine disrupters and interfere with hormonal systems.
6. **Aggregates of FMNs**: Large aggregates of FMNs may lead to obstruction in the capillaries and blood vessels.

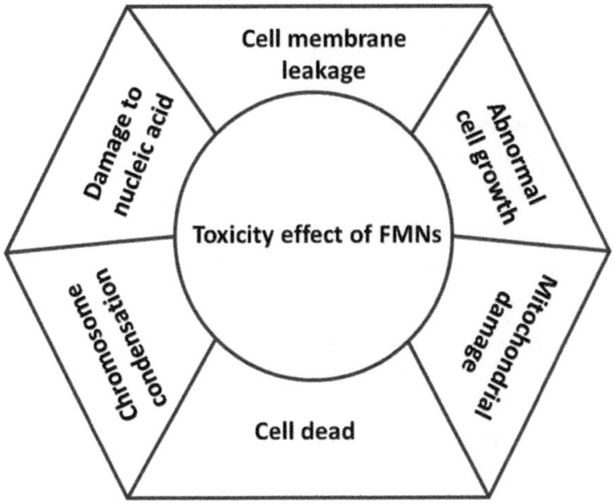

Fig. 12.3 Toxicity effects of FMNs when used for biological applications

While studying the toxicity of FMNs, it is necessary to study it in in vivo and in vitro.

1. **In Vitro toxicity of FMNs**: In Vitro toxicity evaluation is a simple and inexpensive method because of the controlled environment [40, 41]. In this, assays are analyzed for cell viability, differentiation, and proliferation. Some of the techniques used are optical microscopy, electron microscopy, and atomic force microscopy, which are used to form images of the cell before and after the effect of the MNPs, and can be analyzed through different software. Some of the new avenues for the study of the mechanism of toxicity are gene expression analysis, which studies how genes are transcribed to synthesize proteins and other products, metabolomics, which deals with the study of cell metabolism products, and proteomics, which is the study of proteins on a large scale. These studies done in vitro will help analyse the biocompatibility of MNPs with the cell, their interaction with the cell membrane, accumulation in the organs, or any other physiological effects.
2. **In Vivo toxicity of FMNs**: In vivo interaction of FMNs is very complex and dynamic [42–44]. Factors that affect the toxicity and accumulation inside the body are shape, size, stability, surface charge, and functionalization group. It has been observed that FMNs less than 10 nm are removed by the body through excretion. However, FMNs larger than 200 nm get captured by the spleen. Therefore, for practical application, size is limited between 10 and 100 nm. Once inside, FMNs interact with cells and other components, and get absorbed, and reach various organs, and are metabolized. Therefore, the design of the FNM should consider aspects such as absorption, metabolism, distribution in the body, and excretion so as to understand the toxicity inside the body. One of the most important concerns is the accumulation of intercellular iron after the degradation of FNM, which can damage proteins and nucleic acids.

Along with health hazards, FMNs also pose a grave danger to the environment. Some of the ways these nanoparticles can disrupt the environment are listed (Fig. 12.4) [45].

1. **Aquatic life toxicity**: FMNs, when discharged into water sources, can harm aquatic organisms by affecting their reproduction, growth, and survival.
2. **Soil toxicity**: By releasing toxic compounds into the soil, these compounds may reduce soil fertility by interacting with the soil microbes.
3. **Bioaccumulation in the food chain**: FMNs may enter the food chain through microbes in the soil to plants and smaller animals and eventually to higher-order species like humans.
4. **Accumulation in the environment**: Since FMNs do not degrade easily, they accumulate in the environment, leading to long-term environmental effects.
5. **Contamination due to leaching of toxic chemicals**: Degradation of FMNs may release toxic metals such as iron, nickel, cobalt, etc., and functionalization groups leading to toxicity in the environment.

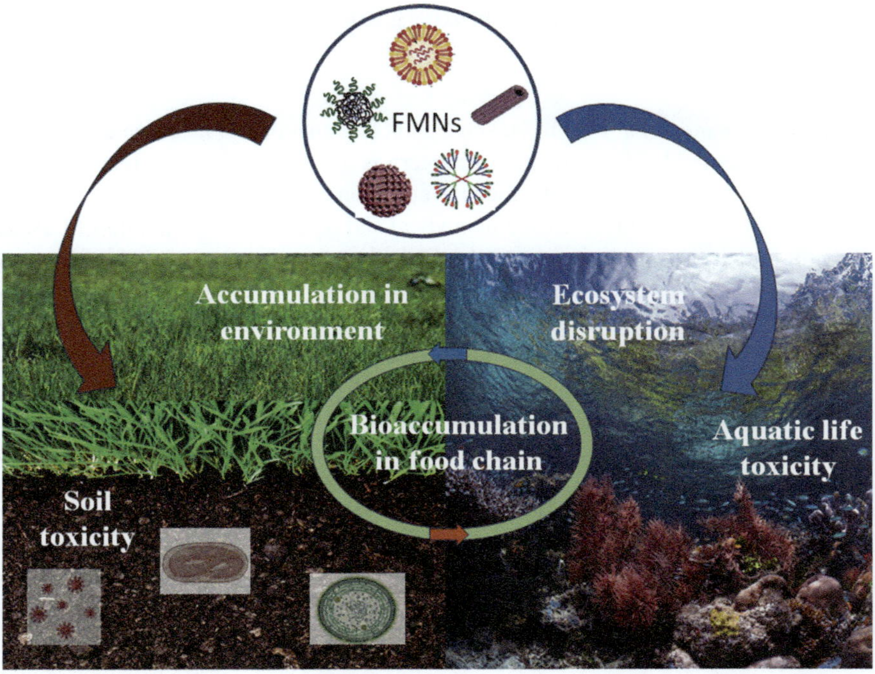

Fig. 12.4 Environmental hazards of functionalized magnetic nanoparticles

6. **Ecosystem disruption**: Interaction of FMNs with different ecosystems may change the ecological balance and biodiversity.

Since exposure to FMNs is considered a great hazard for humans, animals, and the environment, exposure assessment should be made mandatory. One of the greatest risks of exposure is workplace exposure. Synthesis of nanomaterials or their use in gas or liquid form poses the most risk in terms of exposure level. Cleaning and disposal of FMNs can lead to exposure. Also, exposure may occur due to the wastewater containing FMNs.

Therefore, it is extremely important to assess the exposure and risk analysis by particle sampling in the workplace. Personal sampling will provide accurate data on the exposure to an individual, and area sampling and real-time sampling can be used to assess the need for better practices at work. Furthermore, it is also necessary to measure the exposure before production or processing of FMNs, so that background data can be obtained and compared to the actual scenario after the exposure. There are various techniques used for exposure monitoring [46, 47] such as Electron microscopy analysis (SEM and TEM), Condensation particle counter (CPC), Scanning mobility particle sizer (SMPS), and Electrical low-pressure impactor (ELPI) can be used alone or in unison to determine the mass, number, and surface area of the FMNs, which are the most important exposure metrics.

12.4 Guidelines for Safe Handling and Disposal, and Other Safety Measures

Even though nanotechnology has taken a giant leap in terms of its usage and application in various fields of life, the main concern is the potential after-effects of its use and its impact on human life, of which the common man has very limited knowledge. Various nanoparticles that are used in medical and other fields, especially FMNs, are difficult to detect once they enter the biological system. The amount of information on its toxicity and its environmental impact is also limited. Hence, it becomes necessary to establish a way to control the exposure during the production stage and also the application stage. The problem doesn't end with the application, we also have to study the degradation pathway taken by the FMNs after their use. So, safe handling during the use and disposal of the waste also has to be taken into consideration.

12.4.1 Strategies to Control Exposure

Exposure to FMNs should be avoided as far as possible. Some of the practices that can be followed to control exposure are [48–53]:

1. By substituting harmful chemicals with less hazardous materials for functionalization, the exposure to harmful chemicals can be reduced.
2. Engineering control techniques, such as providing proper ventilation, can reduce airborne exposure to nanoaerosols.
3. By using physical separation, such as barriers, isolation booths, and closed containers, during the processing of FMNs.
4. Biological safety cabinets (BSCs) may be developed so that the person working and the environment may be protected from possible spills and aerosol exposure.
5. Remote control of nanomaterial production equipment.
6. Proper training and instructions should be given to the personnel for safe handling and standard operating procedures (SOPs).
7. All the surfaces that have been exposed to nanomaterials should be cleaned thoroughly using an industrial vacuum cleaner with filters.
8. Proper work ethics should be formulated to deal with nanomaterials in various forms, as for liquids, gloves should be worn and proper care taken so as not to spill it, for gas phase nanomaterial, proper ventilation with filters should be used, and for nanomaterial powder, face mask having filters should be used.
9. Proper health monitoring of persons dealing with FMNs should be conducted regularly to avoid long-term effects.
10. Proper induction programmes, training, risk assessment, and health monitoring should be recorded on a daily or monthly basis.
11. Storage of FMNs is also to be considered, as they may corrode and need to be protected from water and air.

12. Proper use of Personal protective equipment (PPE) should be encouraged to reduce exposure.

Exposure to FMNs may result in various long-lasting implications both for life and for the environment, as discussed in earlier sections. Therefore, it is necessary to reduce the exposure limit during and after the processing. However, another equally important part is the safe handling of FMNs [54].

12.4.2 Effective Handling of FMNs

1. Knowledge of the proper use and location of the emergency equipment that might be required while handling the FMNs.
2. Proper hygiene should be followed, such as washing hands frequently, not consuming food near the processing of FMNs, and taking proper care not to contaminate the nearby surroundings and common areas.
3. Proper containers should be used to store the FMNs and should be marked properly to avoid any confusion.
4. In case of spill and contamination, proper SOPs are to be followed, and contaminated materials should be discarded as chemical waste.
5. Nanoparticle spill kits can be made ready in the laboratories for taking rapid action in case of such a scenario.
6. Any experiments related to FMNs should be conducted in a closed chamber, and if a closed chamber is not possible, proper extraction of vapor should be made viable.
7. Waste and by-products released during and after production should be taken care of with minimum exposure to humans.

12.4.3 Treatment of FNM Waste

It has been studied that FNM waste can be treated in three different ways, namely, waste treatment plants, waste incineration plants, and landfills. The steps followed in all the cases are different and are discussed further (Fig. 12.5) [55].

4.3.1 Treatment in waste treatment plant: The steps followed are:

   a. Collection and segregation: FMNs are identified based on size and segregation.
   b. Analysis: Risk analysis is done through characterization.
   c. Transport and storage: Specialized containers are used to store and transport FMNs.

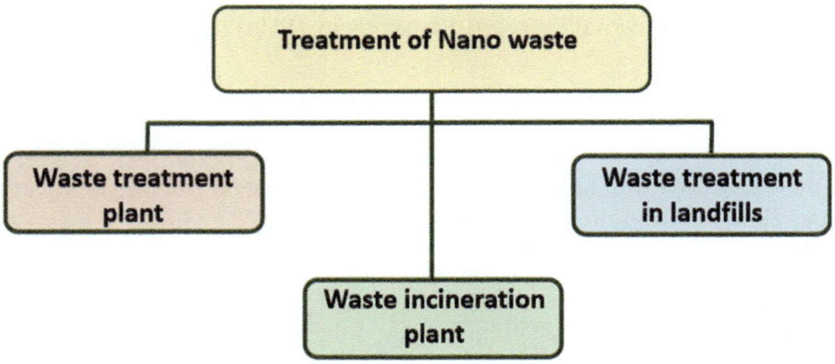

Fig. 12.5 Various methods of treatment of nano waste

d. Treatment of FMNs: By using physical methods like sedimentation, filtration, and centrifugation, or chemical methods such as reduction, oxidation, or biological methods like phytoremediation, the nanomaterial is pretreated.
e. The treated FMNs can be burned and discharged into the environment if it comply with the safety regulations.

4.3.2 Treatment in waste incineration plant: the steps followed are:

a. Collection and segregation: FMNs are identified on the basis of size and segregated.
b. Analysis: Risk analysis is done through characterization.
c. Transport and storage: Specialized containers are used to store and transport FMNs.
d. Encapsulation and stabilization: Nano waste is encapsulated and stabilized so that harmful releases during incineration can be prevented.
e. Incineration: Nanowaste is treated at a very high temperature to break down the nanoparticle.
f. Filtration: Scrubbers and filters are used to extract any traces of FMNs in the effluent gases.
g. Advanced gas treatment: to eliminate even the traces of nanoparticles left.

4.3.3 Treatment in landfills: the steps followed are:

a. Collection and segregation: FNs are identified on the basis of size and segregation.
b. Analysis: Risk analysis is done through characterization.
c. Transport and storage: Specialized containers are used to store and transport FMNs.
d. Designing the landfill: Landfills are designed in a linear manner. A multilayer linear system consisting of clay and other synthetic liners is arranged such that nothing leaks through into the surrounding area. If it does seep through, the system is developed to collect any liquid.

e. Nano trash is compressed to reduce its volume and covered with non-hazardous waste, and kept in a specially made container to check seeping.
f. Continuous monitoring is done to ascertain whether there is any leakage, and appropriate authorities are also informed of the same.

12.5 Regulatory Guidelines and Policies for FMNs' Safety

Regulatory guidelines are particularly important as they ensure the safe usage of FMNs in various products used in the medical, cosmetic, and food industries, thus ensuring the safety of the customers from health risks that may occur due to exposure to FMNs. Policies also take care of the environmental implications by addressing the impact of FMNs on biodiversity, the ecosystem, and, as a whole, the environment. These guidelines are especially for the companies that deal with the development and manufacturing of FMNs so that they adhere to these industry standards of safety protocols and practices. The guidelines will help in furthering the new developments in the FNP industry so that it may benefit a larger crowd while at the same time minimizing the adverse effects it may have on an individual and the environment. The main aim of these policies is to communicate the risk factors to the general public and stakeholders by promoting transparency, and all the while building trust and confidence in the use of FMNs. As the technology is rapidly advancing, the regulatory framework has to keep pace with it so that safety is not compromised. There is no doubt that policies and frameworks for the development and application of FNPs play a crucial role in the safety of human health as well as the environment. Some of the main aims of the regulatory guidelines are [56, 57]:

1. Implement regulations according to the site so that employees and the environment are safeguarded.
2. Create guidelines to be followed by the employees to safeguard against any unpredictable event while managing the FMNs.
3. Some of the functional groups that are found to be hazardous should be avoided, all the while retaining the benefits.
4. Provide guidelines for the safe disposal of FMNs and their storage over the full lifetime of these nanoparticles.

Many countries, including India, have developed various regulations and guidelines in this regard. In India, the regulations are developed by various agencies such as the Department of Biotechnology (DBT), the Ministry of Environment, Forest, and Climate Change (MoEFCC), the Food Safety and Standards Authority of India (FSSAI), and the Indian Council of Agricultural Research (ICAR) [58]. In the USA, the Environmental Protection Agency and various other federal government agencies, like as the FDA, make the regulations for the use of FMNs [59]. In Europe, since there are no separate regulations for FMNs, but are governed by the same rules as for other chemicals, which are governed by the Registration, Evaluation, Authorization, and Restriction of Chemicals (REACH) law [60].

The regulatory process developed for guidelines pertaining to the safe handling of FNPs and the waste it creates is:

1. Identification of waste, whether it be of FNM origin.
2. Toxicity evaluation is done to determine if it poses any danger to human life and the environment.
3. Segregation of FMNs from other non-hazardous waste.
4. Storage in dedicated containers and labeling of these containers with potential risks.
5. Transportation of waste according to the SOP.
6. Waste disposal using suitable methods as discussed earlier.
7. Inspection of storage and disposal of FMNs on a regular basis to see if all the regulations and guidelines are followed.
8. Create emergency protocols that have to be followed in case of accidental spills.
9. Routine compliance checks are to be done to ensure that all the safety regulations are being followed.

12.6 Conclusion and Future Perspectives

It is evident that although nanotechnology on magnetic nanoparticles is being investigated more and more in the fields of science, a growing number of researchers, working on projects and different applications, are also exposing the numerous harmful consequences of magnetic nanomaterials on both people and the environment. This implies that in order to quantify exposure among employees, customers, and the environment, we must first identify and evaluate the risks related to the production and handling of nanoparticles. We must then focus on risk management through control bands and risk categorization. Therefore, it is required to have strict quality control of the goods made possible by nanotechnology, regulatory compliance, and giving consumers peace of mind about their health and safety.

Acknowledgements Dr. Yukti Monga expresses gratitude to Shyamlal College, University of Delhi, Delhi, India, for the motivation and resources that made this collaborative work possible.

Abbreviations

BSCs	Biological Safety Cabinets
CPC	Condensation Particle Counter
DBT	Department of Biotechnology
ELPI	Electrical Low-Pressure Impactor
FSSAI	Food Safety and Standards Authority of India
ICAR	Indian Council of Agricultural Research
MNPs	Functionalized Magnetic Nanoparticles

MoEFCC	Ministry of Environment, Forest, and Climate Change
MRI	Magnetic Resonance Imaging
REACH	Registration, Evaluation, Authorization, and Restriction of Chemicals
ROS	Reactive Oxygen Species
SEM	Scanning Electron Microscopy
SMPS	Scanning Mobility Particle Sizer
SOPs	Standard Operating Procedures
TEM	Transmission Electron Microscopy

References

1. Malik S, Muhammad K, Waheed Y (2023) Nanotechnology: a revolution in modern industry. Molecules 28(2):661. https://doi.org/10.3390/molecules28020661.PMID:36677717;PMCID: PMC9865684
2. Nanotechnologies—health and safety practices in occupational settings relevant to nanotechnologies, ISO/TR 12885:2008(E), First edition 2008-10-01
3. Senthilkumar N, Sharma PK, Sood N, Bhalla N (2021) Designing magnetic nanoparticles for in vivo applications and understanding their fate inside human body. Coord Chem Rev 445:214082
4. Angelakeris M (2017) Magnetic nanoparticles: a multifunctional vehicle for modern theranostics. Biochimica et Biophysica Acta (BBA) Gen Subj 1861(6):1642–1651
5. Villalonga A, Villalonga R, Vilela D (2021) Hybrid magnetic nanoparticles for electrochemical biosensors. In: Ehrmann A, Nguyen TA, Ahmadi M, Farmani A, Nguyen-Tri P (eds) Magnetic nanoparticle-based hybrid materials. Woodhead Publishing Series in Electronic and Optical Materials. Woodhead Publishing
6. Zhu K, Ju Y, Xu J, Yang Z, Gao S, Hou Y (2018) Magnetic nanomaterials: chemical design, synthesis, and potential applications. Acc Chem Res 51(2):404–413
7. Berry CC, Curtis ASG (2003) Functionalisation of magnetic nanoparticles for applications in biomedicine. J Phys D Appl Phys 36:R198–R206
8. Neamtu M, Nadejde C, Hodoroaba VD et al (2018) Functionalized magnetic nanoparticles: Synthesis, characterization, catalytic application and assessment of toxicity. Sci Rep 8:6278. https://doi.org/10.1038/s41598-018-24721-4
9. Keçili R, Ghorbani-Bidkorbeh F, Dolak İ, Canpolat G, Karabörk M, Hussain CM (2021) Functionalized magnetic nanoparticles as powerful sorbents and stationary phases for the extraction and chromatographic applications. Trends Anal Chem (TrAC) 143:116380
10. Chen Z, Wu C, Zhang Z, Wu W, Wang X, Yu Z (2018) Synthesis, functionalization, and nanomedical applications of functional magnetic nanoparticles. Chin Chem Lett 29(11):1601–1608
11. MIT. Potential risks of nanomaterials and how to safely handle materials of uncertain toxicity. http://web.mit.edu/environment/ehs/topic/nanomaterial.html
12. Work place exposure to nanoparticle—a report on European risk observatory, European agency for safety and health at work (2009). http://osha.europa.eu/en/publications/literature_reviews/workplace_exposure_to_nanoparticles
13. Explosion hazards associated with nanopowders—a literature review, health and safety laboratory, HSL/2004/12. http://www.hse.gov.uk/research/hsl_pdf/2004/hsl04-12.pdf
14. Key guidance for managing the potential risks of nanotechnology. http://www.safenano.org/KnowledgeBase/Guidance.aspx
15. Malhotra N, Lee JS, Liman RAD, Ruallo JMS, Villaflores OB, Ger TR, Hsiao CD (2020) Potential toxicity of iron oxide magnetic nanoparticles: a review. Molecules 25:3159

16. Shete PB, Patil RM, Tiwale BM, Pawar SH (2015) Water dispersible oleic acid-coated Fe_3O_4 nanoparticles for biomedical applications. J Magn Magn Mater 377:406–410
17. Tran TT-D, Van Vo T, Tran PH-L (2015) Design of iron oxide nanoparticles decorated oleic acid and bovine serum albumin for drug delivery. Chem Eng Res Des 94:112–118
18. Coricovac D-E, Moacă E-A, Pinzaru I, Cîtu C, Soica C, Mihali C-V, Păcurariu C, Tutelyan VA, Tsatsakis A, Dehelean C-A (2017) Biocompatible colloidal suspensions based on magnetic iron oxide nanoparticles: synthesis, characterization and toxicological profile. Front Pharmacol 8:154
19. Soares PI, Laia CA, Carvalho A, Pereira LC, Coutinho JT, Ferreira IM, Novo CM, Borges JP (2016) Iron oxide nanoparticles stabilized with a bilayer of oleic acid for magnetic hyperthermia and MRI applications. Appl Surf Sci 383:240–247
20. Medeiros SF, Filizzola JO, Fonseca VF, Oliveira PF, Silva TM, Elaissari A, Santos AM (2015) Synthesis and characterization of stable aqueous dispersion of functionalized double-coated iron oxide nanoparticles. Mater Lett 160:522–525
21. Valdiglesias V, Fernandez-Bertolez N, Kiliç G, Costa C, Costa S, Fraga S, Bessa MJ, Pasaro E, Teixeira JP, Laffon B (2016) Are iron oxide nanoparticles safe? Current knowledge and future perspectives. J Trace Elem Med Biol 38:53–63
22. Józefczak A, Hornowski T, Skumiel A, Závišová V, Koneracká M, Tomašovičová N, Timko M, Kopčanský P, Kelani H (2012) Effect of the molecular weight of poly (ethylene glycol) on the properties of biocompatible magnetic fluids. Int J Thermophys 33:640–652
23. Soleymani M, Edrissi M (2016) Synthesis of bilayer surfactant-coated magnetic nanoparticles for application in magnetic fluid hyperthermia. J Dispers Sci Technol 37:693–698
24. Arruebo M, Fernández-Pacheco R, Ibarra MR, Santamaría J (2007) Magnetic nanoparticles for drug delivery. Nano Today 2:22–32
25. Tran N, Webster TJ (2010) Magnetic nanoparticles: biomedical applications and challenges. J Mater Chem 20:8760–8767
26. Markides H, Rotherham M, El Haj A (2012) Biocompatibility and toxicity of magnetic nanoparticles in regenerative medicine. J Nanomater 2012:13
27. Correia Carreira S, Walker L, Paul K, Saunders M (2015) The toxicity, transport and uptake of nanoparticles in the in vitro BeWo b30 placental cell barrier model used within nanotest. Nanotoxicology 9:66–78
28. Shukla S, Jadaun A, Arora V, Sinha RK, Biyani N, Jain V (2015) In vitro toxicity assessment of chitosan oligosaccharide coated iron oxide nanoparticles. Toxicol Rep 2:27–39
29. Gholami A, Rasoul-amini S, Ebrahiminezhad A, Seradj SH, Ghasemi Y (2015) Lipoamino acid coated superparamagnetic iron oxide nanoparticles concentration and time dependently enhanced growth of human hepatocarcinoma cell line (Hep-G2). J Nanomater 16:150
30. Di Bona K, Xu Y, Gray M, Fair D, Hayles H, Milad L, Montes A, Sherwood J, Bao Y, Rasco J (2015) Short-and long-term effects of prenatal exposure to iron oxide nanoparticles: influence of surface charge and dose on developmental and reproductive toxicity. Int J Mol Sci 16:30251–30268
31. E 2535 (2007) Standard guide for handling unbound engineered nanoscale particles in occupational settings. ASTM, USA
32. U.S. National Institute for Occupational Safety and Health, Centers for Disease Control and Prevention (2006) Approaches to safe nanotechnology—an information exchange with NIOSH
33. Stern ST, McNeil SE (2008) Toxicol Sci 101(1):4 [5] University of New Hampshire (2007) Nanomaterials Safety Program
34. ISO/TR27628 (2007) Workplace atmospheres—ultrafine, nanoparticle and nano-structured aerosols—exposure characterization and assessment. 1st edn, Switzerland
35. British Standard Institute (BSI) (2007) Nanotechnologies—part 2: guide to safe handling and disposal of manufactured nanomaterials. PD 6699-2:2007, UK
36. Ostiguy C et al (2006) Health effects of nanoparticles. IRSST report 469, Montréal (Québec)
37. European Commision, Health & Consumer Protection Diretorate, Scientific Committee on Emerging and Newly Identified Health Risks (SCENIHR) (2006) The appropriateness of existing methodologies to assess the potential risks associated with engineered and adventitious products of nanotechnologies

38. Ostiguy C et al (2006) Nanoparticles—actual knowledge about occupational health and safety risks and prevention measures. IRSST report 470, Montréal (Québec)
39. Soenen SJ, De Cuyper M (2011) How to assess cytotoxicity of (iron oxide-based) nanoparticles: a technical note using cationic magnetoliposomes. Contrast Media Mol Imaging 6:153–164
40. Schrand AM, Dai L , Schlager JJ, Hussain SM (2012) Toxicity testing of nanomaterials. In: New technologies for toxicity testing. Springer, Toledo, OH, USA, pp 58–75
41. Shen C-C, Wang C-C, Liao M-H, Jan T-R (2011) A single exposure to iron oxide nanoparticles attenuates antigen-specific antibody production and t-cell reactivity in ovalbumin-sensitized balb/c mice. Int J Nanomed 6:1229
42. Schlachter EK, Widmer HR, Bregy A, Lönnfors-Weitzel T, Vajtai I, Corazza N, Bernau VJ, Weitzel T, Mordasini P, Slotboom J (2011) Metabolic pathway and distribution of superparamagnetic iron oxide nanoparticles: In vivo study. Int J Nanomed 6:1793
43. Malindretos P, Sarafidis PA, Rudenco I, Raptis V, Makedou K, Makedou A, Grekas DM (2007) Slow intravenous iron administration does not aggravate oxidative stress and inflammatory biomarkers during hemodialysis: a comparative study between iron sucrose and iron dextran. Am J Nephrol 27:572–579
44. Pritchard DK (2004) Literature review—explosion hazards associated with nanopowders. Health & Safety Laboratory (HSL), UK
45. Aitken RJ, Creely KS, Tran CL (2004) Nanoparticles: an occupational hygiene review. HSE, UK
46. VCI & BauA (2007) Guidance for handling and use of nanomaterials in the workplace. Germany
47. Senn E et al (2000) Controlling chemical exposure. Concise guidance on 16 components of industrial hygiene controls. New Jersey Department of Health and Senior Services, Division of Epidemiology, Environmental and Occupational Health Service
48. Gerritzen G et al (2006) A review of current practices in the nanotechnology industry. In: Phase two report: survey of current practices in the nanotechnology workplace. The International Council on Nanotechnology
49. Mark D (2004) Nanomaterials: a risk to health at work?. In: First international symposium on occupational health implications of nanomaterials, UK, 12–14 Oct 2004
50. Texas A&M University (2005) Interim guideline for working safely with nanotechnology. USA
51. World Health Organization (2004) Laboratory biosafety manual, 3rd edn. Geneva
52. Conti JA et al (2008) Health and safety practices in the nanomaterials workplace: results from an international survey. American Chemical Society, Environmental Science & Technology
53. Guidelines and best, practices for safe handling of nanomaterials in research laboratories and industries, compiled by Nano mission, DST, Govt of India (with inputs from Nano regulatory task, force (2024)
54. Suman TY, Pei DS (2021) Nanomaterial waste management. In: Nanomaterials recycling. Elsevier, pp 21–36
55. Groso A, Petri-Fink A, Magrez A, Riediker M, Meyer T (2010) Management of nanomaterials safety in research environment. Part Fibre Toxicol 7:1–8
56. Schulte PA, Salamanca-Buentello F (2007) Ethical and scientific issues of nanotechnology in the workplace. Environ Health Perspect 115:5–12
57. Amenta V, Aschberger K, Arena M, Bouwmeester H, Moniz FB, Brandhoff P, Gottardo S, Marvin HJ, Mech A, Pesudo LQ, Rauscher H (2015) Regulatory aspects of nanotechnology in the Agri/feed/food sector in EU and non-EU countries. Regul Toxicol Pharmacol 73:463–476
58. Schwirn K, Voelker D, Galert W, Quik J, Tietjen L (2020) Environmental risk assessment of nanomaterials in the light of new obligations under the REACH regulation: which challenges remain and how to approach them? Integr Environ Assess Manag 16:706–717
59. Franco A, Hansen SF, Olsen SI, Butti L (2007) Limits and prospects of the "incremental approach" and the European legislation on the management of risks related to nanomaterials. Regul Toxicol Pharmacol 48:171–183
60. Rasmussen K, Rauscher H, Kearns P, González M, Riego J (2019) Developing OECD test guidelines for regulatory testing of nanomaterials to ensure mutual acceptance of test data. Regul Toxicol Pharmacol 104:74–83

Chapter 13
Conclusions and Future Outlook of Functionalized Magnetic Nanomaterials (FMNs) in Advanced Applications

Rakshita Yadav and Shikha Gulati

Abstract Functionalized magnetic nanomaterials (FMNs) have emerged as transformative materials with diverse applications spanning biomedicine, industry, and energy storage. This chapter provides a comprehensive conclusion to the advancements and challenges in FMNs research, summarizing their pivotal role in drug delivery, wastewater treatment, catalysis, and energy storage. Despite their remarkable potential, FMNs face challenges such as scalability, functionalization limitations, and concerns regarding toxicity and environmental safety. Future research aims to address these challenges through improved synthesis techniques, eco-friendly modifications, and enhanced biocompatibility. The integration of FMNs with artificial intelligence and smart materials is expected to drive innovation, revolutionizing industries ranging from healthcare to sustainable energy. As FMNs continue to evolve, interdisciplinary collaboration and technological advancements will play a crucial role in unlocking their full potential for next-generation applications.

Keywords Functionalization · Magnetic · Nanomaterials · Applications · Biomedical · Energy storage

13.1 Introduction

Functionalized magnetic nanomaterials (FMNs) have garnered significant attention due to their exceptional properties, including high surface area, tunable magnetic behavior, and versatile functionalization capabilities. These unique characteristics make FMNs invaluable across a wide spectrum of applications, spanning

---

R. Yadav
Department of Biochemistry, Sri Venkateswara College, University of Delhi, Delhi, India

S. Gulati (✉)
Department of Chemistry, Sri Venkateswara College, University of Delhi, Delhi, India
e-mail: shikhagulati@svc.ac.in

S. Gulati (ed.), *Functionalized Magnetic Nanomaterials*, Nanostructure Science and Technology, https://doi.org/10.1007/978-3-031-97199-0_13

biomedicine, environmental remediation, industrial catalysis, and energy storage. Over the past few decades, extensive research has been conducted to harness the full potential of FMNs, leading to groundbreaking advancements that have enhanced their efficiency, stability, and applicability in real-world scenarios.

In biomedical applications, FMNs have revolutionized targeted drug delivery, magnetic resonance imaging (MRI) [1], and hyperthermia treatment for cancer [2]. Their ability to respond to external magnetic fields allows for precise localization and controlled release of therapeutic agents, minimizing side effects and improving treatment efficacy. Figure 13.1 illustrates some of these applications of FMNs[3]. Similarly, in environmental applications, FMNs have demonstrated remarkable efficiency in wastewater treatment and pollutant removal due to their high adsorption capacity and ease of separation from treated water [4]. Industrial sectors have also benefited from FMNs, particularly in catalysis, where their magnetic properties facilitate catalyst recovery and reuse, improving process efficiency and sustainability [5]. Furthermore, the integration of FMNs in energy storage technologies, such as batteries and supercapacitors, has opened new avenues for enhancing energy efficiency and sustainability [6, 7]. Despite these advancements, challenges remain in optimizing FMN synthesis, improving functionalization techniques, and addressing concerns related to toxicity and environmental safety. As research continues to evolve, interdisciplinary approaches integrating nanotechnology, materials science, and artificial intelligence are expected to drive the next phase of FMN innovation [8, 9].

This book volume provides a comprehensive exploration of functionalized magnetic nanomaterials (FMNs), covering their fundamental principles, synthesis techniques, characterization methods [11], and advanced applications across multiple scientific and industrial domains. It begins with an in-depth discussion of the structural and physicochemical properties of FMNs, detailing various functionalization strategies that enhance their stability, reactivity, and application-specific performance. The book extensively examines their role in biomedical fields, including drug delivery, magnetic resonance imaging (MRI), hyperthermia therapy, and biosensing [11]. Additionally, it highlights FMNs' significance in environmental applications such as wastewater treatment, heavy metal removal, and pollutant degradation [4]. Their use in energy storage and conversion, particularly in supercapacitors, batteries [7], and catalysis [12], is also discussed, showcasing their potential for sustainable energy solutions. Furthermore, the book addresses the challenges associated with FMN development, including scalability, stability, toxicity, and regulatory considerations, offering insights into potential solutions and future research directions. By integrating contributions from leading experts worldwide, this volume serves as a valuable resource for researchers, academicians, and industry professionals, providing a holistic understanding of FMNs and their transformative impact on science and technology.

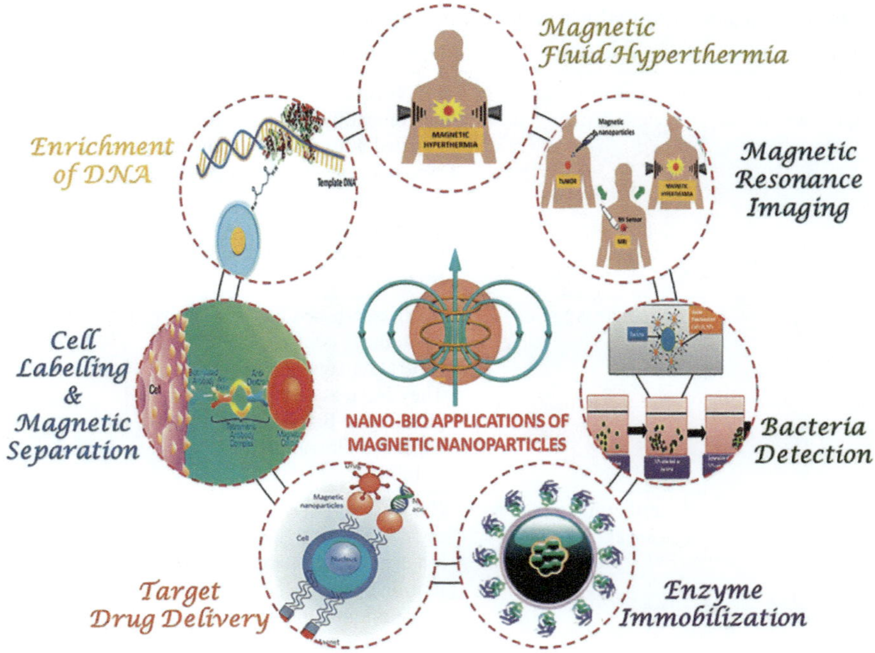

Fig. 13.1 A schematic representation of various applications of magnetic nanoparticles. Reproduced from [10] originally published under a CC BY license

13.2 Summary of FMNs in Advanced Applications

Functionalized magnetic nanomaterials (FMNs) have demonstrated immense potential across a wide range of applications, revolutionizing fields such as biomedicine, environmental science, and energy storage (Table 13.1). Their ability to be tailored for specific functionalities has enabled groundbreaking advancements in targeted drug delivery, pollutant remediation, and sustainable energy solutions. This concluding chapter summarizes the key insights gained throughout this book and discusses the future directions that could further enhance the capabilities of FMNs in advanced scientific and industrial applications.

13.2.1 Biomedical Applications

FMNs have emerged as powerful tools in modern medicine, offering innovative solutions in diagnosis and treatment. Their role in drug delivery and targeted therapy has shown promising results, particularly in enhancing therapeutic efficiency while minimizing systemic side effects [3, 22]. FMNs functionalized with biocompatible coatings enable precise delivery of drugs to diseased tissues, such as tumors,

Table 13.1 Summary of FMNs in advanced applications

Application area	Sub-application	Description	References
Biomedical applications	Drug delivery and targeted therapy	FMNs functionalized with biocompatible coatings enhance targeted drug delivery, reducing systemic toxicity and improving bioavailability. These nanomaterials can cross biological barriers, such as the blood–brain barrier	[13]
	MRI contrast agents	Iron oxide-based FMNs improve MRI image contrast due to their superparamagnetic properties. They serve as safer alternatives to gadolinium-based agents with enhanced biocompatibility and stability	[14, 15]
	Hyperthermia treatment for cancer	FMNs generate localized heat upon exposure to an alternating magnetic field, enabling non-invasive cancer treatment while minimizing damage to surrounding tissues	[16]
Industrial and environmental applications	Wastewater treatment and pollutant removal	FMNs selectively adsorb heavy metals and organic pollutants, enabling efficient removal and magnetic separation. They are functionalized with ligands for enhanced selectivity	[17, 18]
	Catalysis and industrial processing	FMNs act as reusable catalysts in organic synthesis and industrial reactions, improving efficiency and sustainability in chemical processes	[15, 19]
Energy storage and conversion	Batteries and supercapacitors	FMNs improve energy density and charge–discharge efficiency in lithium-ion batteries and supercapacitors. Functionalized FMNs enable superior electrochemical performance	[20]
	Sustainable energy solutions	FMNs serve as catalysts in hydrogen storage, fuel cells, and thermoelectric applications, contributing to clean energy advancements	[21]

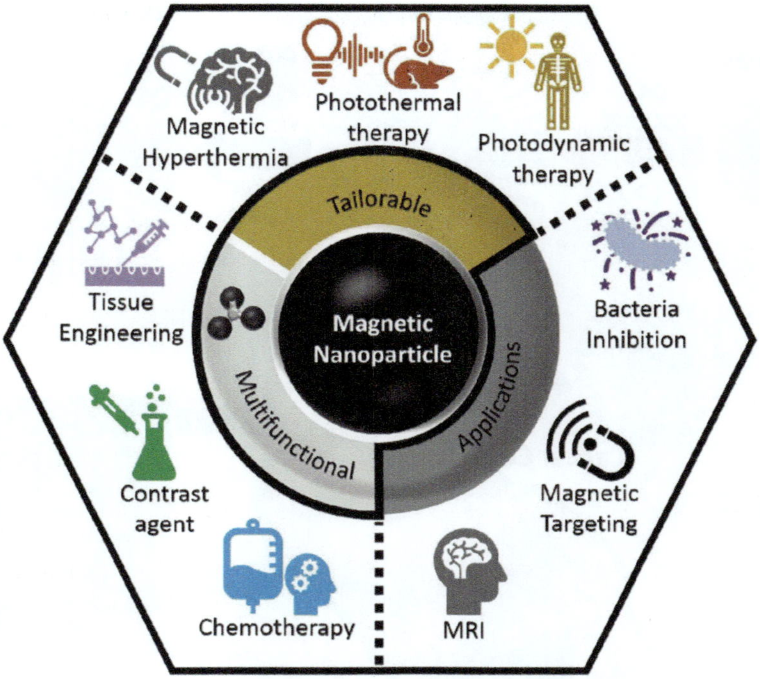

Fig. 13.2 Biomedical applications of MNPs. Reproduced with permission from [11]

under external magnetic field control. Additionally, FMNs serve as highly effective MRI contrast agents, improving the accuracy of medical imaging [23]. Their superior magnetic properties enhance contrast resolution, enabling early disease detection. FMNs are also widely studied for hyperthermia treatment, where they generate localized heat to destroy cancer cells upon exposure to an alternating magnetic field, providing a non-invasive therapeutic approach as illustrated in Fig. 13.2 [24].

13.2.1.1 Drug Delivery and Targeted Therapy

One of the most significant applications of FMNs in biomedicine is their role in targeted drug delivery. Traditional drug administration methods often suffer from poor bioavailability, systemic toxicity, and nonspecific distribution, which limit their therapeutic efficiency [25, 26]. FMNs, when functionalized with biocompatible coatings and drug molecules, provide an effective strategy for site-specific drug delivery(Fig. 13.3). These nanomaterials can be surface-functionalized with drugs, peptides, or antibodies and guided to the target site using an external magnetic field [27, 28]. Upon reaching the designated location, such as a tumor or infection site, the drug can be released in a controlled manner through pH-sensitive, temperature-sensitive, or enzymatic degradation mechanisms [28]. This method significantly

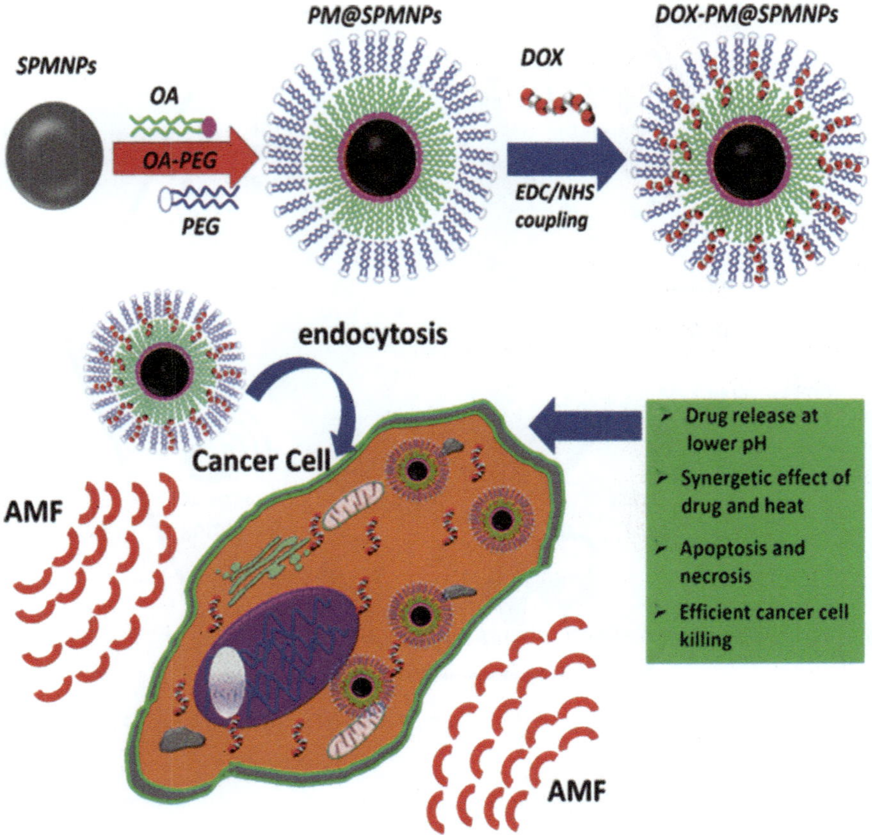

Fig. 13.3 Schematic illustration showing the overall mechanism of anticancer drug (DOX) conjugation into polymer-encapsulated SPMNPs and the concept of cancer cell killing through synergetic therapy using an alternating magnetic field (AMF). Reproduced with permission from [31]

enhances drug bioavailability while minimizing side effects [29]. It also allows FMNs to overcome biological barriers, such as the blood–brain barrier, which is crucial for treating neurological disorders [30]. Despite these advantages, challenges such as biocompatibility concerns, potential toxicity, and large-scale production must be addressed to ensure their clinical viability.

13.2.1.2 Magnetic Resonance Imaging (MRI) Contrast Agents

Magnetic nanoparticles, particularly iron oxide-based FMNs, have emerged as superior MRI contrast agents due to their superparamagnetic properties, high relaxivity, and ability to enhance image contrast [23]. Conventional contrast agents, such as gadolinium-based compounds, pose limitations, including potential toxicity and short circulation times [14]. FMNs provide a safer and more efficient alternative by altering

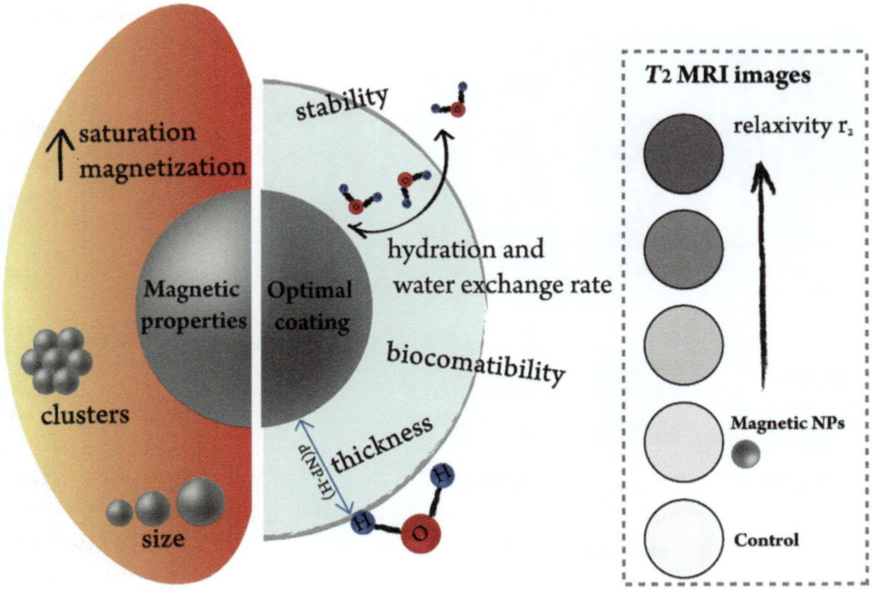

Fig. 13.4 Magnetic nanoparticle-based T_2 MRI Contrast Agents. Reproduced from [34] originally published under a CC BY License

the local magnetic field within tissues, thereby improving MRI image contrast as illustrated in Fig. 13.4.[32, 33]. Functionalization with biocompatible coatings such as polyethylene glycol (PEG), [32] dextran, or silica enhances their stability, circulation time, and targeting capabilities. These characteristics make FMNs particularly useful for detecting tumors and neurodegenerative diseases with greater accuracy. Although FMNs offer significant improvements in MRI diagnostics, issues related to aggregation, immune system clearance, and long-term safety must be carefully examined before widespread clinical adoption [23].

13.2.1.3 Hyperthermia Treatment for Cancer

FMNs have also demonstrated remarkable potential in cancer treatment through magnetic hyperthermia. This innovative and minimally invasive approach involves the localized heating of tumor tissues using FMNs exposed to an alternating magnetic field (AMF) [13, 35]. As these nanoparticles convert electromagnetic energy into heat, the temperature of cancerous tissues rises to a range of 42–45 °C, triggering irreversible cellular damage, disrupting tumor metabolism, and enhancing the effects of chemotherapy and radiotherapy [16]. The selective heating ability of FMNs ensures minimal damage to surrounding healthy tissues, making hyperthermia an attractive treatment option. Additionally, FMN-based hyperthermia can be used in combination with other therapies, such as immunotherapy, for improved outcomes (Fig. 13.5)

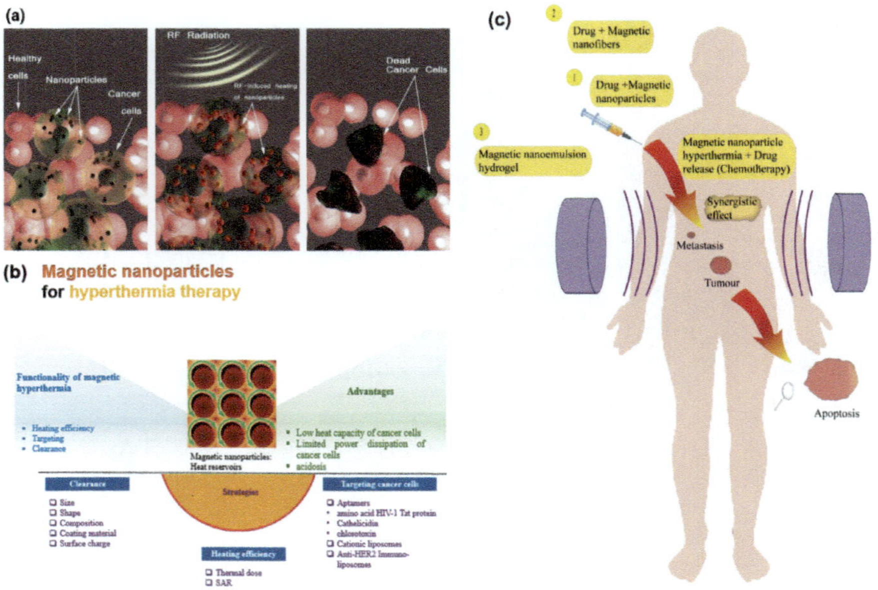

Fig. 13.5 **a** Schematic representation of the magnetic hyperthermia treatment procedure. Reproduced with permission from [37] **b** Factors affecting the functionality of magnetic hyperthermia. Reproduced from [35] originally published under a BCC BY License. **c** Combined MIH and chemotherapy for the treatment of metastatic cancer. Reproduced with permission from [16]

[16]. However, the success of this technique depends on factors such as nanoparticle stability, efficient heat generation, and controlled accumulation at tumor sites. Research efforts are currently focused on optimizing FMN compositions, improving biodistribution, and minimizing potential systemic toxicity to make this therapy more effective and widely accessible [36].

13.2.2 Industrial and Environmental Applications of FMNs

In the industrial and environmental sectors, FMNs have been extensively applied in wastewater treatment and pollutant removal due to their high adsorption capacity and magnetic separation capabilities. They effectively remove heavy metals, dyes, and organic pollutants, contributing to cleaner water resources. Moreover, FMNs play a critical role in catalysis and industrial processing, where they enhance reaction efficiencies, reduce energy consumption, and facilitate easy catalyst recovery and reuse, making chemical processes more sustainable.

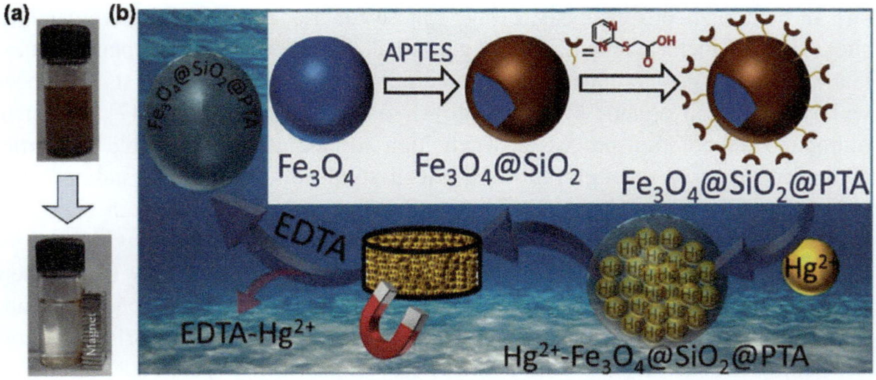

Fig. 13.6 **a** Photographic image and **b** a schematic illustration of the separation of magnetic nanoparticles with the help of a magnet for mercury metal removal from water. Reproduced with permission from [38]

13.2.2.1 Wastewater Treatment and Pollutant Removal

One of the most critical environmental challenges today is water pollution caused by heavy metals, organic dyes, and other hazardous contaminants [17]. FMNs have emerged as highly effective materials for wastewater treatment due to their ability to selectively adsorb pollutants and be easily separated from treated water using an external magnetic field [38]. The functionalization of FMNs with specific ligands, such as carboxyl, amine, or thiol groups, enhances their affinity for heavy metals like lead, mercury, and arsenic, allowing for efficient removal even at low concentrations (Fig. 13.6) [38, 39]. In addition to heavy metal removal, FMNs have proven effective in treating organic pollutants, such as dyes and pharmaceutical residues, which are commonly found in industrial wastewater [4, 40, 41]. Functionalized magnetic nanocatalysts can accelerate degradation reactions, breaking down complex organic pollutants into non-toxic byproducts. Moreover, FMNs can be regenerated and reused multiple times, [18] making them a cost-effective and sustainable solution for large-scale water purification. Despite these advantages, challenges such as nanoparticle aggregation, stability in different water conditions, and large-scale implementation need to be addressed to maximize their real-world application.

13.2.2.2 Catalysis and Industrial Processing

FMNs have also demonstrated significant potential in catalysis and industrial processing, where they serve as efficient, reusable, and easily recoverable catalysts [12, 42]. Traditional catalysts often suffer from issues such as low stability, difficulty in separation, and high operational costs. FMNs provide a viable alternative by offering high catalytic activity combined with easy magnetic recovery, eliminating the need for complex and energy-intensive separation techniques [43,

44]. In heterogeneous catalysis, FMNs can be functionalized with active catalytic sites, such as metal nanoparticles (e.g., palladium, platinum, or copper) or enzymatic biomolecules, to enhance reaction efficiency [45]. These catalysts have been widely applied in organic synthesis [46], hydrogenation reactions [47], and environmental remediation processes. Their high surface area and tunable properties allow for increased reaction rates and selectivity, leading to more sustainable and efficient industrial processes. FMNs are also being explored in green chemistry applications, where they facilitate environmentally friendly reactions with minimal waste production [12, 48]. Their ability to be recovered magnetically makes them highly desirable for catalytic processes in the pharmaceutical, petrochemical, and food industries. However, challenges such as catalyst deactivation, surface fouling, and long-term stability must be addressed to improve their practical usability in industrial-scale reactions [44]. Figure 13.7 illustrates some common heterogeneous catalytic applications of FMNs.

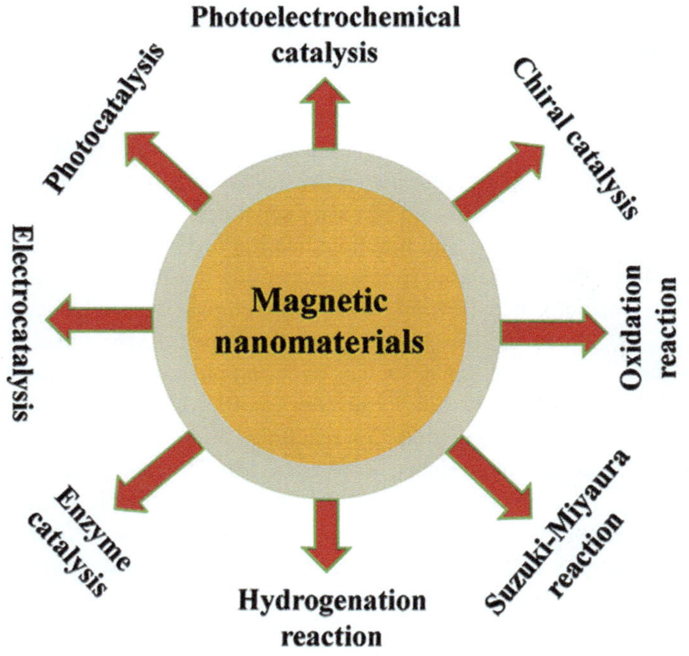

Fig. 13.7 aVarious applications of FMNs in heterogeneous catalysis. Reproduced with permission from [43]

13.2.3 Energy Storage and Conversion

The growing global demand for efficient and sustainable energy storage systems has driven significant interest in functionalized magnetic nanomaterials (FMNs). Their unique magnetic properties, high surface area, tunable functionality, and excellent conductivity make them promising candidates for advanced energy storage and conversion applications, as depicted in Fig. 13.8 [49]. FMNs play a crucial role in enhancing the performance of batteries, supercapacitors, and other energy storage devices while also contributing to sustainable energy solutions [50]. Their integration into modern energy systems offers new pathways for developing high-performance, eco-friendly, and durable energy storage materials [51].

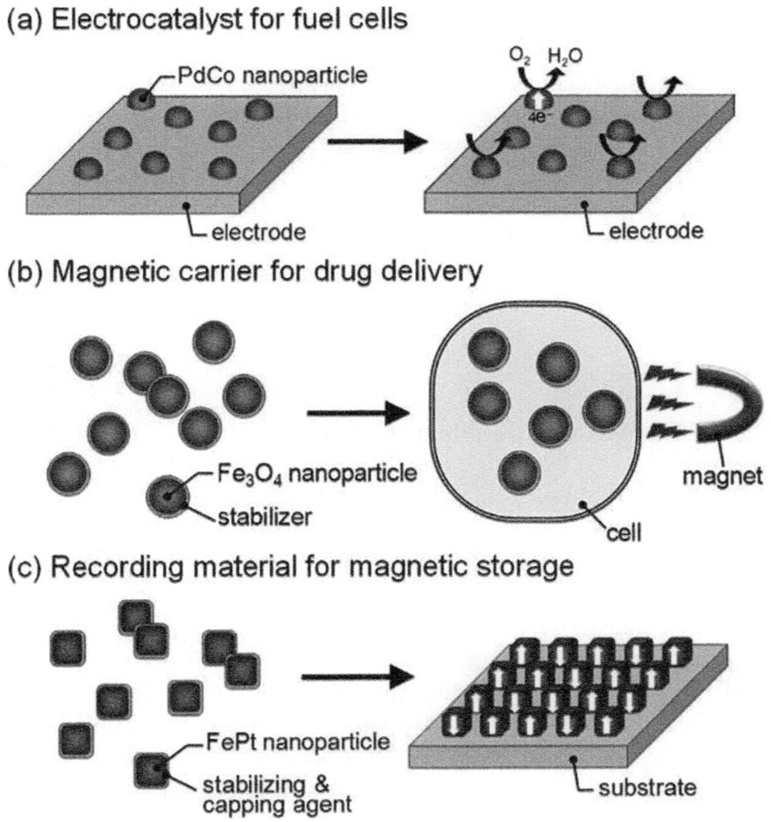

Fig. 13.8 **a** electrocatalyst (PdCo alloy) for fuel cells synthesized by electrodeposition, **b** magnetic carrier (Fe_3O_4) for drug delivery synthesized by hydrolysis, and **c** recording material (FePt alloy) for magnetic storage synthesized by thermal decomposition of organometallic compounds. Reproduced with permission from [50]

13.2.3.1 FMNs in Batteries and Supercapacitors

Rechargeable batteries and supercapacitors are at the forefront of energy storage technologies, enabling the transition to renewable energy systems, portable electronics, and electric vehicles. FMNs have emerged as promising materials for improving battery performance by enhancing energy density, charge–discharge efficiency, and long-term stability. In lithium-ion batteries (LIBs), FMNs serve as efficient electrode materials due to their excellent electrical conductivity and large surface area, which facilitates fast ion transport and charge storage [20]. Functionalized magnetic nanoparticles, such as Fe_3O_4 and $CoFe_2O_4$, have been investigated as potential anode materials, offering high theoretical capacities and improved cycling stability[7]. The incorporation of FMNs into battery components also enhances electrochemical performance by reducing polarization effects and increasing the overall efficiency of charge storage mechanisms [52].

Supercapacitors, which are essential for high-power energy storage and rapid charge–discharge applications, also benefit from FMNs. The incorporation of FMNs into electrode materials improves charge transfer kinetics and enhances capacitance, resulting in superior energy storage capabilities. By functionalizing FMNs with conductive polymers or carbon-based materials (e.g., graphene or carbon nanotubes), [53] Researchers have developed high-performance supercapacitors with improved cycling stability, mechanical flexibility, and charge retention. However, challenges such as nanoparticle aggregation, electrode degradation, and scalability issues must be addressed to enable their widespread adoption in commercial energy storage devices [54].

13.2.3.2 Role of FMNs in Sustainable Energy Solutions

The global shift toward renewable energy sources necessitates the development of efficient energy storage and conversion technologies. FMNs play a significant role in advancing sustainable energy solutions, including hydrogen storage, fuel cells, and thermoelectric applications [21, 55]. Their ability to enhance catalytic activity, improve charge transfer efficiency, and enable magnetic separation makes them valuable in next-generation energy systems.

One of the most promising applications of FMNs is in hydrogen storage and fuel cells, where they serve as catalysts for hydrogen evolution reactions (HER) and oxygen reduction reactions (ORR) [56]. FMNs functionalized with transition metal catalysts, such as iron, cobalt, or nickel, exhibit enhanced electrocatalytic performance, making them suitable for green hydrogen production and fuel cell applications [57]. Their magnetic properties also allow for easy recovery and reuse, contributing to the development of cost-effective and sustainable hydrogen energy systems. FMNs have also been explored in thermoelectric applications, where they enhance the efficiency of waste heat recovery and energy conversion [58]. Their ability to improve thermal conductivity and charge carrier mobility enables the development of high-performance thermoelectric materials, which convert waste heat into usable electricity [59]. These advancements have the potential to improve energy efficiency in industrial processes, automotive systems, and electronic devices.

13.3 Current Challenges and Limitations

While functionalized magnetic nanomaterials (FMNs) have demonstrated immense potential across biomedical, environmental, and industrial applications, several critical challenges and limitations continue to hinder their widespread adoption. These challenges include stability and scalability issues, functionalization complexities, and concerns related to toxicity and environmental safety. Addressing these limitations is crucial for advancing FMNs toward commercial viability and sustainable applications.

13.3.1 Stability and Scalability Issues

One of the primary challenges in FMN development is ensuring long-term stability, particularly under varying environmental conditions. Magnetic nanoparticles tend to aggregate due to van der Waals forces and magnetic dipole interactions, which can significantly reduce their functionality in applications such as drug delivery, catalysis, and water purification (Fig. 13.9) [60]. Maintaining the structural and functional integrity of FMNs in different media, including biological fluids and industrial solvents, remains a significant hurdle. Scalability is another pressing concern. Many synthesis methods for FMNs, such as co-precipitation, sol–gel, and thermal decomposition, are optimized for small-scale laboratory production [58]. However, translating these methods into large-scale, cost-effective manufacturing while maintaining particle uniformity, functionalization efficiency, and magnetic properties is challenging [61]. Industrial-scale production demands consistent quality control, reproducibility, and energy-efficient processes, all of which require further technological advancements.

13.3.2 Functionalization Challenges

The surface functionalization of FMNs plays a pivotal role in enhancing their stability, selectivity, and biocompatibility for specific applications. However, achieving uniform, reproducible, and durable functionalization remains complex. The effectiveness of FMNs often depends on the chemical nature of the surface ligands, polymers, or coatings, which must be carefully selected to maintain colloidal stability and

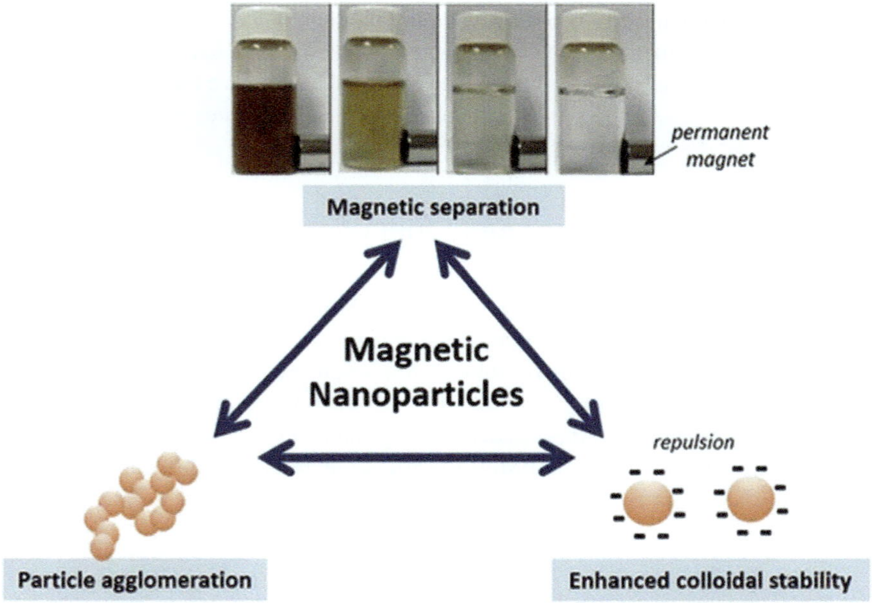

Fig. 13.9 Colloidal stability and particle aggregation of MNPs affect their long-term stability under different environments. Reproduced with permission from [60]

avoid undesired interactions with biological or environmental systems. Another key challenge is ensuring that functionalization does not negatively impact the magnetic properties of the nanomaterials. Some coatings or modifications may reduce magnetic responsiveness, making the FMNs less effective for applications like hyperthermia treatment, MRI contrast enhancement, or targeted drug delivery. Additionally, large-scale functionalization processes must be developed to ensure uniform coating and stability across bulk production [62].

13.3.3 Toxicity and Environmental Safety Concerns

As FMNs gain traction in biomedical and environmental applications, concerns regarding their potential toxicity and long-term environmental impact have become increasingly significant. Due to their small size and high surface reactivity, FMNs can penetrate biological membranes, accumulate in organs, and induce oxidative stress, inflammation, or genotoxic effects in living organisms [62, 63]. These toxicological risks pose challenges in applications such as drug delivery, imaging, and cancer therapy, where FMNs come into direct contact with human tissues [64]. In environmental applications, the fate and transport of FMNs in ecosystems are still not fully understood. Their potential to persist in soil and water bodies, interact with microbial communities, or leach toxic components raises concerns about their long-term ecological impact [65]. Developing biodegradable or eco-friendly FMNs and conducting comprehensive life-cycle assessments will be crucial for mitigating these risks. Moreover, regulatory frameworks governing the safe use, disposal, and long-term effects of FMNs remain underdeveloped in many regions [65, 66]. A standardized approach to risk assessment, toxicity testing, and environmental monitoring is necessary to ensure FMNs can be safely integrated into commercial and industrial applications without causing unintended harm [67].

13.4 Future Directions in FMNs Research

As Functionalized Magnetic Nanomaterials (FMNs) continue to evolve, ongoing research is directed toward overcoming existing challenges and unlocking new possibilities for their applications. The future of FMN's research lies in advancements in synthesis and functionalization techniques, improving biocompatibility, developing sustainable nanomaterials, and integrating FMNs with artificial intelligence (AI) and smart materials. These advancements will not only enhance their efficiency and safety but also broaden their applicability across diverse scientific and industrial fields.

13.4.1 Advancements in Synthesis and Functionalization Techniques

The precise control of FMNs' size, morphology, and surface chemistry is crucial for their functionality. Future research will focus on refining scalable and reproducible synthesis methods that ensure uniform particle size and controlled functionalization[68]. Advanced techniques such as microfluidic synthesis [69], electrochemical deposition [70], and atomic layer deposition [71] could offer greater precision and scalability. Additionally, plasma-based and microwave-assisted synthesis methods are expected to improve energy efficiency while maintaining high material quality [72]. Innovations in functionalization strategies will further enhance FMNs' stability, solubility, and selectivity for specific applications. Research will emphasize multifunctional coatings, where FMNs are engineered with responsive polymers, bioinspired ligands, and self-assembling nanostructures to create smart materials that adapt to environmental stimuli [62, 73].

13.4.2 Enhancing Biocompatibility and Reducing Toxicity

While FMNs hold promise for biomedical applications, concerns about toxicity and biocompatibility remain a major challenge. Future research will explore biodegradable coatings, natural polymer-based surface modifications, and bioinspired functionalization to minimize adverse biological interactions [74]. Zwitterionic and hydrophilic surface coatings could significantly reduce protein adsorption and immune responses, improving their compatibility for in vivo applications [75]. Additionally, systematic in vitro and in vivo toxicity assessments will be crucial in understanding FMNs' long-term effects on biological systems [76]. The development of safe-by-design FMNs, where toxicity mitigation is integrated into the early stages of material design, will be a key focus to facilitate their clinical and environmental applications.

13.4.3 Development of Sustainable and Eco-Friendly FMNs

The sustainability of FMN production and application is a growing concern, particularly in large-scale industrial and environmental applications. Researchers are shifting toward green synthesis approaches, utilizing plant extracts, microorganisms, and biomolecules to synthesize FMNs in an environmentally friendly manner, as illustrated in Fig. 13.10 [77, 78]. These methods not only reduce toxic chemical usage but also improve cost-effectiveness and eco-compatibility. Moreover, efforts to develop biodegradable and recyclable FMNs will drive sustainability further [79]. Future studies will aim to create FMNs that degrade into non-toxic components after their intended use, reducing environmental accumulation and potential ecological risks. Additionally, research into magnetic nanocomposites derived from renewable sources will support circular economy principles, promoting resource-efficient nanotechnology [80, 81].

13.4.4 Integration of FMNs with Artificial Intelligence (AI) and Smart Materials

The convergence of FMNs with artificial intelligence (AI) and smart materials will revolutionize their application landscape. AI-driven material discovery and computational modeling will facilitate the design of high-performance FMNs with tailored properties for specific applications, reducing trial-and-error experimental approaches (Fig. 13.11) [82]. Machine learning algorithms can predict synthesis conditions, optimize functionalization, and enhance process scalability, accelerating the development of next-generation FMNs[83]. In addition, smart FMNs that respond to external stimuli such as pH, temperature, magnetic fields, and light will open new frontiers

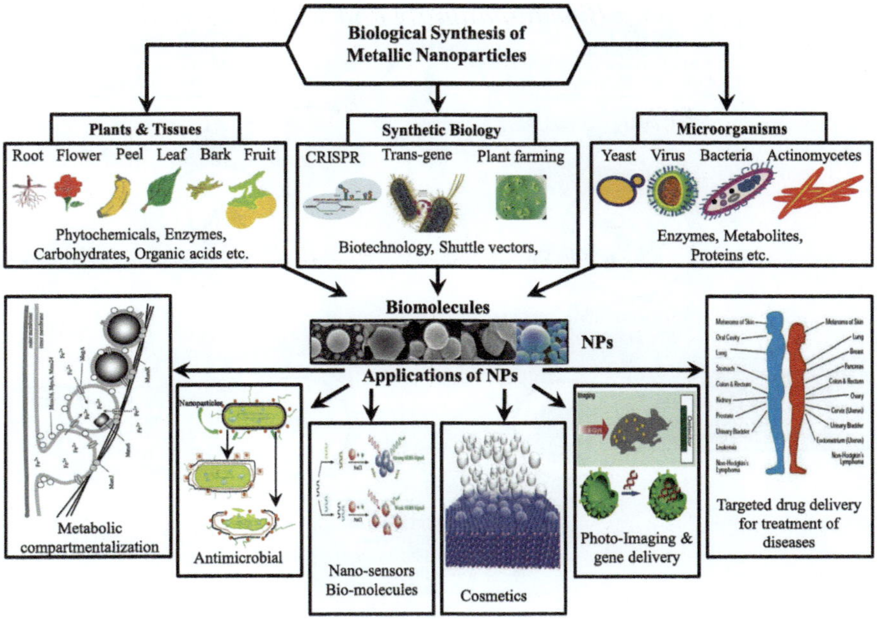

Fig. 13.10 Metallic nanoparticles (MNPs) produced biologically by plants and microbes: cellular absorption, biocompatibility, and biomedical applications. Reproduced with permission from [77]

in drug delivery, biosensing, and environmental remediation[84]. The integration of FMNs with nanorobotics, wearable biosensors, and self-healing materials will create advanced technologies with real-time responsiveness, precision control, and enhanced efficiency in healthcare, diagnostics, and industrial processes [85, 86].

13.5 Concluding Remarks

Functionalized magnetic nanomaterials (FMNs) have emerged as a transformative class of materials with applications spanning biomedical, environmental, industrial, and energy sectors. Over the years, advancements in synthesis techniques, surface functionalization, and application-specific modifications have propelled FMNs from fundamental research to real-world implementation. Their unique combination of magnetic properties, high surface area, tunable surface chemistry, and functional versatility has enabled their integration into cutting-edge technologies such as targeted drug delivery, wastewater treatment, catalysis, and energy storage systems. Despite these advancements, challenges such as scalability, stability, toxicity, and environmental sustainability remain critical hurdles in FMN development. Addressing these concerns through green synthesis, advanced characterization techniques, and safe-by-design approaches will be pivotal in ensuring the

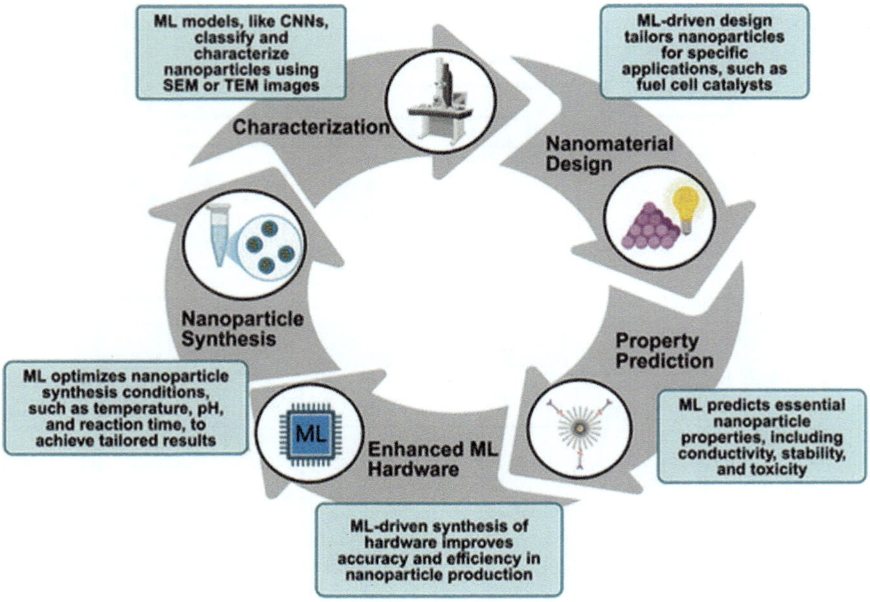

Fig. 13.11 Convergence and collaboration between ML and nanotechnology. Reproduced from [83]. Originally published under a CC BY 4.0 license

long-term viability and acceptance of FMNs across industries. Moreover, regulatory frameworks must evolve to ensure the responsible and ethical deployment of these nanomaterials in healthcare and environmental applications. Looking ahead, FMNs are expected to play a crucial role in next-generation smart materials, AI-driven nanotechnology, and sustainable solutions, revolutionizing sectors that rely on precision materials and nanotechnology-driven innovations. With continued advancements, FMNs hold immense potential to shape the future of scientific and industrial progress.

Abbreviations

AI	Artificial intelligence
AMF	Alternating magnetic field
FMNs	Functionalized magnetic nanoparticles
HER	Hydrogen evolution reaction
LIBs	Lithium-ion batteries
MRI	Magnetic resonance imaging
ORR	Oxygen reduction reaction
PEG	Polyethylene glycol

References

1. Sun C, Lee JSH, Zhang M (2008) Magnetic nanoparticles in MR imaging and drug delivery. Adv Drug Deliv Rev 60:1252–1265. https://doi.org/10.1016/j.addr.2008.03.018
2. Magnetic nanoparticles: biomedical applications and challenges. J Mater Chem. RSC Publishing. https://pubs.rsc.org/en/content/articlelanding/2010/jm/c0jm00994f/unauth. Accessed 20 Mar 2025
3. Emerging trends in the nanomedicine applications of functionalized magnetic nanoparticles as novel therapies for acute and chronic diseases. J Nanobiotechnol. https://doi.org/10.1186/s12951-022-01595-3. Accessed 23 Mar 2025
4. Ramu S, Kainthla I, Chandrappa L et al (2024) Recent advances in metal organic frameworks–based magnetic nanomaterials for waste water treatment. Environ Sci Pollut Res 31:167–190. https://doi.org/10.1007/s11356-023-31162-8
5. Fuentes-García JA, Sanz B, Mallada R et al (2023) Magnetic nanofibers for remotely triggered catalytic activity applied to the degradation of organic pollutants. Mater Des 226:111615. https://doi.org/10.1016/j.matdes.2023.111615
6. Inamuddin, Boddula R, Ahmer MF, Asiri AM (2020) Lithium-ion batteries: materials and applications. Materials Research Forum LLC
7. Synthesis, characterization, and applications of nanomaterials for energy conversion and storage. https://www.mdpi.com/1420-3049/28/21/7383. Accessed 28 Mar 2025
8. Nanotoxicology and nanosafety: safety-by-design and testing at a glance. https://www.mdpi.com/1660-4601/17/13/4657. Accessed 23 Mar 2025
9. Assessing the in vitro and in vivo toxicity of superparamagnetic iron oxide nanoparticles. Chem Rev. https://doi.org/10.1021/cr2002596. Accessed 23 Mar 2025
10. Katz E (2020) Magnetic nanoparticles. Magnetochemistry 6:6. https://doi.org/10.3390/magnetochemistry6010006
11. Lu et al (2007) Magnetic nanoparticles: synthesis, protection, functionalization, and application. Angewandte Chemie Int Edn. Wiley Online Library. https://doi.org/10.1002/anie.200602866. Accessed 22 Mar 2025
12. Polshettiwar V, Varma RS (2010) Green chemistry by nano-catalysis. Green Chem 12:743. https://doi.org/10.1039/b921171c
13. Bañobre-López M, Teijeiro A, Rivas J (2013) Magnetic nanoparticle-based hyperthermia for cancer treatment. Rep Pract Oncol Radiother 18:397–400. https://doi.org/10.1016/j.rpor.2013.09.011
14. Ranga A, Agarwal Y, Garg KJ (2017) Gadolinium based contrast agents in current practice: risks of accumulation and toxicity in patients with normal renal function. Indian J Radiol Imaging 27:141–147. https://doi.org/10.4103/0971-3026.209212
15. Avasthi A, Caro C, Pozo-Torres E et al (2020) Magnetic nanoparticles as MRI contrast agents. Top Curr Chem (Cham) 378:40. https://doi.org/10.1007/s41061-020-00302-w
16. Jose J, Kumar R, Harilal S et al (2020) Magnetic nanoparticles for hyperthermia in cancer treatment: an emerging tool. Environ Sci Pollut Res 27:19214–19225. https://doi.org/10.1007/s11356-019-07231-2
17. Toxicity of metal oxides, dyes, and dissolved organic matter in water: implications for the environment and human health. https://www.mdpi.com/2305-6304/12/2/111. Accessed 29 Mar 2025
18. Gómez-Pastora J, Bringas E, Ortiz I (2014) Recent progress and future challenges on the use of high performance magnetic nano-adsorbents in environmental applications. Chem Eng J 256:187–204. https://doi.org/10.1016/j.cej.2014.06.119
19. Parra-Arroyo L, González-González RB, Chavez-Santoscoy RA et al (2023) Magnetic nanomaterials assisted nanobiocatalysis to abate groundwater pollution. MethodsX 10:102161. https://doi.org/10.1016/j.mex.2023.102161
20. Yu S, Ng VMH, Wang F et al (2018) Synthesis and application of iron-based nanomaterials as anodes of lithium-ion batteries and supercapacitors. J Mater Chem A 6:9332–9367. https://doi.org/10.1039/C8TA01683F

21. Magnetic-based nanocomposites for hydrogen generation as renewable energy. ScienceDirect. https://www.sciencedirect.com/science/article/abs/pii/B9780323857482000190. Accessed 10 Mar 2025
22. Tiwari R, Tiwari G, Parashar P (2023) Theranostics applications of functionalized magnetic nanoparticles. In: Jain K, Jain NK (eds) Multifunctional and targeted theranostic nanomedicines: formulation, design and applications. Springer Nature, Singapore, pp 361–382
23. Avasthi A, Caro C, Pozo-Torres E et al (2020) Magnetic nanoparticles as MRI contrast agents. Top Curr Chem (Z) 378:40. https://doi.org/10.1007/s41061-020-00302-w
24. Kumar CSSR, Mohammad F (2011) Magnetic nanomaterials for hyperthermia-based therapy and controlled drug delivery. Adv Drug Deliv Rev 63:789–808. https://doi.org/10.1016/j.addr.2011.03.008
25. Jain KK (2008) Drug delivery systems—an overview. In: Jain KK (ed) Drug delivery systems. Humana Press, Totowa, NJ, pp 1–50
26. Adepu S, Ramakrishna S (2021) Controlled drug delivery systems: current status and future directions. Molecules 26:5905. https://doi.org/10.3390/molecules26195905
27. Berry CC, Curtis ASG (2003) Functionalisation of magnetic nanoparticles for applications in biomedicine. J Phys D: Appl Phys 36:R198. https://doi.org/10.1088/0022-3727/36/13/203
28. Liu JF, Jang B, Issadore D, Tsourkas A (2019) Use of magnetic fields and nanoparticles to trigger drug release and improve tumor targeting. Wiley Interdiscip Rev Nanomed Nanobiotechnol 11:e1571. https://doi.org/10.1002/wnan.1571
29. Chamundeeswari M, Jeslin J, Verma ML (2019) Nanocarriers for drug delivery applications. Environ Chem Lett 17:849–865. https://doi.org/10.1007/s10311-018-00841-1
30. Namdeo M, Saxena S, Tankhiwale R et al (2008) Magnetic nanoparticles for drug delivery applications. J Nanosci Nanotechnol 8:3247–3271. https://doi.org/10.1166/jnn.2008.399
31. Thorat ND, Bohara RA, Noor MR et al (2017) Effective cancer theranostics with polymer encapsulated superparamagnetic nanoparticles: combined effects of magnetic hyperthermia and controlled drug release. ACS Biomater Sci Eng 3:1332–1340. https://doi.org/10.1021/acsbiomaterials.6b00420
32. Nitin N, LaConte LEW, Zurkiya O et al (2004) Functionalization and peptide-based delivery of magnetic nanoparticles as an intracellular MRI contrast agent. J Biol Inorg Chem 9:706–712. https://doi.org/10.1007/s00775-004-0560-1
33. Mao X, Xu J, Cui H (2016) Functional nanoparticles for magnetic resonance imaging. WIREs Nanomed Nanobiotechnol 8:814–841. https://doi.org/10.1002/wnan.1400
34. Kostevšek N (2020) A review on the optimal design of magnetic nanoparticle-based T_2 MRI contrast agents. Magnetochemistry 6:11. https://doi.org/10.3390/magnetochemistry6010011
35. Fundamentals to apply magnetic nanoparticles for hyperthermia therapy. https://www.mdpi.com/2079-4991/11/5/1203. Accessed 29 Mar 2025
36. Peiravi M, Eslami H, Ansari M, Zare-Zardini H (2022) Magnetic hyperthermia: potentials and limitations. J Indian Chem Soc 99:100269. https://doi.org/10.1016/j.jics.2021.100269
37. Tamarov KP, Osminkina LA, Zinovyev SV et al (2014) Radio frequency radiation-induced hyperthermia using Si nanoparticle-based sensitizers for mild cancer therapy. Sci Rep 4:7034. https://doi.org/10.1038/srep07034
38. Keshta BE, Gemeay AH, Kumar Sinha D et al (2024) State of the art on the magnetic iron oxide nanoparticles: synthesis, functionalization, and applications in wastewater treatment. Results Chem 7:101388. https://doi.org/10.1016/j.rechem.2024.101388
39. Marimón-Bolívar W, Tejeda-Benítez L, Herrera AP (2018) Removal of mercury (II) from water using magnetic nanoparticles coated with amino organic ligands and yam peel biomass. Environ Nanotechnol Monit Manag 10:486–493. https://doi.org/10.1016/j.enmm.2018.10.001
40. Gutierrez AM, Dziubla TD, Hilt JZ (2017) Recent advances on iron oxide magnetic nanoparticles as sorbents of organic pollutants in water and wastewater treatment. Rev Environ Health 32:111–117. https://doi.org/10.1515/reveh-2016-0063
41. Abu-Dief AM, Hamdan SK (2016) Functionalization of magnetic nano particles: synthesis, characterization and their application in water purification

42. Haqjow H, Raoufi F (2021) Silica-coated magnetic nanoparticles functionalized cobalt complex: a recyclable and efficient catalyst for the C–C bond formation. Res Chem Intermed 47:4113–4128. https://doi.org/10.1007/s11164-021-04522-7
43. Applications of magnetic nanomaterials in heterogeneous catalysis. ACS Appl Nano Mater. https://doi.org/10.1021/acsanm.9b00976. Accessed 19 Feb 2025
44. Faez Sead F, Jain V, Kumar A et al (2025) Magnetically recoverable catalysts for efficient multicomponent synthesis of organosulfur compounds. RSC Adv 15:3928–3953. https://doi.org/10.1039/D4RA08769K
45. Mokhtary M (2016) Recent advances in catalysts immobilized on magnetic nanoparticles. J Iran Chem SOC 13:1827–1845. https://doi.org/10.1007/s13738-016-0900-4
46. Abu-Dief AM, Abdel-Fatah SM (2018) Development and functionalization of magnetic nanoparticles as powerful and green catalysts for organic synthesis. Beni-Suef Univ J Basic Appl Sci 7:55–67. https://doi.org/10.1016/j.bjbas.2017.05.008
47. Abu-Reziq R, Wang D, Post M, Alper H (2007) Platinum nanoparticles supported on ionic liquid-modified magnetic nanoparticles: selective hydrogenation catalysts. Adv Synth Catal 349:2145–2150. https://doi.org/10.1002/adsc.200700129
48. Das C, Sen S, Singh T et al (2020) Green synthesis, characterization and application of natural product coated magnetite nanoparticles for wastewater treatment. Nanomaterials 10:1615. https://doi.org/10.3390/nano10081615
49. Hemanth NR, Mohili RD, Patel M et al (2022) Metallic nanosponges for energy storage and conversion applications. J Mater Chem A 10:14221–14246. https://doi.org/10.1039/D2TA02057B
50. Osaka T, Iida H, Tominaka S, Hachisu T (2008) New trends in nanoparticles: syntheses and their applications to fuel cells, health care, and magnetic storage. Isr J Chem 48:333–347. https://doi.org/10.1560/IJC.48.3-4.333
51. Review on the recent developments in magnetic nanocomposites for energy storage applications. Semiconductors. https://doi.org/10.1134/S106378262460219X. Accessed 29 Mar 2025
52. Chandra S, Patel MD, Lang H, Bahadur D (2015) Dendrimer-functionalized magnetic nanoparticles: a new electrode material for electrochemical energy storage devices. J Power Sources 280:217–226. https://doi.org/10.1016/j.jpowsour.2015.01.075
53. Abellán G, Coronado E, Martí-Gastaldo C et al (2013) Magnetic nanocomposites formed by $FeNi_3$ nanoparticles embedded in graphene. application as supercapacitors. Part Part Syst Charact 30:853–863. https://doi.org/10.1002/ppsc.201300186
54. Functionalized graphene-based nanocomposites for supercapacitor application. J Phys Chem C. https://doi.org/10.1021/jp201673e. Accessed 29 Mar 2025
55. Wang J, Musameh M, Laocharoensuk R et al (2006) Pt/Ru-functionalized magnetic spheres for a magnetic-field stimulated methanol and oxygen redox processes: towards on-demand activation of fuel cells. Electrochem Commun 8:1106–1110. https://doi.org/10.1016/j.elecom.2006.04.019
56. Wang R, Jia J, Wang H et al (2013) CNx-modified Fe_3O_4 as Pt nanoparticle support for the oxygen reduction reaction. J Solid State Electrochem 17:1021–1028. https://doi.org/10.1007/s10008-012-1948-4
57. Zhong C-J, Luo J, Njoki PN et al (2008) Fuel cell technology: nano-engineered multimetallic catalysts. Energy Environ Sci 1:454–466. https://doi.org/10.1039/B810734N
58. Review on recent progress in magnetic nanoparticles: synthesis, characterization, and diverse applications. Frontiers. https://doi.org/10.3389/fchem.2021.629054/full. Accessed 29 Mar 2025
59. Vasilakaki M, Chikina I, Shikin VB et al (2020) Towards high-performance electrochemical thermal energy harvester based on ferrofluids. Appl Mater Today 19:100587. https://doi.org/10.1016/j.apmt.2020.100587
60. Yeap SP, Lim J, Ooi BS, Ahmad AL (2017) Agglomeration, colloidal stability, and magnetic separation of magnetic nanoparticles: collective influences on environmental engineering applications. J Nanopart Res 19:368. https://doi.org/10.1007/s11051-017-4065-6

61. Multifunctional magnetic iron oxide nanoparticles: diverse synthetic approaches, surface modifications, cytotoxicity towards biomedical and industrial applications. BMC Mater. https://doi.org/10.1186/s42833-019-0002-6. Accessed 29 Mar 2025
62. Comanescu C (2023) Recent advances in surface functionalization of magnetic nanoparticles. Coatings 13:1772. https://doi.org/10.3390/coatings13101772
63. Patankar KK, Hussain CM (2022) Fundamentals and industrial applications of magnetic nanoparticles. Woodhead Publishing
64. Magnetic nanoparticles: an update of application for drug delivery and possible toxic effects. Arch Toxicol. https://doi.org/10.1007/s00204-011-0773-3. Accessed 29 Mar 2025
65. Biocompatibility, toxicity concerns, environmental and safety considerations, and legal aspects of functionalized magnetic nanoparticles—functionalized magnetic nanoparticles for theranostic applications. Wiley Online Library. https://doi.org/10.1002/9781394172917.ch17. Accessed 29 Mar 2025
66. Bushra R, Ahmad M, Alam K et al (2024) Recent advances in magnetic nanoparticles: key applications, environmental insights, and future strategies. Sustain Mater Technol 40:e00985. https://doi.org/10.1016/j.susmat.2024.e00985
67. Antonio Chávez-Hernández J, Jimena Velarde-Salcedo A, Navarro-Tovar G, Gonzalez C (2024) Safe nanomaterials: from their use, application, and disposal to regulations. Nanoscale Adv 6:1583–1610. https://doi.org/10.1039/D3NA01097J
68. Araújo EV, Carneiro SV, Neto DMA et al (2024) Advances in surface design and biomedical applications of magnetic nanoparticles. Adv Coll Interface Sci 328:103166. https://doi.org/10.1016/j.cis.2024.103166
69. Microfluidic synthesis, control, and sensing of magnetic nanoparticles: a review. https://www.mdpi.com/2072-666X/12/7/768. Accessed 29 Mar 2025
70. Ramimoghadam D, Bagheri S, Hamid SBA (2014) Progress in electrochemical synthesis of magnetic iron oxide nanoparticles. J Magn Magn Mater 368:207–229. https://doi.org/10.1016/j.jmmm.2014.05.015
71. High densities of magnetic nanoparticles supported on graphene fabricated by atomic layer deposition and their use as efficient synergistic microwave absorbers. Nano Res. https://doi.org/10.1007/s12274-014-0432-0. Accessed 29 Mar 2025
72. Rahman TU, Roy H, Fariha A et al (2023) Progress in plasma doping semiconductor photocatalysts for efficient pollutant remediation and hydrogen generation. Sep Purif Technol 320:124141. https://doi.org/10.1016/j.seppur.2023.124141
73. Baig N, Kammakakam I, Falath W (2021) Nanomaterials: a review of synthesis methods, properties, recent progress, and challenges. Mater Adv 2:1821–1871. https://doi.org/10.1039/D0MA00807A
74. Markides H, Rotherham M, El Haj AJ (2012) Biocompatibility and toxicity of magnetic nanoparticles in regenerative medicine. J Nanomater 2012:614094. https://doi.org/10.1155/2012/614094
75. New insight about biocompatibility and biodegradability of iron oxide magnetic nanoparticles: stereological and in vivo MRI monitor. Sci Rep. https://www.nature.com/articles/s41598-019-43650-4. Accessed 29 Mar 2025
76. In vitro and in vivo biocompatibility of apatite-coated magnetite nanoparticles for cancer therapy. J Mater Sci Mater Med. https://doi.org/10.1007/s10856-013-4862-0. Accessed 29 Mar 2025
77. Verma C, Verma DK, Berdimurodov E et al (2024) Green magnetic nanoparticles: a comprehensive review of recent progress in biomedical and environmental applications. J Mater Sci 59:325–358. https://doi.org/10.1007/s10853-023-08914-5
78. Kazemi S, Hosseingholian A, Gohari SD et al (2023) Recent advances in green synthesized nanoparticles: from production to application. Mater Today Sustain 24:100500. https://doi.org/10.1016/j.mtsust.2023.100500
79. Virkutyte J, Varma RS (2011) Green synthesis of metal nanoparticles : biodegradable polymers and enzymes in stabilization and surface functionalization. Chem Sci 2:837–846. https://doi.org/10.1039/C0SC00338G

80. Veluswamy R, Balasubramaniam G, Natarajan M et al (2024) Multifunctional and sustainable hydroxyapatite from natural products for biomedical and industrial applications—a comprehensive review. Sustain Chem Pharm 41:101653. https://doi.org/10.1016/j.scp.2024.101653
81. Liu C et al (2021) Biopolymers derived from trees as sustainable multifunctional materials: a review. Adv Mater. Wiley Online Library. https://doi.org/10.1002/adma.202001654. Accessed 29 Mar 2025
82. AI-driven innovation and designing of nanomaterials for a better tomor. https://doi.org/10.1201/9781003479239-24/ai-driven-innovation-designing-nanomaterials-better-tomorrow-subbulakshmi-padma-priya-thiyagaraj-bhoomika. Accessed 29 Mar 2025
83. Tripathy A, Patne AY, Mohapatra S, Mohapatra SS (2024) Convergence of nanotechnology and machine learning: the state of the art, challenges, and perspectives. Int J Mol Sci 25:12368. https://doi.org/10.3390/ijms252212368
84. Yang HY, Li Y, Lee DS (2018) Multifunctional and stimuli-responsive magnetic nanoparticle-based delivery systems for biomedical applications. Adv Ther 1:1800011. https://doi.org/10.1002/adtp.201800011
85. Goga AS (2025) Integrating artificial intelligence in nanomaterials science: pathways to revolutionary materials discovery and design: ethics and risks. In: Sandu AV, Vizureanu P, Abdullah MMAB et al (eds) Selected papers from the international conference on innovative research. Springer Nature Switzerland, Cham, pp 1–11
86. Advancements in micro/nanorobots in medicine: design, actuation, and transformative application. ACS Omega. https://doi.org/10.1021/acsomega.4c09806. Accessed 29 Mar 2025

Index

A

Adsorption, 24, 25, 28, 47, 106, 137, 141, 143–147, 151, 153, 161, 163–168, 170, 171, 176, 183, 213, 234, 237, 266, 272, 280
Advanced nanomaterials, viii
Advanced oxidation, 147, 153
Alternative magnetic field, 40, 51, 79, 106, 108, 113, 236, 268–271
Antibacterial applications, 19
Antibodies, 9, 11, 12, 18, 29, 39, 41, 47, 59, 65, 70, 71, 73, 80, 89, 91, 92, 105, 108, 109, 114, 120, 123, 126, 127, 143, 233, 269
Antiferromagnetism, 5, 6
Antiviral applications, 39, 113

B

Bioaccumulation, 21, 152, 249, 253, 254
Biocompatibility, 3, 5, 7, 9, 10, 12–16, 19–21, 30, 41, 42, 44, 48, 50, 53, 54, 59, 60, 64, 67, 68, 79, 80, 84–86, 90, 93, 96, 98, 105, 106, 109, 110, 114, 127, 130, 140, 141, 143, 178, 180, 191, 199, 213, 227, 230–236, 239, 254, 265, 268, 270, 277, 279–281
Biological imaging, 219, 223
Biomarkers, 13, 119–123, 126–130
Biomedical applications, 2, 3, 5, 7–10, 12, 23, 26, 39, 41–44, 54, 60, 61, 66, 80, 81, 84, 85, 89, 90, 96, 97, 106, 112–114, 124, 125, 169, 199, 210–212, 233, 250, 251, 266–269, 280, 281
Biosensing, 4, 10, 12, 29, 30, 39, 42, 52–54, 80, 89, 90, 96, 99, 119–124, 127–130, 193, 223, 235, 266, 281
Blood-brain barrier, 59, 69–71, 268, 270
Bone regeneration, 19, 50, 211

C

Cancer therapy, 11, 40, 41, 51, 64, 68–70, 79–81, 83, 84, 89, 97, 100, 236, 250, 278
Cancer treatment, 15–17, 39, 40, 50, 51, 54, 63, 66, 68, 69, 79–83, 86, 89, 93, 96, 97, 99, 125, 211, 232, 268, 271
Cardiac regeneration, 50
Catalysis, 1–3, 5–9, 11, 141, 147, 148, 175, 177, 178, 180, 189, 193, 199, 200, 203, 219, 223, 231–234, 240, 265, 266, 268, 272–274, 277, 281
Chemical vapor deposition, 205, 231
Controlled release, 22, 28, 59, 62, 64, 74, 90, 91, 97, 232, 266

D

Dendrimers, 16, 79, 90, 93, 97, 184, 211, 232
Diagnostic devices, 7, 41, 60
Diagnostics, 1, 4, 7, 12–14, 28, 30, 31, 39–41, 45–47, 54, 60, 62, 67, 71–74, 80, 105, 106, 108–110, 112, 114, 119–121, 123–130, 177, 199, 220, 235, 239, 271, 281
Dispersion, 87, 109, 112, 122, 150, 164, 169, 176, 180, 188, 190, 199, 201–203, 206, 207, 222, 229, 240

S. Gulati (ed.), *Functionalized Magnetic Nanomaterials*, Nanostructure Science and Technology, https://doi.org/10.1007/978-3-031-97199-0

Drug delivery, 1, 3, 5–7, 9–12, 15–17, 19, 20, 22, 28, 30, 31, 39–42, 50, 52, 54, 59–71, 74, 80, 89, 106, 124, 169, 178, 199, 212, 219, 223, 224, 232–236, 240, 250, 265–269, 275, 277, 278, 281
Dye extraction, 165

E
Electrochemical sensors, 3, 28, 175, 190, 191
Electrospinning, 20, 203, 204
Energy, 3, 10, 11, 18, 23, 26, 30, 31, 45, 51, 81, 84, 87, 88, 100, 112, 139, 148, 167, 168, 170, 175, 177, 178, 181, 183–188, 191, 193, 199, 200, 206, 208, 212, 213, 219, 221–223, 225, 226, 229, 232, 237, 238, 240–242, 265–268, 271–273, 275–277, 279, 281
Energy storage, 11, 24, 41, 175, 178, 181–185, 193, 203, 212, 213, 219, 220, 232, 238, 265–268, 275, 276, 281
Environmental applications, 9, 10, 160, 166, 167, 177, 250, 266, 268, 272, 278, 280, 282
Environmental impact, 5, 153, 213, 249, 256, 278
Environmental monitoring, 28, 119, 127–130, 190, 191, 223, 279
Environmental persistence, 249
Environmental remediation, 1–3, 5, 8–11, 30, 41, 159, 168, 171, 180, 213, 219, 232, 236, 240, 266, 274, 281
Environmental safety, 137, 265, 266, 277, 278
Exposure limits, 249, 257

F
Ferromagnetism, 5, 6
Ferroptosis, 79, 80
FMN applications, 1–12, 24–26, 28–31, 39, 41, 51, 54, 66, 105, 112, 114, 119–125, 127–129, 143, 161, 168, 175, 181, 191, 193, 208, 219, 235, 240, 249, 265–269, 272, 274, 276–280, 282
Fuel cells, 3, 10, 23, 25, 31, 41, 193, 212, 241, 242, 268, 275, 276
Functionalization, 2, 3, 5, 8–13, 15, 16, 18, 25, 39, 40, 44, 48, 54, 59, 64, 67, 71, 80, 89, 90, 97–99, 105, 109, 119, 120, 123, 141, 145, 152–154, 164, 169, 170, 175, 177–180, 206, 212, 219, 222, 223, 229–233, 240–242, 251, 252, 254, 256, 265, 266, 271, 273, 277–280
Functionalized Magnetic Nanomaterials (FMNs), 1–13, 23–31, 39–41, 45–54, 59, 60, 66, 68–72, 74, 105, 108, 109, 112, 114, 119–130, 140, 143–145, 148, 150–154, 159, 161, 164, 165, 167, 168, 171, 175, 181, 182, 191, 199–204, 206–210, 212, 213, 219, 222–224, 229, 231, 233–236, 240–242, 249–260, 265–282
Functionalized nanomaterials, 41, 171, 222–224, 230, 232, 235, 238–242

G
Genetic material detection, 120
Graphene, 12, 20, 23, 25, 26, 28, 53, 69, 169, 171, 188, 199, 210, 212, 213, 220, 222, 223, 231, 232, 235, 238, 276
Graphene oxide, 53, 79, 90, 92, 93, 97, 127, 146, 147, 163, 164, 167–169, 188, 199, 212, 222

H
Hazardous chemicals, 161, 192
Health risks, 145, 249, 253, 259
Heavy metal removal, 145, 161, 162, 266, 273
Hydrogen production, 3, 23, 25, 31, 178, 276
Hyperthermia, 1, 3, 7, 12, 15, 17, 18, 39, 40, 42, 44, 51, 69, 79–86, 97, 100, 106, 113, 233, 266, 268, 269, 271, 278

I
Imaging, 3, 5, 7, 9–14, 20, 23, 30, 39–42, 45–47, 50, 54, 60, 63, 66, 67, 80, 85, 97, 100, 105, 106, 108–112, 114, 124, 208, 210, 212, 222, 223, 226, 229, 232, 233, 235, 236, 239, 250, 266, 270, 278
Industrial applications, 5, 7, 31, 177, 191, 249, 267, 277, 279

L
Ligand exchange, 79, 89, 90, 98, 145, 180, 227
Ligands, 9, 10, 16, 40, 41, 59, 64–66, 71, 80, 89, 96, 98, 105, 108, 109, 114, 119, 120, 123–126, 140, 153, 159, 179, 180, 223, 226, 227, 230–233, 235, 241, 252, 268, 273, 277, 279

M
Magnetic anisotropy, 7, 84–86
Magnetic clusters, 44
Magnetic hyperthermia, 17, 18, 50–52, 68, 79, 80, 82–86, 89, 96, 97, 99, 203, 233, 235, 236, 271, 272
Magnetic nanocarriers, 62, 69, 79, 99, 100
Magnetic nanomaterials, 13, 39, 43, 53, 54, 60, 106, 128, 138, 141, 144, 153, 168, 171, 175–178, 188, 189, 191, 193, 201, 206, 207, 211, 229, 230, 233, 260
Magnetic nanoparticles, 1, 3, 7, 13–18, 20, 22, 26, 39, 40, 42, 43, 47–50, 53, 59–64, 66–69, 71–74, 79–81, 84–87, 96–98, 100, 106, 113, 138, 140–143, 145, 151–153, 160, 162–165, 168, 171, 179, 181, 185–192, 203, 206, 208, 210–212, 231–233, 238, 249–251, 255, 260, 267, 270, 271, 273, 276, 277
Magnetic properties, 1, 3, 5–7, 10, 12, 15, 23, 25, 26, 39, 41, 42, 44, 48, 54, 68, 80, 84–86, 88, 89, 97, 105, 108, 119–124, 140, 144, 148, 152, 175, 176, 190, 193, 200, 208, 212, 213, 238, 250, 266, 269, 275–278, 281
Magneto-liposomes, 39
Medical diagnostics, 9, 10, 54, 219
Medical implants, 39
Multifunctional materials, 11, 30, 199, 200, 219, 220

N
Nanocomposites, 14, 19, 22, 25, 26, 49, 53, 68, 146–149, 160, 163, 164, 166, 167, 169, 171, 177, 178, 185, 199–213, 220, 222, 229, 234, 237, 238, 280
Nanomaterial functionalization, 170
Nanomaterials, 13, 14, 19–23, 39–41, 48, 63, 66, 69, 70, 72–74, 79, 80, 106, 108, 125–127, 129, 130, 137, 140, 141, 150, 159–162, 168–171, 175–180, 183, 184, 187, 190, 191, 193, 207, 208, 210, 211, 219–223, 225, 229–234, 238–242, 250, 251, 255, 256, 258, 268, 269, 278, 279, 282
Nanomaterials in energy systems, 3, 11, 23, 26, 30, 31, 41
Nanoparticle stability, 272
Nanoparticle synthesis, 3, 11, 68, 86, 87, 140, 141, 153, 163, 180, 189, 231, 251
Nanostructures, 12, 14, 63, 67, 68, 74, 79, 89, 97, 159, 160, 181, 182, 191, 224, 233–235, 279
Nanotechnology in medicine, 31, 40, 124, 193
Nanotechnology safety, 129, 193, 242, 260, 266

P
Pathogen detection, 120
Pathogen removal, 150, 151
Peptides, 9, 12, 29, 47, 59, 65, 72, 80, 89, 90, 105, 109, 112, 114, 180, 230, 269
Personalization in medicine, 31, 54, 114, 119, 120, 126, 127, 129, 130
Pesticide removal, 167
Pharmaceutical waste, 3, 153, 160, 161, 164, 241, 273
Photocatalysis, 12, 30, 148, 149, 153, 168, 170, 171, 187, 223, 237, 238
Piezoelectric devices, 175, 186
Point-of-care diagnostics, 125–127, 129
Polymer coatings, 44, 89, 90, 96, 147, 233
Polymers, 9, 18–20, 22, 28, 44, 59, 60, 62, 66, 67, 71–73, 79, 85, 86, 90, 96, 97, 109, 141, 151, 153, 161, 168, 170, 177, 180, 183, 186, 187, 199–205, 207–210, 213, 220, 223, 228, 231–233, 238, 240, 241, 252, 270, 276, 277, 279, 280
Precision medicine, 59, 120, 129

Q
Quantum Dots (QDs), 14, 53, 69, 106, 111, 126, 127, 129, 211, 212, 219, 221–228, 233–240

R
Radiotherapy, 18, 40, 52, 79, 81–83, 100, 112, 271
Regenerative medicine, 18, 20, 23, 40, 48, 125, 211
Regulatory frameworks, 249, 251, 259, 279, 282
Risk mitigation strategies, 249

S
Safe handling, 251, 256, 257, 260
Sensor technologies, 14, 26, 29, 30, 153
Silica coating, 9, 11, 44, 63, 89, 90, 96
Stem cell therapy, 21, 49, 50
Supercapacitors, 41, 175, 181, 183–185, 212, 232, 238, 266, 268, 275, 276
Surface coating, 9, 41, 140, 141, 250, 280
Surface engineering, 223, 228
Surface functionalization, 3, 5, 14, 16, 42, 44, 62, 79, 87, 89, 91, 97, 109, 120, 124, 141, 160, 178, 179, 229, 277, 281
Synergistic effects, 13, 15, 17, 52, 199, 219, 223, 233, 235–238

T
Targeted delivery, 3, 9, 41, 60, 69, 70, 72–74, 83, 97, 109, 125
Targeted detection, 4, 30, 31, 54, 60, 65, 66, 73, 114, 119–121, 123, 126, 128, 130, 235, 269
Targeted drug delivery, 5, 7, 15, 40–42, 50, 54, 73, 79, 80, 85, 89–92, 95–97, 99, 200, 203, 220, 233, 266–269, 278, 281
Targeted therapy, 9, 31, 54, 60, 66, 89, 130, 233, 267–269
Technological applications, 3, 4, 30
Toxicity, 3, 5, 9, 10, 12, 14, 16, 20–23, 31, 41, 42, 44, 46, 48, 50, 51, 54, 59, 60, 63, 64, 67, 68, 72, 79, 85, 89, 93, 96, 99, 109, 114, 137, 145, 146, 152–154, 160, 165, 167, 168, 193, 208, 222, 223, 230, 233, 239, 240, 249–254, 256, 260, 265, 266, 268–270, 272, 277–281
Toxicity assessment, 280
Toxicological evaluations, 278

W
Waste management technologies, 161, 168
Wastewater treatment, 3, 7, 10, 31, 138, 141, 143, 153, 159, 161, 162, 166, 169–171, 193, 213, 234, 265, 266, 268, 272, 273, 281
Water purification, 1, 3, 11, 41, 138, 141, 150, 153, 167, 212, 273, 277

Y
Yielding catalysis, 241

Z
Zero-field cooling, 61, 108

GPSR Compliance

The European Union's (EU) General Product Safety Regulation (GPSR) is a set of rules that requires consumer products to be safe and our obligations to ensure this.

If you have any concerns about our products, you can contact us on ProductSafety@springernature.com

In case Publisher is established outside the EU, the EU authorized representative is:

Springer Nature Customer Service Center GmbH
Europaplatz 3
69115 Heidelberg, Germany

Batch number: 10755466

Printed by Printforce, the Netherlands